Elements of Induction Heating

Design, Control, and Applications

Elements of Induction Heating

Design, Control, and Applications

S. Zinn and S. L. Semiatin

Center for Materials Fabrication
Battelle Columbus Division

EPRI Program Managers:

I. L. Harry and R. D. Jeffress

Electric Power Research Institute, Inc.

Palo Alto, California 94303

ASM International

Metals Park, Ohio 44073

First printing, December 1987
Second printing, June 1988
Third printing, September 1991
Fourth printing, May 1995

Library of Congress Cataloging-in-Publication Data

Zinn, S. (Stanley)
Elements of induction heating: design, control, and applications / S. Zinn and S. L. Semiatin; EPRI program managers, I. L. Harry and R. D. Jeffress.
p. cm.
"Bulk of the work was sponsored by the Electric Power Research Institute (EPRI) through its contract with Battelle Columbus Division"—Pref.
Includes index.
ISBN 0-87170-308-4
1. Induction heating. I. Semiatin, S. L. II. Electric Power Research Institute. III. Battelle Memorial Institute. Columbus Laboratories. IV. Title.
TK4601.Z56 1987
671—dc19
87-22384
CIP

SAN: 204-7586

Editorial and production coordination by
Carnes Publication Services, Inc.

PRINTED IN THE UNITED STATES OF AMERICA

TO

J.C. Dăkauskăs

Preface

Induction heating is utilized in a large and ever-increasing number of applications. The most prominent of these are billet heating (prior to hot working), heat treating, metals joining, and metal melting. There are also many special uses of induction heating for both nonmetals and metals. Among these uses are curing of coatings, adhesive bonding, semiconductor fabrication, tin reflow, and sintering of powder metallurgy parts.

The objective of this book is to provide an overview of the range of applications of induction heating technology and methods around the special capabilities of which conventional as well as special process heating jobs can be designed. To this end, the book is divided into 12 chapters. Following an introductory chapter in which the history, applications, and advantages of induction heating are overviewed, the theory of induction heating and induction heating circuits are discussed in Chapters 2 and 3. Major equipment considerations in designing induction heating systems (e.g., power supplies, cooling systems) are summarized in Chapters 4 and 5. With this as background, process design and control guidelines and coil geometry for a variety of applications are described in Chapters 6, 7, 8, and 9. The final chapters of the book address materials handling (Chapter 10), special applications (Chapter 11), and economics (Chapter 12).

The book is written at a somewhat basic level and is intended for those who do not necessarily have a background in electrical engineering or process heating. It may be used as an introductory textbook for undergraduate college students as well as a reference for practicing engineers or shop-floor personnel.

The authors wish to express their gratitude to the organizations that made possible the writing and publication of this book. The bulk of the work was sponsored by the Electric Power Research Institute (EPRI) through its contract with Battelle Columbus Division in establishing a Center for Metals (now, Materials) Fabrication (CMF). The CMF conducts programs that promote the use of efficient electrotechnologies in industry. One of the major objectives of the Center is to encourage the efficient use of electricity for industrial process heating.

Thanks are also due to the publisher, ASM International. ASM has a long history of disseminating information to those in the metals processing industries. The authors appreciate the efforts taken by the ASM staff—in particular, Ms. Sunniva Refsnes—as well as by Ms. Laura Cahill, CMF manager of marketing and communications, in the publication of this book. We would also like to acknowledge the outstanding efforts of Mr. William Carnes, Carnes Publication Services, Inc., and Mr. Craig Kirkpatrick in the production of this book.

A number of the authors' colleagues have made substantial contributions through discussions and comments on the initial drafts of this volume. Chief among these have been G. Bobart, T. Bogan, P. Capolongo, and N. Ross. A number of companies have also supplied photographs used throughout the book; these are acknowledged in the figure captions.

The authors extend a special thanks to Mr. Tom Byrer, CMF director, Mr. Larry Kirkbride, CMF associate director, and Messrs. Les Harry and Bob Jeffress, the EPRI program managers for the project under which this work was conducted. Their ever-present support and encouragement have lightened the task of putting together this volume.

Lastly, the authors express their sincere appreciation to their families and friends, without whose understanding and moral support the writing of this book would not have been possible.

Stanley Zinn
Lee Semiatin

Contents

Chapter 1

Introduction

Electromagnetic induction, or simply "induction," is a method of heating electrically conductive materials such as metals. It is commonly used in process heating prior to metalworking, and in heat treating, welding, and melting (Table 1.1). This technique also lends itself to various other applications involving packaging and curing. The number of industrial and consumer items which undergo induction heating during some stage of their production is very large and rapidly expanding.

As its name implies, induction heating relies on electrical currents that are induced internally in the material to be heated—i.e., the workpiece. These

Table 1.1. Induction heating applications and typical products

Preheating prior to metalworking	Heat treating	Welding	Melting
Forging Gears Shafts Hand tools Ordnance **Extrusion** Structural members Shafts **Heading** Bolts Other fasteners **Rolling** Slab Sheet (can, appliance, and automotive industries)	**Surface Hardening, Tempering** Gears Shafts Valves Machine tools Hand tools **Through Hardening, Tempering** Structural members Spring steel Chain links **Annealing** Aluminum strip Steel strip	**Seam Welding** Oil-country tubular products Refrigeration tubing Line pipe	**Air Melting of Steels** Ingots Billets Castings **Vacuum Induction Melting** Ingots Billets Castings "Clean" steels Nickel-base superalloys Titanium alloys

so-called eddy currents dissipate energy and bring about heating. The basic components of an induction heating system are an induction coil, an alternating-current (ac) power supply, and the workpiece itself. The coil, which may take different shapes depending on the required heating pattern, is connected to the power supply. The flow of ac current through the coil generates an alternating magnetic field which cuts through the workpiece. It is this alternating magnetic field which induces the eddy currents that heat the workpiece.

Because the magnitude of the eddy currents decreases with distance from the workpiece surface, induction can be used for surface heating and heat treating. In contrast, if sufficient time is allowed for heat conduction, relatively uniform heating patterns can be obtained for purposes of through heat treating, heating prior to metalworking, and so forth. Careful attention to coil design and selection of power-supply frequency and rating ensures close control of the heating rate and pattern.

A common analogy used to explain the phenomenon of electromagnetic induction makes use of the transformer effect. A transformer consists of two coils placed in close proximity to each other. When a voltage is impressed across one of the coils, known as the primary winding or simply the "primary," an ac voltage is induced across the other coil, known as the "secondary." In induction heating, the induction coil, which is energized by the ac power supply, serves as the primary, and the workpiece is analogous to the secondary.

The mathematical analysis of induction heating processes can be quite complex for all but the simplest of workpiece geometries. This is because of the coupled effects of nonuniform heat generation through the workpiece, heat transfer, and the fact that the electrical, thermal, and metallurgical properties of most materials exhibit a strong dependence on temperature. For this reason, quantitative solutions exist for the most part only for the heating of round bars or tubes and rectangular slabs and sheets. Nevertheless, such treatments do provide useful insights into the effects of coil design and equipment characteristics on heating patterns in irregularly shaped parts. This information, coupled with knowledge generated through years of experimentation in both laboratory and production environments, serves as the basis for the practical design of induction heating processes.

This book focuses on the practical aspects of process design and control, an understanding of which is required for the implementation of actual induction heating operations. The treatment here is by and large of the "hands-on" type as opposed to an extended theoretical discussion of induction heating or equipment design. Chapters 2 and 3 deal with the basics of induction heating and circuit theory only to the degree that is required in design work. With this as a background, subsequent chapters address the questions of equipment selection (Chapter 4), auxiliary equipment (Chapter 5), process design for common applications (Chapter 6), control systems (Chapter 7), and coil design and fabrication (Chapter 8). The concluding chapters address the questions of special design features (Chapter 9), materials-handling systems (Chapter 10), process design for special applications (Chapter 11), and economic considerations (Chapter 12).

To introduce the subject, a brief review of the history, applications, and advantages of induction heating is given next.

HISTORY

The birth of electromagnetic induction technology dates back to 1831. In November of that year, Michael Faraday wound two coils of wire onto an iron ring and noted that when an alternating current was passed through one of the coils, a voltage was induced in the other. Recognizing the potential applications of transformers based on this effect, researchers working over the next several decades concentrated on the development of equipment for generating high-frequency alternating current.

It was not until the latter part of the 19th century that the practical application of induction to heating of electrical conductors was realized. The first major application was melting of metals. Initially, this was done using metal or electrically conducting crucibles. Later, Ferranti, Colby, and Kjellin developed induction melting furnaces which made use of nonconducting crucibles. In these designs, electric currents were induced directly into the charge, usually at simple line frequency, or 60 Hz. It should be noted that these early induction melting furnaces all utilized hearths that held the melt in the form of a ring. This fundamental practice had inherent difficulties brought about by the mechanical forces set up in the molten charge due to the interaction between the eddy currents in the charge and the currents flowing in the primary, or induction coil. In extreme cases, a "pinch" effect caused the melt to separate and thus break the complete electrical path required for induction, and induction heating, to occur. Problems of this type were most severe in melting of nonferrous metals.

Ring melting furnaces were all but superseded in the early 1900's by the work of Northrup, who designed and built equipment consisting of a cylindrical crucible and a high-frequency spark-gap power supply. This equipment was first used by Baker and Company to melt platinum and by American Brass Company to melt other nonferrous alloys. However, extensive application of such "coreless" induction furnaces was limited by the power attainable from spark-gap generators. This limitation was alleviated to a certain extent in 1922 by the development of motor-generator sets which could supply power levels of several hundred kilowatts at frequencies up to 960 Hz. It was not until the late 1960's that motor-generators were replaced by solid-state converters for frequencies now considered to be in the "medium-frequency" rather than the high-frequency range.*

---

*Modern induction power supplies are classified as low frequency (less than approximately 1 kHz), medium frequency (1 to 50 kHz), or high or radio frequency (greater than 50 kHz).

Following the acceptance of induction heating for metal melting, other applications of this promising technology were vigorously sought and developed. These included induction surface hardening of steels, introduced by Midvale Steel (1927) and the Ohio Crankshaft Company (mid-1930's). The former company used a motor-generator for surface heating and hardening of rolling-mill rolls, a practice still followed almost universally today to enhance the wear and fatigue resistance of such parts. The Ohio Crankshaft Company, one of the largest manufacturers of diesel-engine crankshafts, also took advantage of the surface-heating effect of high ac frequencies and used motor-generators at 1920 and 3000 Hz in surface hardening of crankshaft bearings. This was the first high-production application of induction heating for surface heat treating of metals. The wider application to a multiplicity of other parts was an obvious step. For example, the Budd Wheel Company became interested in induction surface hardening of the internal bores of tubular sections and applied this technique to automotive axle hubs and later to cylinder liners.

World War II provided a great impetus to the use of induction heating technology, particularly in heat treating of ordnance components such as armor-piercing projectiles and shot. The ability to use induction for *local* as well as surface hardening was also called upon to salvage over a million projectiles which had been improperly heat treated, yielding local soft spots. In addition, it was found that tank-track components, pins, links, and sprockets could be hardened in large quantities most effectively by high-frequency induction. In a different area, induction heating was applied to preheating of steel blanks prior to hot forging of parts such as gun barrels.

In recent years, the application of induction heating and melting has increased to the point where most engineers in the metalworking industries are familiar with existing applications and have some ideas for potential uses. In addition, various nonmetals industries are now beginning to develop a familiarity with induction heating principles as they find and develop uses in making their products.

Many of the recent developments have been promoted by the development of high-efficiency solid-state power supplies, introduced in 1967. Over the last several decades, the efficiency of these units has increased to almost 95% in terms of the percentage of line-frequency energy converted to the higher output frequency (Fig. 1.1). In terms of equipment cost per kilowatt available for heating, this has actually resulted in a *decrease* in cost after adjustment for inflation (Fig. 1.2).

APPLICATIONS OF INDUCTION HEATING

As can be surmised from the above discussion, induction heating finds its greatest application in the metals-processing industries (Table 1.1). Primary

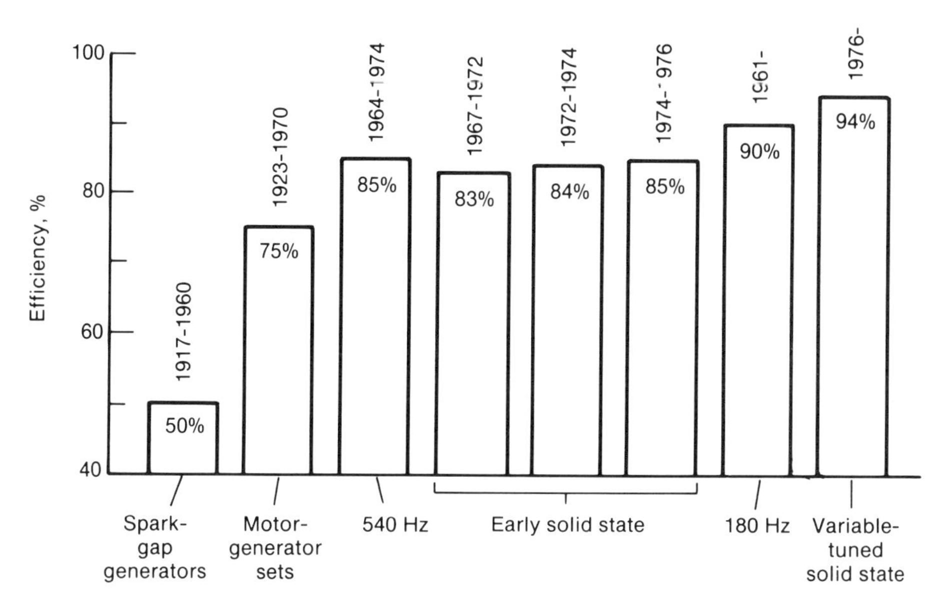

Fig. 1.1. Conversion efficiency of induction heating power supplies (from R. W. Sundeen, *Proceedings, 39th Electric Furnace Conference*, Houston, TX, AIME, New York, 1982, p. 8)

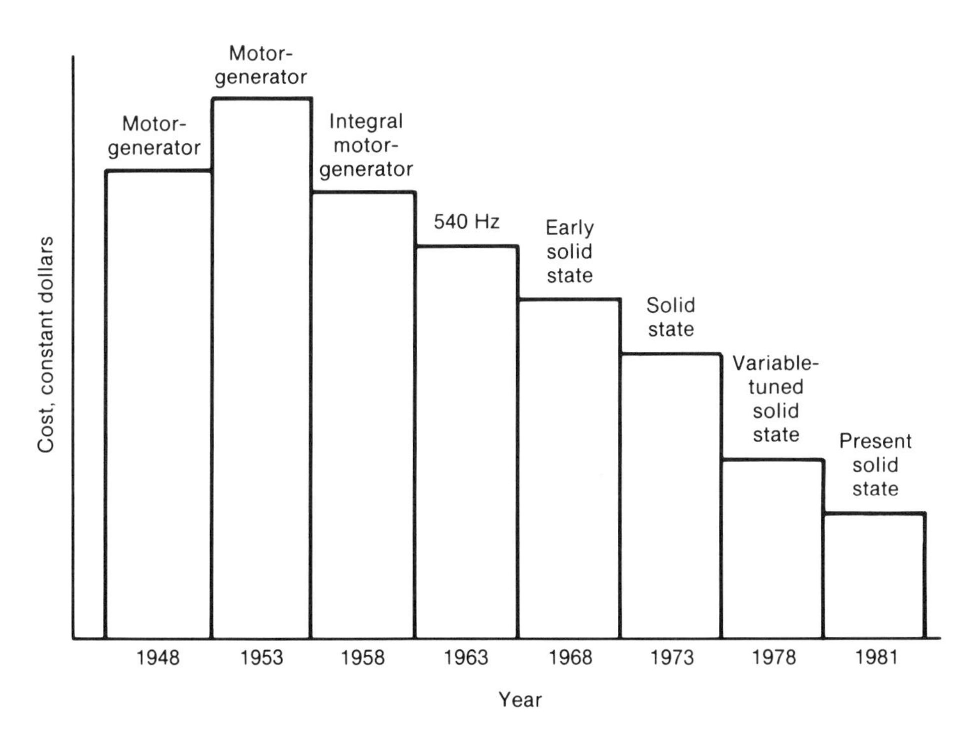

Fig. 1.2. Change in cost of induction heating power supplies since 1948 (from R. W. Sundeen, *Proceedings, 39th Electric Furnace Conference*, Houston, TX, AIME, New York, 1982, p. 8)

uses fall into the major categories of heating prior to metalworking, heat treating, welding, and metal melting. While these are the most common uses, a variety of other operations, such as paint curing, adhesive bonding, and zone refining of semiconductors, are also amenable to induction heating methods. Each of these applications is briefly discussed below.

Preheating Prior to Metalworking. Induction heating prior to metalworking is well accepted in the forging and extrusion industries. It is readily adapted to through preheating of steels, aluminum alloys, and specialty metals such as titanium and nickel-base alloys. Frequently, the workpieces in these types of applications consist of round, square, or round-cornered square bar stock. For steels, the high heating rates of induction processes minimize scale and hence material losses. The rapid heating boosts production rates. Induction heating is also useful for selectively preheating bar stock for forming operations such as heading.

Heat Treating. Induction heating is used in surface and through hardening, tempering, and annealing of steels. A primary advantage is the ability to control the area that is heat treated. Induction hardening, the most common induction heat treating operation, improves the strength, wear, and fatigue properties of steels. Steel tubular products, for example, lend themselves quite readily to hardening by induction in continuous-line operations. Tempering of steel by induction, although not as common as induction hardening of steels, restores ductility and improves fracture resistance. Also less commonly applied is induction annealing, which restores softness and ductility—important properties for forming of steels, aluminum alloys, and other metals.

Melting. Induction processes are frequently used to melt high-quality steels and nonferrous alloys (e.g., aluminum and copper alloys). Advantages specific to induction melting as compared with other melting processes include a natural stirring action (giving a more uniform melt) and long crucible life.

Welding, Brazing, and Soldering. High-frequency induction welding offers substantial energy savings because heat is localized at the weld joint. The most common application of induction welding is welded tube or pipe products that lend themselves to high-speed, high-production automated processing. Induction brazing and soldering also rely on the local heating and control capabilities inherent in the induction heating process.

Curing of Organic Coatings. Induction is used to cure organic coatings such as paints on metallic substrates by generating heat within the substrate. By this means, curing occurs from within, minimizing the tendency for formation of coating defects. A typical application is the drying of paint on sheet metal.

Adhesive Bonding. Certain automobile parts, such as clutch plates and brake shoes, make use of thermosetting adhesives. As in paint curing, induction heating of the metal parts to curing temperatures can be an excellent means of achieving rapid bonding. Metal-to-nonmetal seals, widely used in vacuum devices, also rely heavily on induction heating.

Semiconductor Fabrication. The growing of single crystals of germanium and silicon often relies on induction heating. Zone refining, zone leveling, doping, and epitaxial deposition of semiconductor materials also make use of the induction process.

Tin Reflow. Electrolytically deposited tin coatings on steel sheet have a dull, matte, nonuniform finish. Heating of the sheet to 230 °C (450 °F) by induction causes reflow of the tin coating and results in a bright appearance and uniform coverage.

Sintering. Induction heating is widely used in sintering of carbide preforms because it can provide the necessary high temperature (2550 °C, or 4620 °F) in a graphite retort or susceptor with atmosphere control. Other preforms of ferrous and nonferrous metals can be sintered in a similar manner with or without atmosphere protection.

ADVANTAGES OF INDUCTION HEATING

Prior to the development of induction heating, gas- and oil-fired furnaces provided the prime means of heating metals and nonmetals. Induction heating offers a number of advantages over furnace techniques, such as:

- Quick heating. Development of heat within the workpiece by induction provides much higher heating rates than the convection and radiation processes that occur in furnaces (Fig. 1.3).
- Less scale loss. Rapid heating significantly reduces material loss due to scaling (e.g., for steels) relative to slow gas-fired furnace processes.
- Fast start-up. Furnaces contain large amounts of refractory materials that must be heated during start-up, resulting in large thermal inertia. The internal heating of the induction process eliminates this problem and allows much quicker start-up.
- Energy savings. When not in use, the induction power supply can be turned off because restarting is so quick. With furnaces, energy must be supplied continuously to maintain temperature during delays in processing and to avoid long start-ups.
- High production rates. Because heating times are short, induction heating often allows increased production and reduced labor costs.

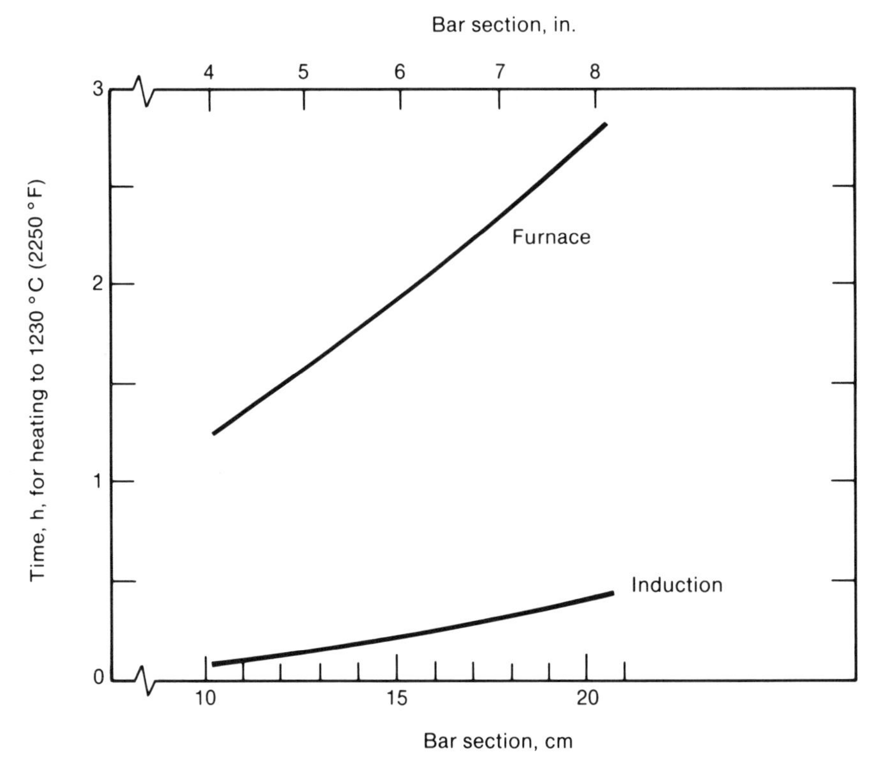

Fig. 1.3. Comparison of times for through heating by induction and gas-fired furnace techniques as a function of bar diameter (from R. Daugherty and A. A. Huchok, *Proceedings, 11th Biennial Conference on Electric Process Heating in Industry*, IEEE, New York, 1973)

In addition to those listed above, other advantages that induction heating systems offer include:

- Ease of automation and control
- Reduced floor-space requirements
- Quiet, safe, and clean working conditions
- Low maintenance requirements.

Chapter 2

Theory of Induction Heating

An induction heating system consists of a source of alternating current (ac), an induction coil, and the workpiece to be heated. The basic phenomena which underlie induction heating are best understood with reference to the interactions between the coil and the workpiece; the role of the power supply in this case is taken into account only in terms of the frequency and magnitude of the ac current which it supplies to the coil. By this means, the electrical and thermal effects which are induced in the workpiece through its coupling with the coil are deduced. The present chapter deals with this interaction. The chapter that follows deals with the interactions between the induction power supply on the one hand and the coil and workpiece on the other.

BASIS FOR INDUCTION HEATING

Induction heating relies on two mechanisms of energy dissipation for the purpose of heating. These are energy losses due to Joule heating and energy losses associated with magnetic hysteresis. The first of these is the sole mechanism of heat generation in nonmagnetic materials (e.g., aluminum, copper, austenitic stainless steels, and carbon steels above the Curie, or magnetic transformation, temperature) and the primary mechanism in ferromagnetic metals (e.g., carbon steels below the Curie temperature). A second, less important means of heat generation by induction for the latter class of materials is hysteresis losses. A simplified but qualitatively useful explanation of hysteresis losses states that it is caused by friction between molecules, or so-called magnetic dipoles, when ferromagnetic metals are magnetized first in one direction and then in the other. The dipoles can be considered as small magnets which turn around with each reversal of the magnetic field associated with an alternating current. The energy required to turn them around is dissipated as heat, the rate of expenditure of which increases with the rate of reversal of the magnetic field—i.e., the frequency of the alternating current.

Eddy-current losses and the Joule heat generation associated with them are described by the same relationships that pertain to other ac circuits or to dc (direct current) circuits. As with other electric currents, eddy currents require a complete electrical path. Associated with a given eddy current is a voltage drop V which, for a pure resistance R, is given by Ohm's Law, $V = IR$, where I denotes current. When a voltage drop occurs, electrical energy is converted into thermal energy or heat. This conversion of energy is analogous to the conversion of potential energy into kinetic energy that occurs in mechanical systems, such as when an object is dropped under the force of gravity from a given height. In the electrical case, the voltage (or potential) drop results in heating at a rate given by $VI = I^2R$. Note that this is a measure of heating *rate*, or power—i.e., it is expressed in units of energy per time.

The question now arises as to how the eddy currents are induced in the workpiece in the first place. An understanding of this mechanism is essential in design of induction coils and in control of heating rates and heating patterns. The basic phenomenon of induction is related to the fact that a magnetic field is associated with any electric current, be it ac or dc.

For an electrical conductor carrying a direct current, the magnetic field (or, more formally, the field of magnetic induction) is aligned at right angles to the current, its strength decreasing with distance from the conductor. The magnitude of the magnetic field varies in proportion to the current; the polarity or direction of the lines of magnetic induction is given by the "right-hand" rule (Fig. 2.1). If a direct current is passed through a solenoid coil, the resulting field strength is greater within the turns of the coil and smaller outside the coil (Fig. 2.2). The magnetic fields between adjacent turns are very small because the lines of magnetic induction for the adjacent turns have different signs and therefore cancel each other. Consider now what happens to the magnetic field if a solid bar is placed in a coil carrying a dc current (Fig. 2.3). If the bar is nonmagnetic, the field is unaffected. On the other hand, if a magnetic steel bar is placed inside the coil, the number of lines of magnetic induction is greatly increased. Because of this, the permeability of the steel is said to be greater than that of the nonmagnetic material. In practice, only the *relative* permeability needs to be known to do electrical calculations. Nonmagnetic materials have permeabilities equivalent to that of air and are taken to have a relative magnetic permeability of unity. In contrast, magnetic materials have relative magnetic permeabilities greater than one.

When a solid bar of electrically conducting material is placed inside a coil carrying a dc current, no eddy currents are induced. If the dc current is replaced with an ac current, however, eddy currents, and heating are induced. As an aid in understanding this, the somewhat simpler example of a solenoid coil surrounding a long, thin sleeve of electrically conducting material is examined first. With ac current, as with dc, a magnetic field surrounds the solenoid coil, but its magnitude and direction vary with time as the magnitude and direction of the ac through the coil vary. This causes the number of lines in the field of magnetic induction, or magnetic flux, which cut through the thin

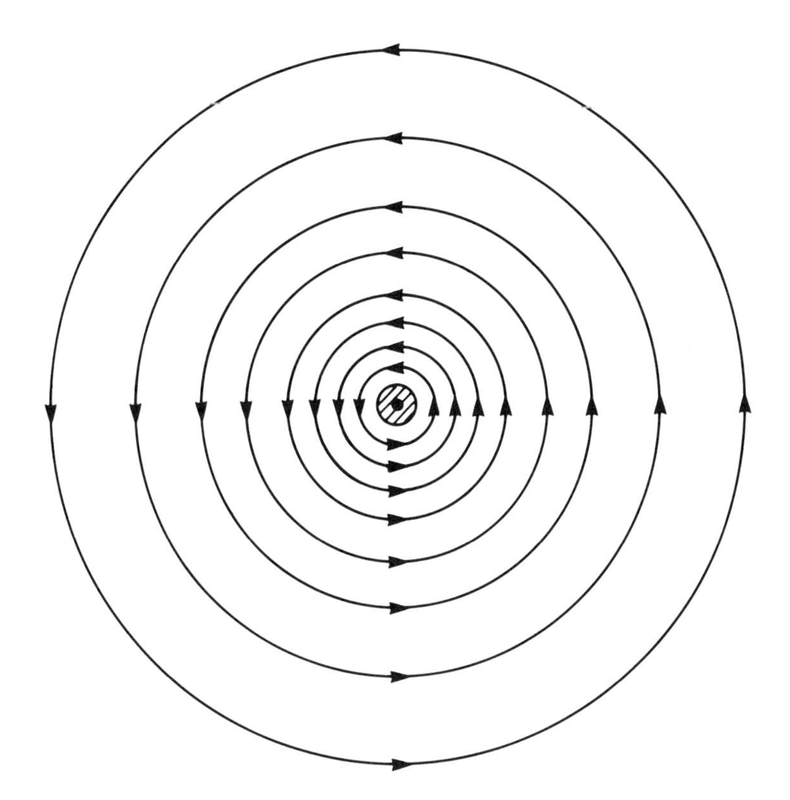

The current is emerging from the page. The relationship between the directions of the magnetic field and the current is expressed by the "right-hand" rule (thumb pointing in direction of current, fingers giving direction of magnetic field).

Fig. 2.1. Magnetic field (lines with arrows) around an electrical conductor (cross-hatched circle at center) carrying a current (from D. Halliday and R. Resnick, *Physics*, Wiley, New York, 1966)

sleeve, to vary. In his experiments in the mid-1800's, Faraday found that such a variation in flux induces a voltage. For the present example, the voltage, or electromagnetic force, induced in the sleeve, E_{sleeve}, is given by

$$E_{sleeve} = -N(\Delta\Phi_\beta/\Delta t)$$

where N is the number of turns in the coil and $\Delta\Phi_\beta/\Delta t$ is the rate at which the flux is changing, in webers (Wb) per second. The above equation is known as Faraday's Law.

It has been pointed out above that the magnetic field strength associated with an electric current varies with the magnitude of the current. Thus, for an ac current in a conducting wire or a solenoid coil, the maximum and minimum values of the magnetic field strength occur at the same times as those of the current (Fig. 2.4). At a peak or a valley of the current or magnetic field, $\Delta\Phi_\beta/\Delta t$ is equal to zero. By examining Fig. 2.4, it is apparent that

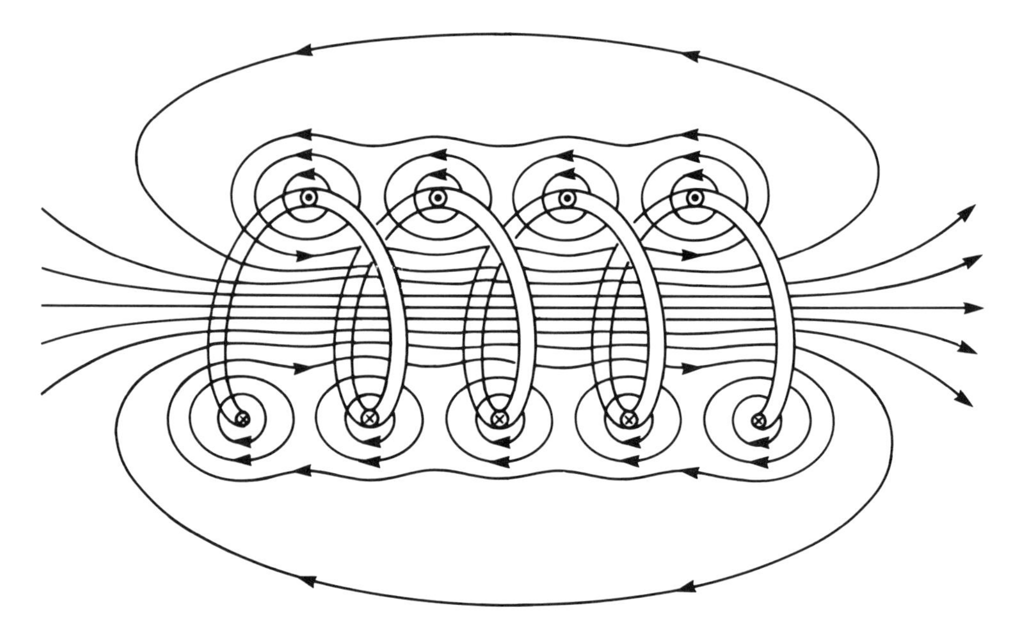

Fig. 2.2. Schematic illustration of the field of magnetic induction associated with a solenoid coil carrying an electric current (from D. Halliday and R. Resnick, *Physics*, Wiley, New York, 1966)

$\Delta\Phi_\beta/\Delta t$ is greatest when the curve for coil current versus time passes through zero. For this reason, the voltage induced in the thin sleeve is greatest when the coil current passes through zero. Moreover, because of the minus sign in Faraday's Law, the induced, or eddy, current associated with the induced voltage is opposite in sign to the coil current. Note also that the induced current paths in the sleeve mirror the coil current paths, as can be deduced from the right-hand rule.

For the thin-sleeve case, it is rather easy to determine the magnitude of the eddy currents and the heating rate from the expression for the induced voltage, E_{sleeve}. The magnetic flux is first found in terms of the coil current, I_c, and the coil geometry in an approximate manner using Ampère's Law. For a solenoid coil, the required relationship* is:

$$\Phi_\beta = (\mu_0 I_c n)(\pi r_0^2)$$

where μ_0 is the permeability constant ($4\pi \times 10^{-7}$ Wb/A·m), n is the number of coil turns per unit length, and r_0 is the mean radius of the coil turns. Next, the resistance of the sleeve is found from $R_{sleeve} = \rho\ell/A$, where ρ is the resistivity of the sleeve material, ℓ is the length of the current path ($= \pi \cdot a$,

*This relationship is based on the flux inside the coil in the *absence* of the sleeve. When the sleeve is placed in the coil, the flux within the sleeve is reduced. Thus, Φ_β calculated from the equation is an upper limit.

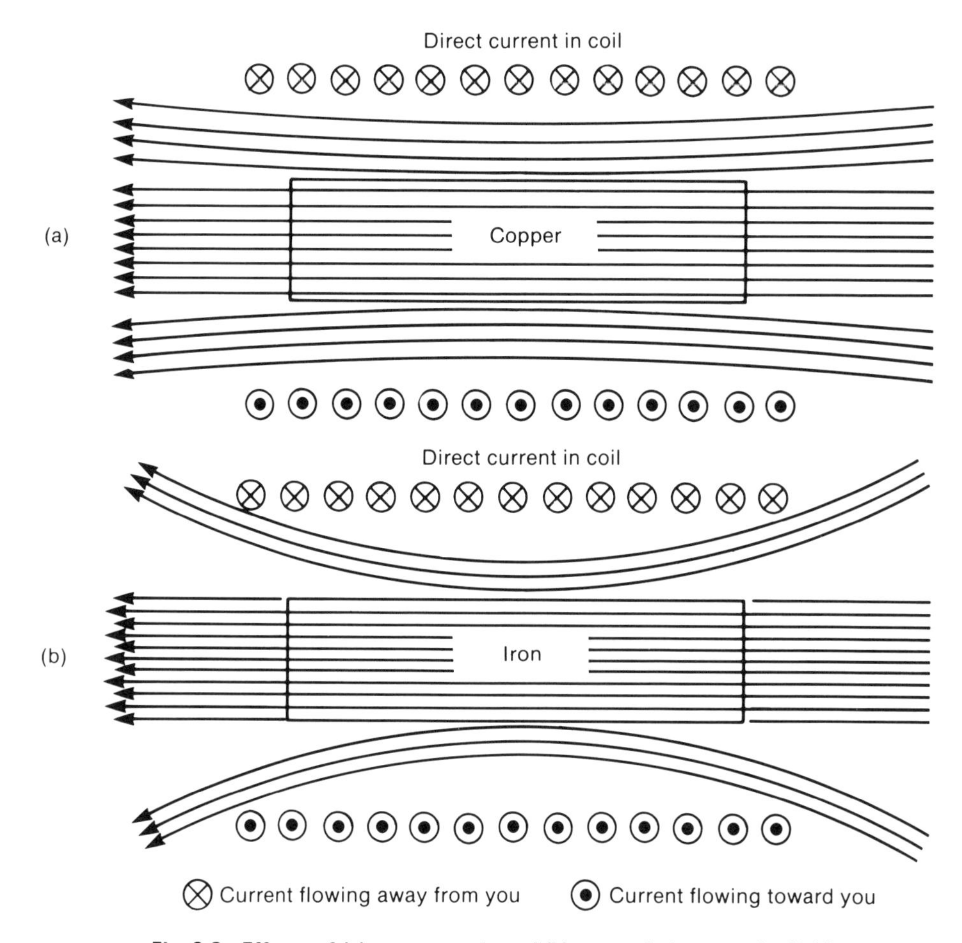

Fig. 2.3. Effects of (a) nonmagnetic and (b) magnetic bars on the field of magnetic induction (i.e., magnetic flux) within a solenoid coil carrying an electric current (from C. A. Tudbury, *Basics of Induction Heating*, Vol 1, John F. Rider, Inc., New York, 1960)

where a is the sleeve diameter), and A is the cross-sectional area of the sleeve (equal to the product of sleeve thickness times sleeve length). The power dissipated by the eddy currents is then equal to $E_{sleeve}\ R_{sleeve}$, or $I^2_{sleeve}\ R_{sleeve}$, where $I_{sleeve} = E_{sleeve}/R_{sleeve}$, from Ohm's Law.

EDDY-CURRENT DISTRIBUTION IN A SOLID BAR

For the thin-sleeve case, the induced eddy currents assume a fixed magnitude depending only on the coil current and geometry. In contrast, when a solid bar is placed inside the coil, the behavior is somewhat more complex.

This situation is best visualized by imaging the solid bar to consist of a number of thin, concentric sleeves (Fig. 2.5). The field of magnetic induction

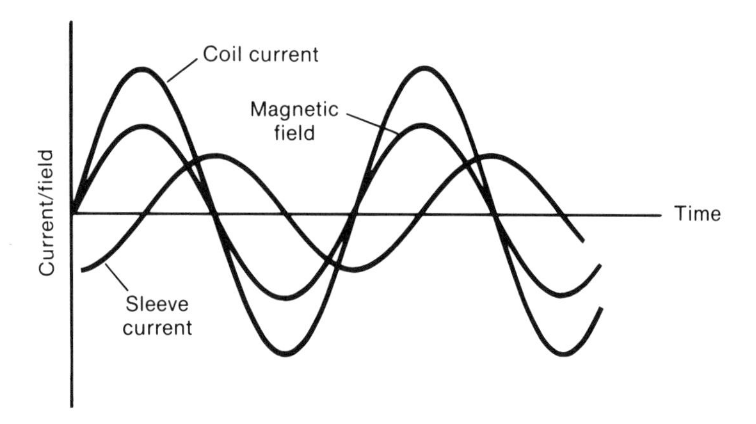

Fig. 2.4. Qualitative variation of the current and the strength of the associated field of magnetic induction with time for a solenoid coil energized by an ac supply. The eddy current induced in an electrically conductive sleeve placed in the induction coil is also shown.

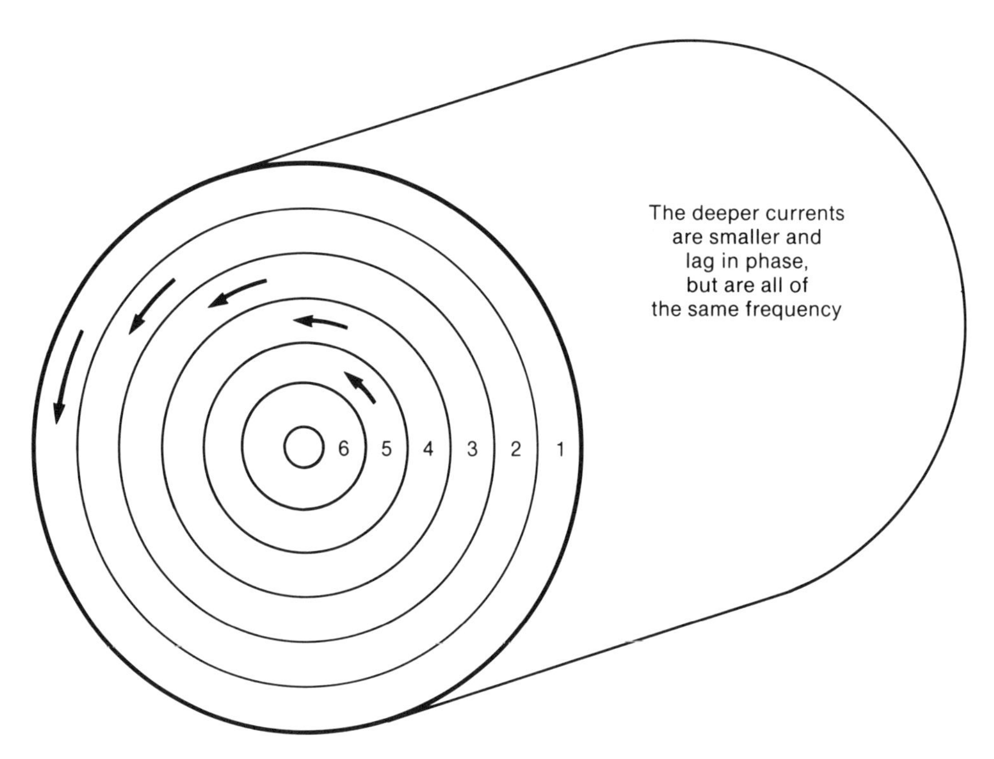

Fig. 2.5. Qualitative variation, as a function of radial position, of the magnitude and phase shift of the eddy currents induced in a solid bar by a solenoid coil carrying an alternating current (from C. A. Tudbury, *Basics of Induction Heating*, Vol 1, John F. Rider, Inc., New York, 1960)

is strongest in the space between the inner diameter of the coil and the outer diameter of the outermost sleeve. A certain amount of flux passes through the outermost sleeve, inducing eddy currents. The question now arises as to whether the magnetic field strength is greater or smaller within the outer sleeve, relative to the external field. This depends on whether the induced current in the outer sleeve tends to reinforce the field or not. If it strengthened the field, a higher voltage would be induced in the sleeve, causing a higher current. This would result in a still stronger field, a higher voltage, etc., a situation which certainly cannot happen. Thus, the magnetic field strength is reduced within the outermost sleeve. Because of this, the current induced in the second sleeve from the surface is smaller than that in the outermost sleeve, the induced current in the third sleeve is smaller than that in the second, and so forth. In general, the magnitudes of the induced currents decrease continuously from the surface irrespective of whether the bar is made of a magnetic or a nonmagnetic material. This phenomenon is known as "skin effect."

Mathematical determination of the current distribution over the cross section of a bar, let alone a more complicated geometry, is a difficult task, a discussion of which is beyond the scope of this book. For the simplest case which can be analyzed—that of a solid round bar—the solutions demonstrate that the induced current decreases exponentially from the surface. The most important result from such solutions is that they allow us to define an effective depth of the current-carrying layers. This depth, known as the reference depth or skin depth, d, depends on the frequency of the alternating current through the coil and the electrical resistivity and relative magnetic permeability of the workpiece; it is very useful in gaging the ability to induction heat various materials as well. The definition of d is:

$$d = 3160\ \sqrt{\rho/\mu f}\ \text{(English units)}$$

or

$$d = 5000\ \sqrt{\rho/\mu f}\ \text{(metric units)}$$

where d is the reference depth, in inches or centimetres; ρ is the resistivity of the workpiece, in ohm-inches or ohm-centimetres; μ is the relative magnetic permeability of the workpiece (dimensionless); and f is the frequency of the ac field of the work coil, in hertz. The reference depth is the distance from the surface of the material to the depth where the induced field strength and current are reduced to $1/e$, or 37% of their surface values. The power density at this point is $1/e^2$, or 14% of its value at the surface.*

Figure 2.6 gives reference depth versus frequency for various common metals. Reference depth varies with temperature, for a fixed frequency, because

*e ≡ base of the natural logarithm = 2.718.

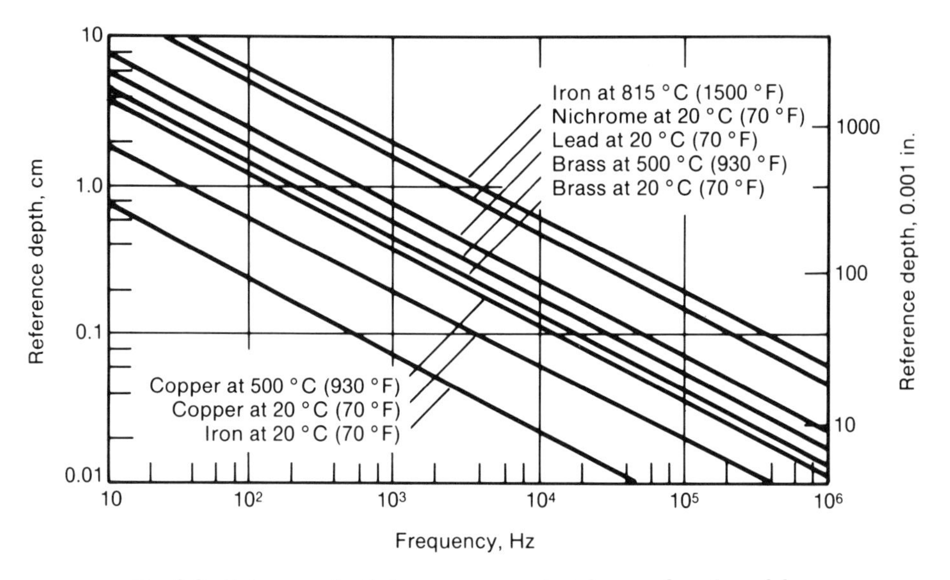

Fig. 2.6. Reference depth for common materials as a function of frequency (from A. F. Leatherman and D. E. Stutz, "Induction Heating Advances: Application to 5800 F," NASA Report SP-5071, National Aeronautics and Space Administration, Washington, 1969)

the resistivities of conductors vary with temperature. Furthermore, for magnetic steels, permeability varies with temperature, decreasing to a value of unity at and above the Curie temperature. Also in these materials, the reference depth below the Curie temperature increases with power density as the steel becomes magnetically saturated and permeability decreases (Fig. 2.7). Because of these effects, the reference depth in nonmagnetic materials may vary by a factor of two or three over a wide heating range, whereas for magnetic steels, it can vary as much as 20 times.

EQUIVALENT RESISTANCE AND EFFICIENCY—SOLID ROUND BAR

The concept of reference depth allows us to define two other very important quantities in the technology of induction heating—namely, equivalent resistance and electrical efficiency. The equivalent resistance R_{eq} is the workpiece resistance which if placed in a series circuit with the induction coil would dissipate as much heat as all the eddy currents in the actual workpiece. In other words, the power dissipated in the workpiece would be equal to $I_c^2 \cdot R_{eq}$. For a solid round bar, it turns out that the equivalent resistance is equal to the product of N^2 and the resistance of a sleeve *one* reference depth thick located at the surface: $R_{eq} = \rho \pi a K_{R_2} N^2/A$, where a is the outer diameter of the sleeve (or solid bar), A is the cross-sectional area of the sleeve ($= d \cdot w$), and K_{R_2} is a factor (Fig. 2.8) to account for the variation of electrical path

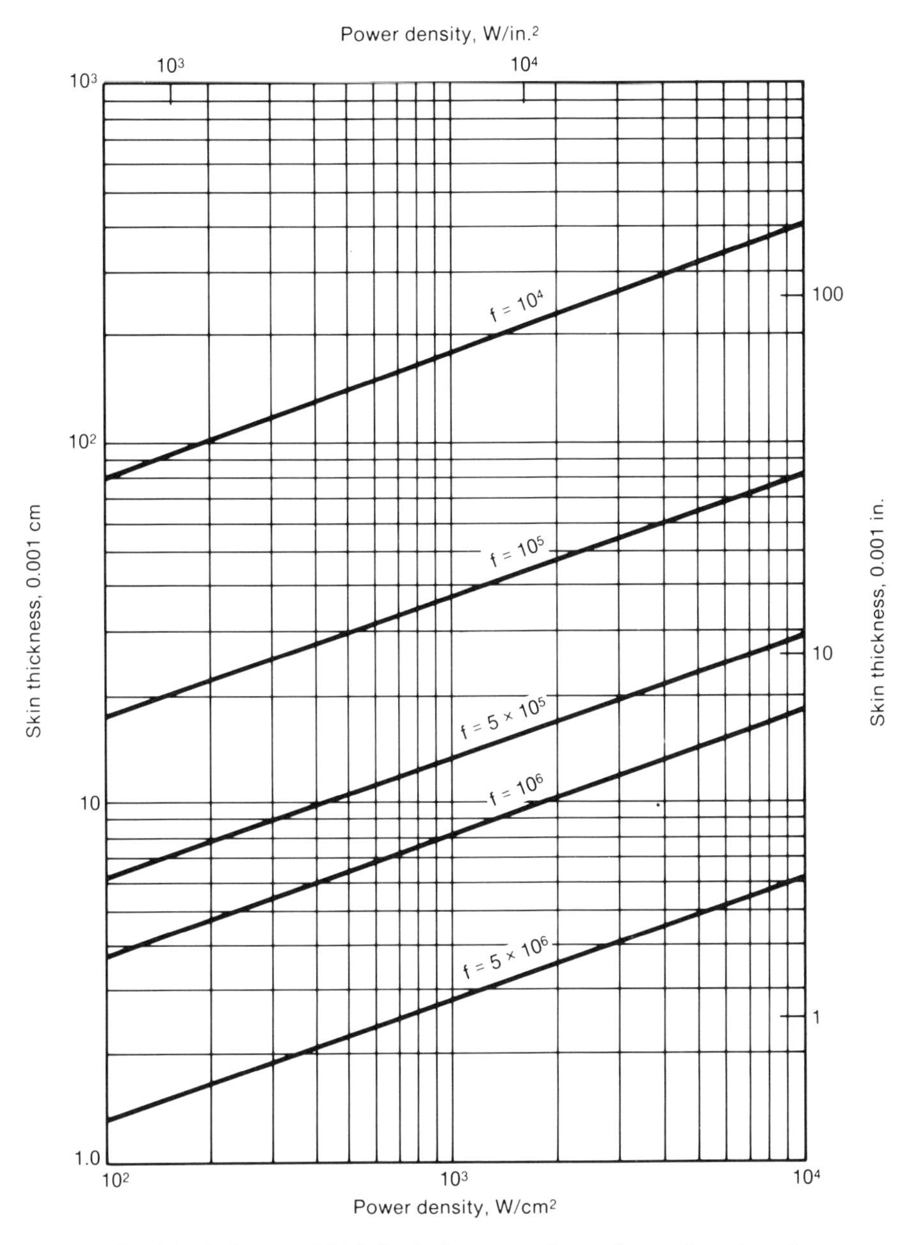

Fig. 2.7. Reference (skin) depth for magnetic steel as a function of power density and frequency (from G. H. Brown, C. N. Hoyler, and R. A. Bierwirth, *Theory and Application of Radio Frequency Heating*, Van Nostrand, New York, 1947)

between the ID and OD of the equivalent sleeve. The dependence of R_{eq} on the number of induction coil turns N is a result of the fact that the induced voltage and induced eddy currents both increase linearly with N according to Faraday's and Ohm's Laws, respectively.

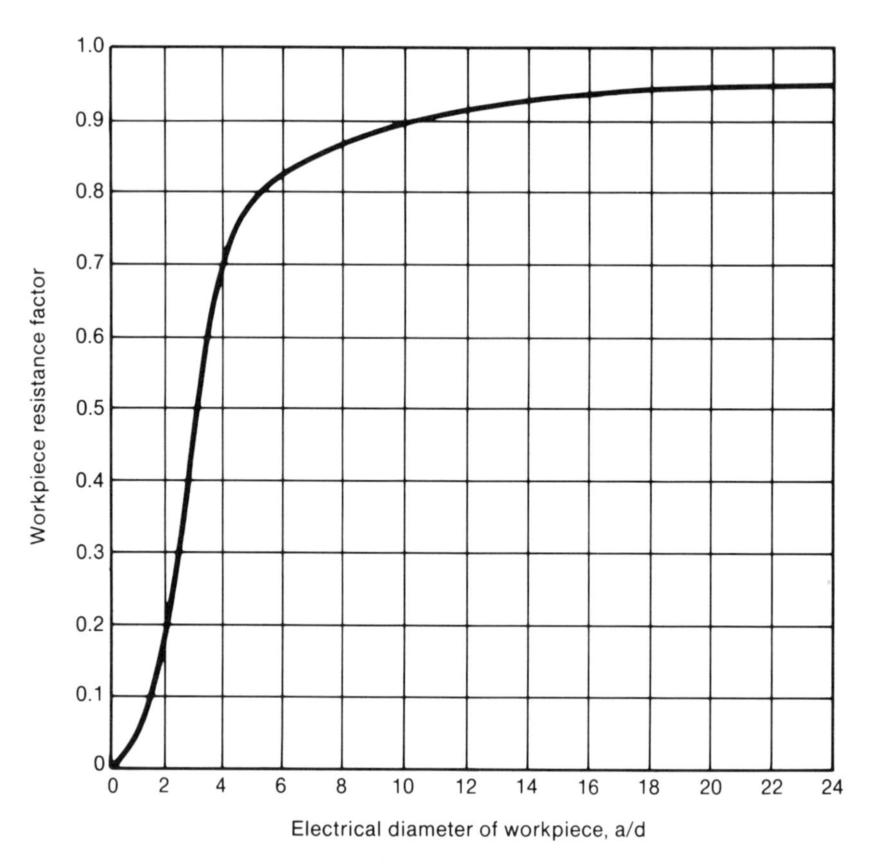

Fig. 2.8. Workpiece resistance correction factor used for calculating workpiece resistance as reflected in induction-coil circuit for solid round bars (from C. A. Tudbury, *Basics of Induction Heating*, Vol 1, John F. Rider, Inc., New York, 1960)

The equivalent resistance, or the workpiece resistance as reflected in the coil circuit as it is often called, allows the electrical efficiency to be calculated. The efficiency η is the ratio of electrical energy dissipated in the workpiece to that dissipated by the coil and workpiece:

$$\eta = \frac{I_c^2 R_{eq}}{I_c^2 R_c + I_c^2 R_{eq}} = \frac{R_{eq}}{R_c + R_{eq}}$$

This equation demonstrates that, for a given equivalent resistance, the efficiency η increases as the coil resistance R_c decreases. Because copper has the lowest resistivity of all common metals, it is the typical choice for the construction of induction coils. On the other hand, for a given value of R_c, efficiency can be increased by increasing R_{eq}. Once the material and workpiece geometry are fixed, all the terms in the relation defining R_{eq}, except K_{R_2}, are specified. Figure 2.8 shows that K_{R_2} is a function of bar diameter over reference depth, a/d. Careful inspection of this plot reveals that K_{R_2} increases rapidly for values of a/d up to approximately 4. Above this value, there is little increase in K_{R_2} for big increases in a/d. Because a is a fixed number for a specific application, the value of a/d is determined by the reference depth d. Thus, an increase in a/d requires a decrease in d. Referring to the equation for reference depth ($d \approx \sqrt{\rho/\mu f}$), it is seen that increased frequencies are needed for higher values of a/d. This has practical significance inasmuch as the cost of induction power supplies increases with frequency.

The trend shown in Fig. 2.8 for K_{R_2} versus a/d forms the basis for the definition of a "critical" frequency. This is the frequency above which relatively little increase in electrical efficiency can be gained by increases in frequency. Viewed from a different perspective, it is the frequency *below* which induction heating efficiency drops rapidly. For a round bar, the critical frequency is that at which the ratio of workpiece diameter to reference depth is approximately 4 to 1. For a sheet or slab heated from both surfaces, the critical ratio of thickness to reference depth is 2.25 to 1. Figure 2.9 shows criti-

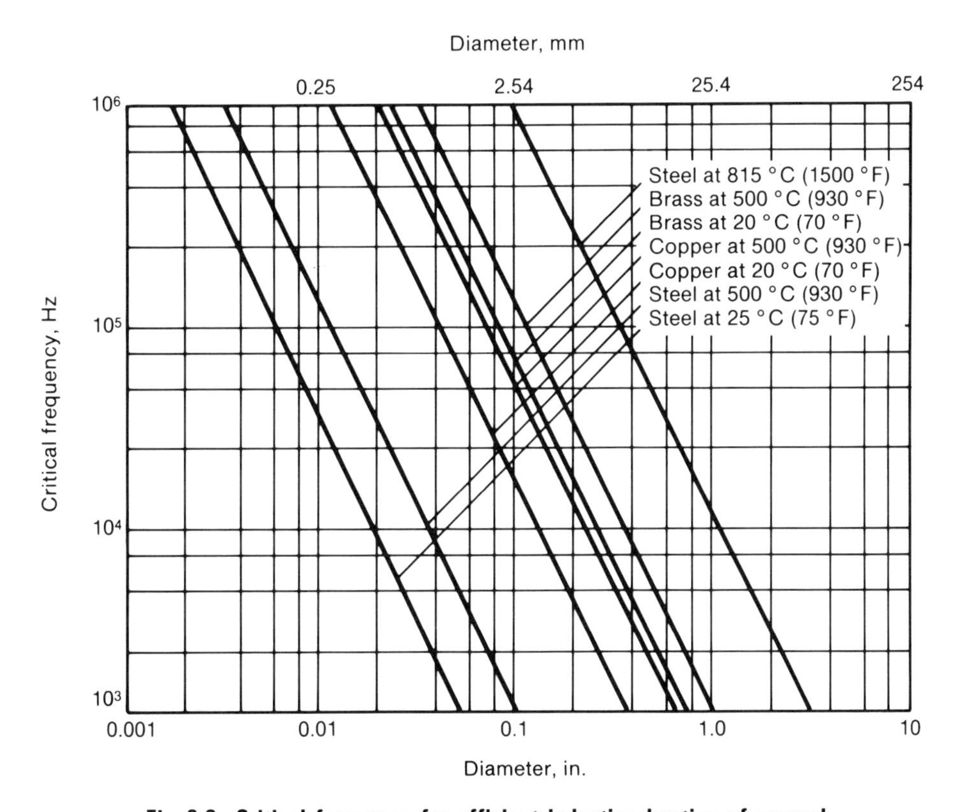

Fig. 2.9. Critical frequency for efficient induction heating of several materials as a function of bar size (from A. F. Leatherman and D. E. Stutz, "Induction Heating Advances: Application to 5800 F," NASA Report SP-5071, National Aeronautics and Space Administration, Washington, 1969)

cal frequency as a function of diameter for round bars. Figure 2.10 shows efficiency of heating as a function of this critical frequency; note the similarity between this plot and the plot in Fig. 2.8. Here, the changes in heating efficiency above and below the critical frequency are apparent. Below the critical frequency, efficiency drops rapidly because less current is induced due to current cancellation. Current cancellation becomes significant when the reference depth is such that eddy currents induced from either side of a workpiece "impinge" upon each other and, being of opposite sign, cancel each other.

For through heating, a frequency close to the critical frequency should be chosen. In contrast, for shallow heating in a large workpiece, a high frequency is selected. This is the situation for surface heat treatment of steels. In these instances, there is no concern about critical frequency because the workpiece diameter will typically be many times the reference depth.

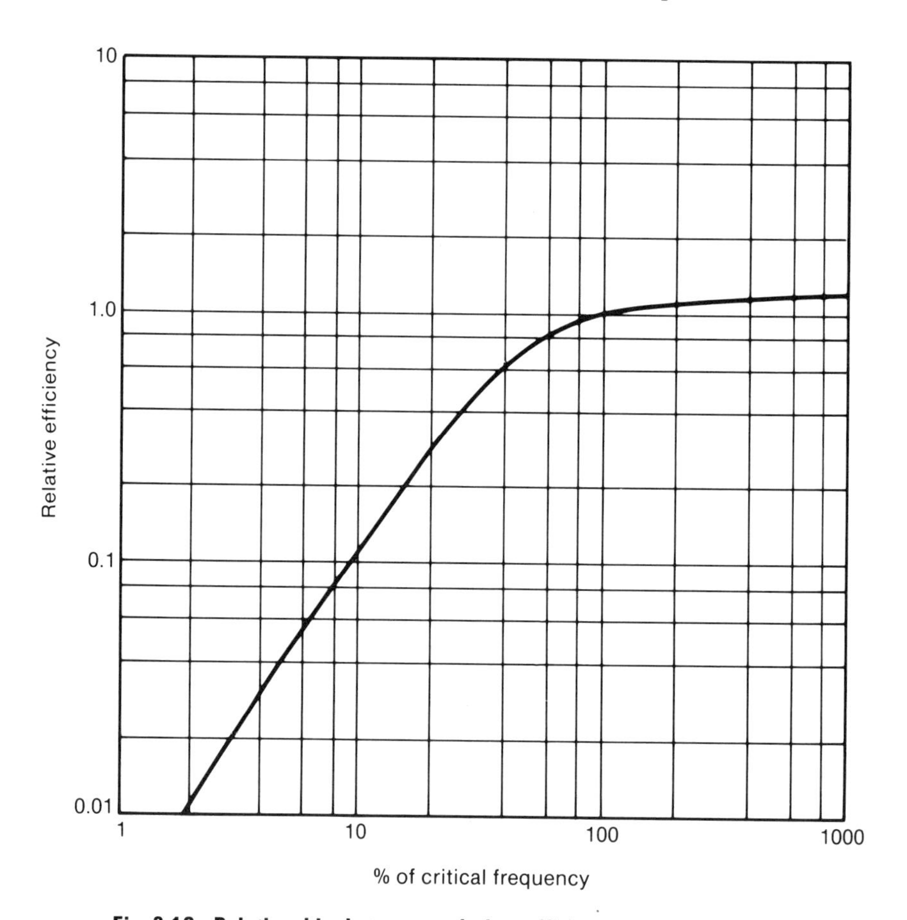

Fig. 2.10. Relationship between relative efficiency and critical frequency (from A. F. Leatherman and D. E. Stutz, "Induction Heating Advances: Application to 5800 F," NASA Report SP-5071, National Aeronautics and Space Administration, Washington, 1969)

EQUIVALENT RESISTANCE AND EFFICIENCY—OTHER GEOMETRIES

The expression given above for electrical efficiency η in terms of the induction-coil resistance and the resistance of the workpiece as reflected in the coil circuit, or equivalent resistance, is a general one for all geometries. The major difficulty in evaluating this expression lies in the determination of the equivalent resistance R_{eq}. Because R_{eq} is a function of geometry, its value can be estimated only for a limited number of particularly simple shapes. These are (*a*) a short, stubby round bar, (*b*) a hollow tube, and (*c*) a rectangular slab. The equation for the equivalent resistance of a short bar is identical to that for a long bar except for a factor K_{S_2} (Fig. 2.11): $R_{eq} = \rho\pi\alpha N^2 K_{R_2} K_{S_2}/d \cdot w$. The K_{S_2} factor takes into account the "fringing" of the magnetic field at the bar ends, which becomes important when the length-to-diameter ratio becomes small. The expressions for a long, hollow tube and a rectangular slab are similar to that for a long, solid bar except for the definition of K_{R_2}, for which the values shown in Fig. 2.12 and 2.13 are employed.

A type of workpiece which is being heated increasingly by induction nowadays is thin sheet. In this case, two pancake-type inductors are used to set up an alternating magnetic field *perpendicular* to the sheet, rather than parallel to it, as it would be for a solenoid coil surrounding the sheet. Known as transverse-flux induction heating, this method allows the use of less-expensive, low-frequency power supplies in order to obtain reasonable heating efficiencies. A precise relationship for R_{eq} for this heating geometry is difficult to derive because of the complex nature of the current paths in the sheet and the dependence of R_{eq} on the size of the air gap. As discussed by Semiatin *et al*,* however, the following expression is thought to give a reasonable first approximation:

$$R_{eq} = (2w + 2\ell)\rho/t^2$$

where w, ℓ, ρ, and t are sheet width, inductor length, workpiece resistivity, and sheet thickness, respectively. The accuracy of this equation was confirmed in a series of low-carbon steel annealing trials using an inductor with an R_c equal to 2.125 Ω. Using an average value of ρ over the temperature range through which the steel was heated for annealing purposes, a coil efficiency η of 83.5% was predicted. This compared favorably with the values between 83 and 86 obtained experimentally through measurements of the power into the coil and the (known) heat capacity of the workpiece material. Further information on the design of transverse-flux induction heating systems is given in Chapter 6.

---

*S. L. Semiatin *et al*, "CMF Reports–Rapid Annealing of Sheet Steels," Center for Metals Fabrication, Columbus, Ohio, 1986.

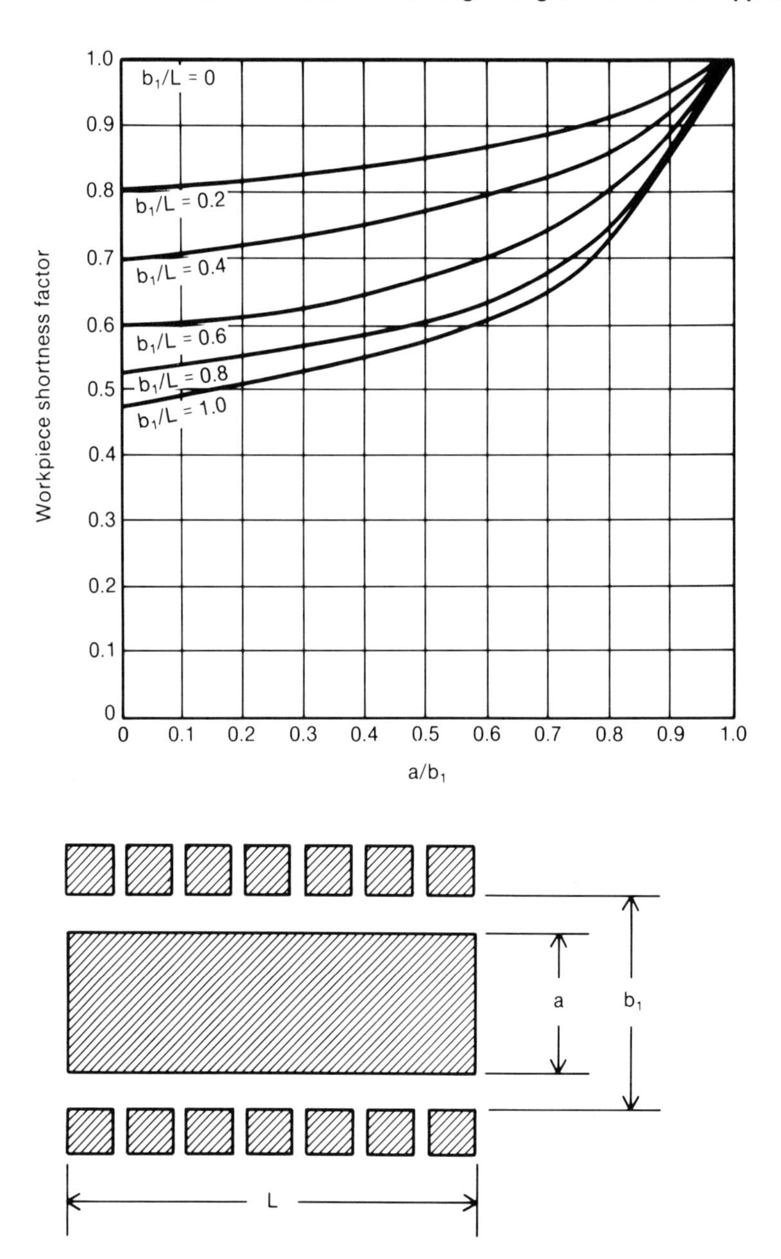

Fig. 2.11. Workpiece shortness correction factor used for calculating workpiece resistance as reflected in induction-coil circuit for short, solid round bars (from C. A. Tudbury, *Basics of Induction Heating*, Vol 1, John F. Rider, Inc., New York, 1960)

For irregular geometries for which efficiency calculations as described above are not possible, minimum frequencies for efficient induction heating are chosen on the basis of the ratio of some characteristic dimension of the workpiece to the reference depth. The frequency is chosen to obtain a ratio of at

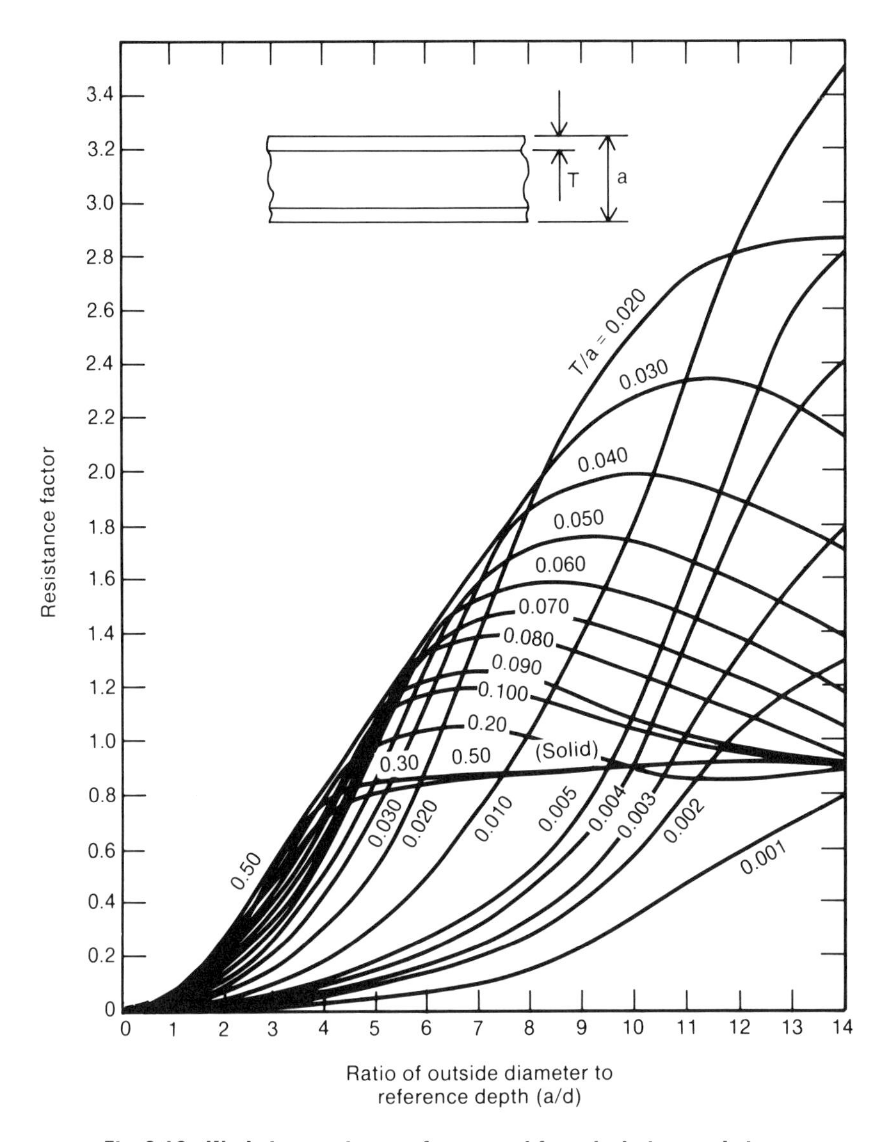

Fig. 2.12. Workpiece resistance factor used for calculating workpiece resistance as reflected in induction-coil circuit for round, hollow loads (from C. A. Tudbury, *Basics of Induction Heating*, Vol 1, John F. Rider, Inc., New York, 1960)

least 4 to 1 or at least 2 to 1 for parts which are basically round or flat, respectively.

DETERMINATION OF POWER REQUIREMENTS

As discussed above, the ac frequency, through its effect on reference depth and efficiency, is one of the important design parameters in induction heat-

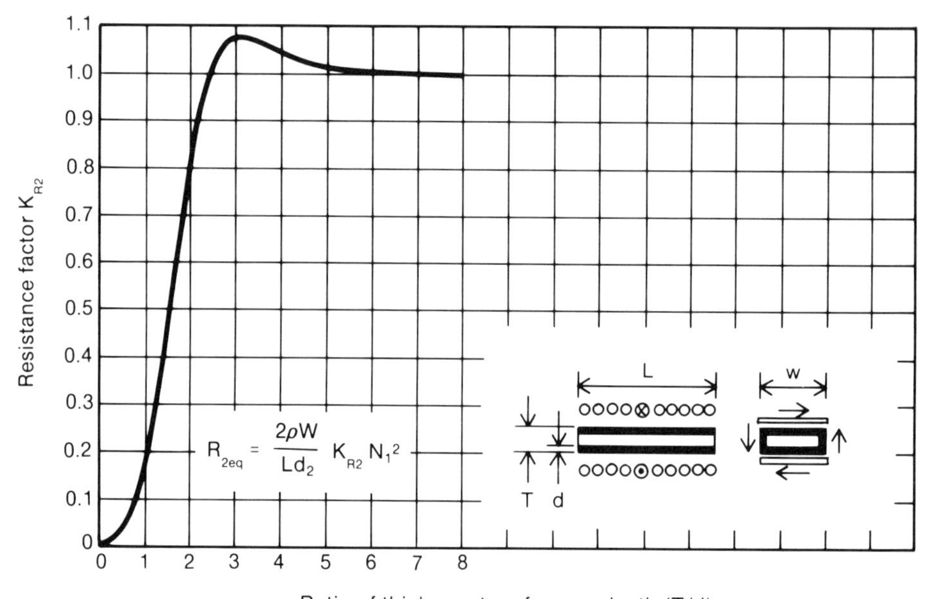

Fig. 2.13. Workpiece resistance factor used for calculating workpiece resistance as reflected in induction-coil circuit for rectangular loads (from C. A. Tudbury, *Basics of Induction Heating*, Vol 1, John F. Rider, Inc., New York, 1960)

ing. The amount of power that is needed for a given application is also an important consideration. If the workpiece is regularly shaped and is to be through heated or melted, the calculation is straightforward. However, if it is to be only selectively heated so that the remainder of the workpiece is a heat sink for the generated heat, calculation of the power needed may be difficult. The latter calculation is not considered here; guidelines for surface heating for the purpose of surface heat treatment of steel are given in Chapter 6.

For through-heating applications, the power density should be kept relatively low to allow conduction from the outer layers (which are heated more rapidly by higher current densities) to the inner layers. There will always be a temperature gradient, but this can be minimized by careful selection of induction heating parameters. Neglecting the temperature gradient, the absorbed power depends on the required temperature rise ΔT, the total weight to be heated per unit time W, and the specific heat of the material c. The power P_1 to be supplied to the load is then given by $P_1 = Wc\Delta T$.

To determine the total input power needed from the power source which supplies ac current to the induction coil, the power lost from the workpiece due to radiation and convection and the loss in the coil itself due to Joule heating must be added to P_1. Heat loss by convection is usually small and is neglected in calculations of power requirements for typical rapid-heating applications.

Radiation losses are calculated by means of the expression $P_2 = Ae\sigma(T_2^4 - T_1^4)$, where e is the emissivity of the workpiece surface, σ is the Stefan-Boltzmann constant, T_1 and T_2 are the workpiece and ambient temperatures (in K), respectively, and A is the surface area of the workpiece. The radiation power loss can vary greatly during the heating cycle because of surface-condition changes dependent on material and temperature. Typical emissivities are 0.1 to 0.2 for aluminum at 200 to 595 °C (390 to 1100 °F) and 0.80 for oxidized steel.

The power lost in the induction coil, $P_3 = I_c^2 R_c$, depends on frequency, coil design, and the size of the air gap between the inductor and the workpiece, among other factors. At low frequencies (e.g., 60 Hz), R_c is about equal to the dc resistance value, and P_3 is calculated directly from knowledge of the coil current. At higher frequencies, a skin effect is also established in the inductor, and this must be taken into account when estimating P_3. The calculation of R_c in these cases is discussed by Tudbury.*

With the above definitions, the total power required is equal to $P_1 + P_2 + P_3$, and the over-all heating efficiency of the system is $P_1/(P_1 + P_2 + P_3)$.

*C. A. Tudbury, *Basics of Induction Heating*, Vol 1, John F. Rider, Inc., New York, 1960.

Chapter 3

Tuning of Induction Heating Circuits and Load Matching

The previous chapter described the fundamentals of induction heating with respect to the transfer of energy between an induction coil and an electrically conductive workpiece. The effects of coil geometry, workpiece properties, and power-supply frequency on the electrical, or coil, efficiency were summarized. With proper designs, coil efficiencies in excess of 90% are readily obtained. However, the over-all system efficiency, defined as the percentage of the electrical energy drawn from the power line that is actually used in workpiece heating, depends on four other factors: (1) the conversion efficiency of the power supply, (2) tuning of the induction heating circuit, (3) matching of the induction heated load and coil to the induction power supply, and (4) coupling of the coil and the workpiece. Power-supply conversion efficiency is a function of the specific type of converter employed. As described in more detail in Chapter 4, solid-state generators tend to have the highest conversion efficiencies and vacuum-tube-type radio-frequency oscillators the lowest.

In this chapter, attention is focused on the transfer of energy between the power supply and the induction heating coil. The most efficient transfer requires that the induction heated load and coil be matched to the power supply and that the electrical circuit containing these elements be properly tuned. Load matching enables the full rated power of the induction generator to be drawn effectively; tuning ensures that this power is used for actual heating. In practice, it is usually best to tune the appropriate induction heating circuit to obtain what is known as a unity "power factor" and then to match the coil and workpiece "impedance" to that of the power supply. These procedures are described below.

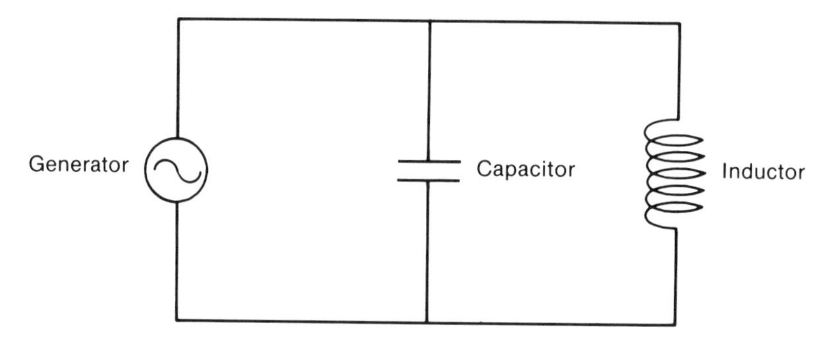

Fig. 3.1. A simple parallel resonant circuit (from P. H. Brace, Induction Heating Circuits and Frequency Generation, in *Induction Heating*, ASM, Metals Park, OH, 1946, p 36)

TUNING OF INDUCTION HEATING CIRCUITS

Most induction heating systems make use of an ac electrical circuit known as a "tank" circuit. Such circuits include a capacitor and an inductor.* The latter may be the induction coil itself or a separate so-called tank coil. The term "tank" is derived from the fact that the capacitor and inductor both serve as storage-tank-type devices for energy—electrostatic in nature for the capacitor and electromagnetic in the case of the inductor. The tank-circuit components may be connected in parallel (Fig. 3.1) or in series (Fig. 3.2).

Tuning in induction heating generally refers to the adjustment of the capacitance or inductance of a tank circuit containing the induction coil and workpiece, so that the "resonant" frequency of the circuit is equal or close to the frequency of the induction power supply. This resonant frequency is defined by the expression f_0 (in hertz) $= 1/2\pi\sqrt{LC}$, where L is inductance (in henries) and C is capacitance (in farads).

Series Resonant Circuits

To understand how power is stored or dissipated in various elements of a series (or parallel) tank circuit, it is important to recognize the nature of the voltage and current passing through each. When an ac voltage is impressed across a resistance R, the voltage drop is in phase with the current and is equal to the product of the current and the resistance, $V_R = IR$. On the other hand, inductors and capacitors offer a different kind of "opposition" to current flow, known as inductive reactance X_L and capacitive reactance X_C, respectively. They are defined as:

$$X_L = 2\pi fL \text{ and } X_C = 1/2\pi fC$$

where f is frequency.

*In actual induction heating installations, the tank-circuit elements (other than the induction coil itself) are frequently contained in a separate enclosure apart from the power supply. This enclosure, most often called the "heat station," may frequently contain a load-matching transformer as well and may be adjacent to or remote from the power supply itself.

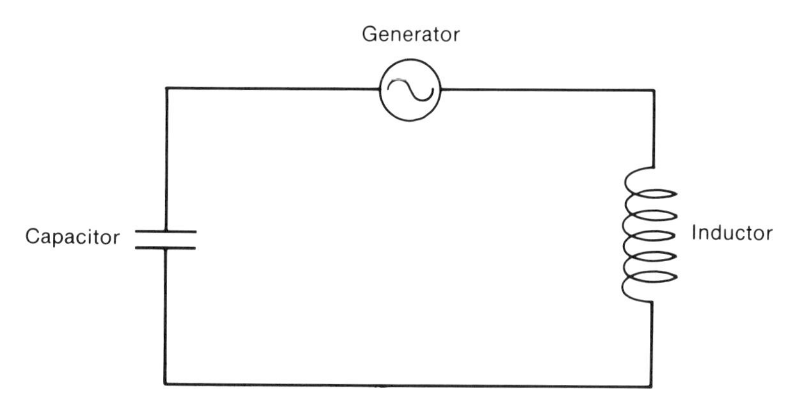

Fig. 3.2. A simple series resonant circuit (from P. H. Brace, Induction Heating Circuits and Frequency Generation, in *Induction Heating*, ASM, Metals Park, OH, 1946, p 36)

In contrast to the current-voltage relationship for a pure resistance, the peaks and valleys of an ac current through an inductor "lag" the corresponding peaks and valleys of the voltage drop across it by 90° (one-fourth of the ac cycle). The current through a capacitor "leads" the voltage by 90°. When an ac current passes through an inductor, a capacitor, and a resistor in *series*, the net opposition to the current flow or "impedance" is obtained by a vectorial sum of the voltages, $V_L = IX_L$, $V_C = IX_C$, and $V_R = IR$ (Fig. 3.3):

$$E = \sqrt{(|V_L| - |V_C|)^2 + V_R^2} = I\sqrt{(X_L - X_C)^2 + R^2}$$

where E is the applied electromotive force and I is the current flowing through the circuit. The impedance Z is defined as $\sqrt{(X_L - X_C)^2 + R^2}$. Thus, $E = IZ$, an equation which expresses the ac equivalent of Ohm's Law for resistors in dc circuits.

At resonance in a series circuit containing an inductor, a capacitor, and a resistor, the inductive and capacitive reactance are equal and the over-all circuit impedance Z appears as a pure resistance. As the resistive component of such a series circuit is decreased, the circulating current will be higher (Fig. 3.4). In addition, when the resistance is low, the current drops rapidly as the operating frequency varies from the resonant frequency.

The ratio of inductive reactance to the resistance of the circuit is sometimes called the quality factor, or $Q(Q = X_L/R = 2\pi fL/R)$. Circuits having higher values of Q therefore produce higher currents in the system. The values of Q for a resonant circuit typically range between 20 and 100. The Q factor also offers a comparison between the total energy in a tuned circuit and the energy dissipated by the resistance in a circuit. Because no energy is dissipated in a pure inductance or capacitance, keeping the resistance as low as possible will reduce the losses. In induction heating, therefore, because nearly all the resistance of a resonant circuit is in the induction coil, its resistance should be kept low, thus ensuring maximum efficiency.

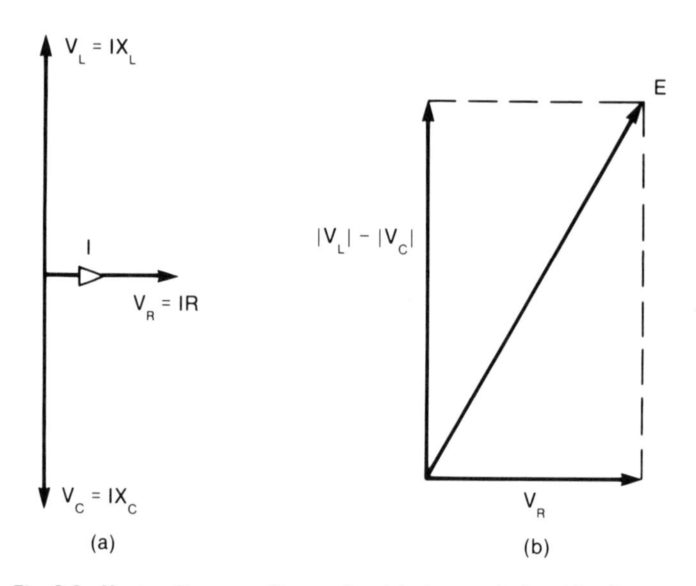

Fig. 3.3. Vector diagrams illustrating (a) phase relationships between current and voltage in the various components of an ac series LCR circuit and (b) relationship between the applied emf E and the voltage drops across the resistor (V_R), the inductor (V_L), and the capacitor (V_C) in such a circuit (from C. A. Tudbury, *Basics of Induction Heating*, Vol 2, John F. Rider, Inc., New York, 1960)

True, Reactive, and Apparent Power

The definitions of reactance and impedance allow several other important quantities in induction heating to be defined. These are the true, reactive, and apparent power. True power, P_T, is the power which is available to the coil and workpiece for actual heating. It is expressed in kilowatts (kW) and is equal to I^2R. Apparent power, P_A, is equal to I^2Z, or total volts times total amperes. To differentiate it from true power, apparent power is expressed in kilovolt-amperes (kVA). The third power quantity is the reactive power, $P_R = I^2(X_L - X_C)$; this is the power which is out of phase. It results in no heating, and its units are expressed as kilovars (kVAR). Because of the relationship among Z, R, and $(X_L - X_C)$, the various power quantities are themselves related through $P_A = \sqrt{P_R^2 + P_T^2}$.

The "power factor" for a given induction heating operation is defined as the ratio of true power to apparent power, P_T/P_A. In induction heating, it is desirable to have a power factor at the power source as close to unity as possible.

The power factor may also be defined as the cosine of the phase angle between the current and voltage in an electrical circuit. For a tuned series circuit containing an inductor, a capacitor, and a resistor, which appears to consist of a pure resistance, the current and voltage are in phase, $P_R = 0$, and the power factor is equal to unity. For both a pure inductance and a pure capacitance, the phase angle is equal to 90°, $P_T = 0$, and the power factor

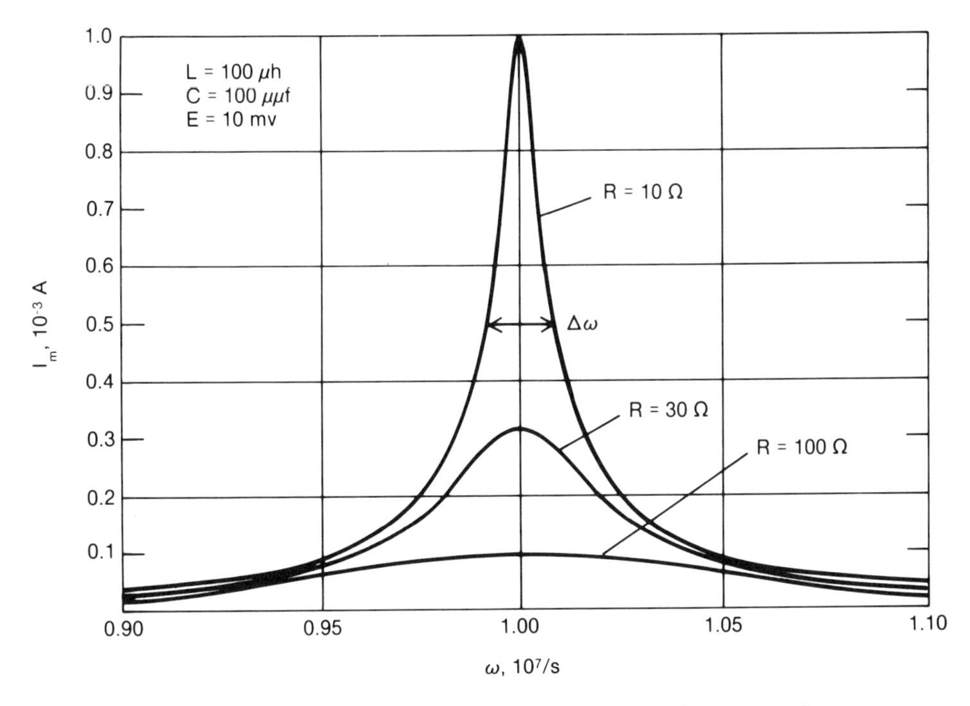

Fig. 3.4. Current amplitude as a function of frequency for an ac series LCR circuit whose components have the values indicated (from D. Halliday and R. Resnick, *Physics*, Wiley, New York, 1966)

viewed across these elements alone is equal to zero. Note also that low power factors correspond to high values of Q.

Parallel Resonant Circuits

The most common tank circuit used in induction heating is the *parallel* resonant circuit. The simplest type, as referred to previously, uses a capacitance in parallel with an inductance. The impedance of a parallel circuit at resonance is equal to $(2\pi fL)^2/R$ or $(2\pi fL)Q$. At resonance, Q will be high and thus impedance will also be high.

The parallel resonant tank (Fig. 3.5) is tuned to the frequency of oscillation or resonance by making the inductive reactance equal to the capacitive reactance at the desired frequency. As previously stated, when these are equal the currents will be equal and opposite in phase. Hence, they cancel each other in the circuit, and the line current I_1 is thus very small. The small line current results from the fact that the inductor has a small but finite resistance causing a slight phase-angle shift so that complete current cancellation cannot take place. On the other hand, the circulating current shown as I_2 is very large, depending on the applied voltage and the reactance of the capacitor at the resonant frequency ($= 1/2\pi\sqrt{LC}$, as in a resonant series LCR circuit). In a parallel resonant, or tank, circuit, therefore, the impedance is at a maximum

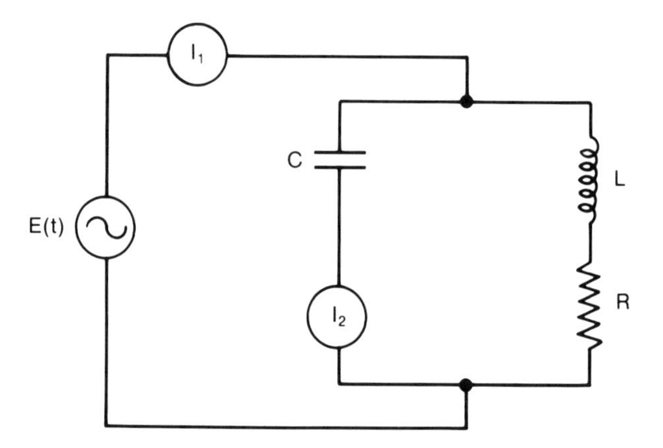

Fig. 3.5. A simple parallel resonant (tank) circuit consisting of a supply voltage E and a capacitor in parallel with an inductor and a resistor

across the L-C circuit, and the line current is very small. Also, the power factor as viewed across the voltage terminals is nearly unity.

Tuning Capacitors

Capacitor banks often are employed to tune induction heating circuits to obtain a resonant frequency close to the power-supply frequency.* Generally, such devices are of two types: oil-filled, multiple-tap units (Fig. 3.6) for operation at 10,000 Hz or lower; and ceramic or solid dielectric capacitors (Fig. 3.7) used in newer equipment operating at frequencies generally above 50 kHz.

Because of their use in tuning induction heating circuits, capacitors are usually rated in terms of the reactive power (kVAR) which they can provide in such applications, rather than in farads. Capacitors are generally tapped for step selection, and are connected in banks of multiple units (Fig. 3.8) for specific required kVAR ratings, at low and medium frequencies. At high frequencies, capacitors generally have fixed ratings, and total tank capacity is adjusted by connecting or removing units as required.

Increasing tank capacitance lowers the resonant frequency, and, with subsequent decreases in capacitive reactance and tank impedance, the current increases. Accordingly, the current-carrying capability of a capacitor must be increased as the frequency decreases. Capacitors for use at 10 kHz are rated as high as 2000 kVAR. It is essential that the proper ratings be used for the specific system. Reactive power is given by the formula $P_R = V_C^2/X_C = 2\pi f C V_C^2$, so the kVAR rating must be reduced for a specific capacitor by the

*Note that at resonance, $X_L = 2\pi fL = 1/2\pi fC = X_C$, thereby defining the resonant frequency $f_0 = 1/2\pi\sqrt{LC}$. If the resonant frequency f_0 is set as the power-supply frequency and L is determined by the inductance of the heating coil or an auxiliary tank coil, then tuning involves selection of the proper capacitance.

Fig. 3.6. Typical water-cooled capacitors used to tune low- to medium-frequency induction heating circuits (from P. H. Brace, Induction Heating Circuits and Frequency Generation, in *Induction Heating*, ASM, Metals Park, OH, 1946, p 36)

square of the voltage, or directly as the frequency, when it is used at a lower frequency. In view of the large kVAR input to capacitors, it is therefore important that they have high efficiencies. Accordingly, most capacitors have power factors less than 0.0003.

Fig. 3.7. Typical ceramic capacitors used to tune induction heating circuits (source: Lindberg Cycle-Dyne, Inc.)

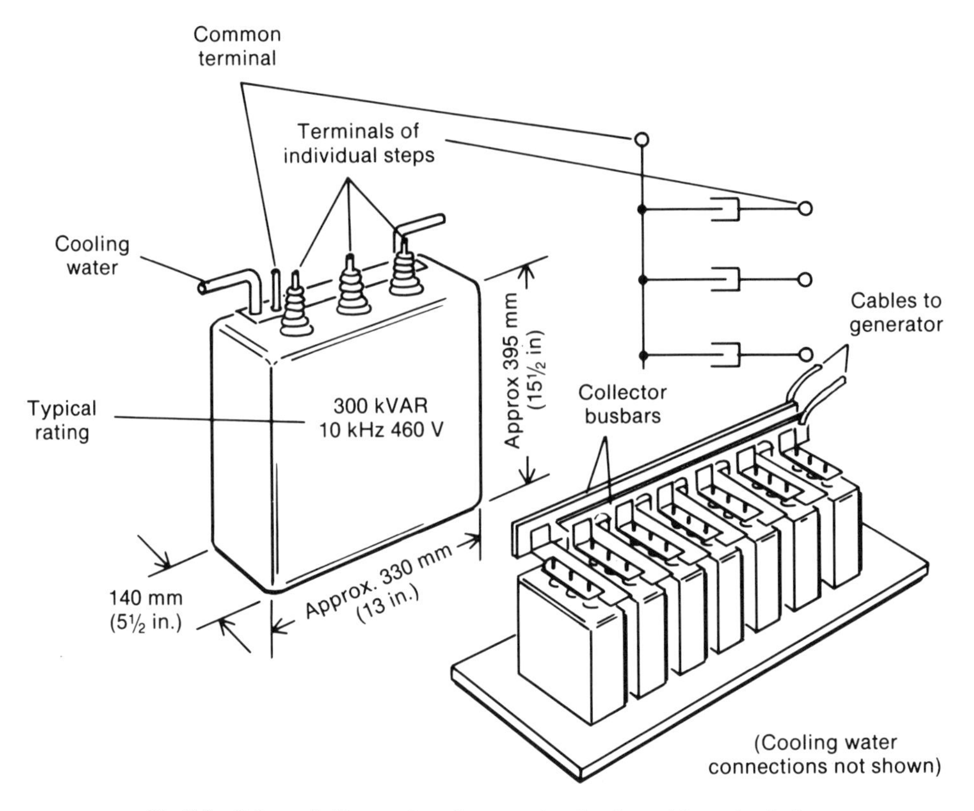

Fig. 3.8. Schematic illustration of a capacitor bank used in tuning induction heating circuits for low- and medium-frequency induction heating power supplies (from C. A. Tudbury, *Basics of Induction Heating*, Vol 2, John F. Rider, Inc., New York, 1960)

The power factor of the coil(s) is generally low, and the current lags in phase. Because the capacitor is used to correct the power factor to unity, it must be capable of drawing a leading current which is equal to the lagging component of the current drawn by the inductance. The capacitors operate at rated output voltage and the resultant reactive power (kVAR) of the capacitor must usually be several times the rated kW power of the supply.

The example shown in Fig. 3.9 will serve to illustrate the use of capacitors in tuning. Consider the case in which the coil-heating power requirement (i.e., true power) is 200 kW. Assume that 2000 kVA are being drawn from the power supply during the initial setup; thus the power factor is only 0.1 (lagging). To estimate the capacitance required to achieve a unity power factor, the reactive power is first estimated as $\sqrt{2000^2 - 200^2} = 1990$ kVAR. Thus, capacitors which provide 1990 kVAR of leading reactive power must be added to the tank circuit.

In power supplies such as motor-generators, it is frequently beneficial to have a slightly *leading* power factor—i.e., a capacitive reactance which slightly exceeds the inductive reactance. In the above example, to obtain a power factor

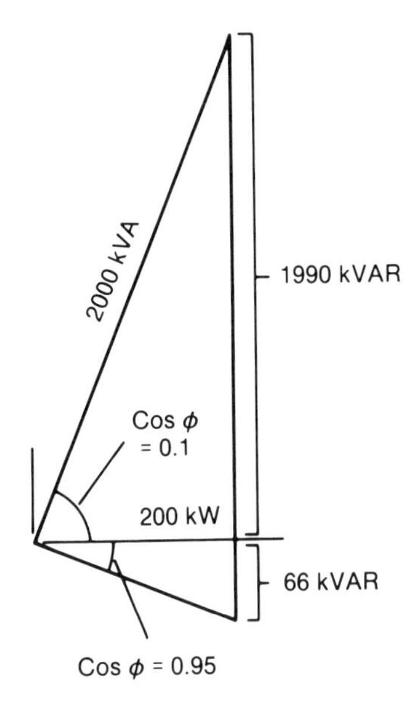

Fig. 3.9. Vector diagram illustrating power-factor correction for a fixed-frequency induction heating power supply

of 0.95 which is leading, an additional capacitance of 200 $\tan(\cos^{-1} 0.95) =$ 66 kVAR must be utilized. Thus, the total kVAR requirement would be $1990 + 66 = 2056$ kVA.

TRANSFORMERS AND IMPEDANCE MATCHING

Induction heating sources have rated current and voltage limits that cannot be exceeded without damage to the source. The ratio of the rated voltage to the rated current is the effective impedance of the source. To obtain the greatest transfer of energy from the source to the load, the impedances of the two should be as close to each other as possible. If they do not match, transformers are then employed.

A transformer is a device consisting of two coils (or windings) which have "mutual" inductance between them (Fig. 3.10). The *primary* winding is connected to the supply, and a voltage is induced in the *secondary* winding, which is separated from the primary by an iron core or air core. A transformer can be used to increase (step up) or decrease (step down) voltages.

The relationship between the voltages in the primary (E_P) and secondary (E_S) is determined by the ratio of the turns in each:

$$\frac{E_P}{E_S} = \frac{N_P}{N_S}$$

Here, N_P and N_S denote the turns in the primary and secondary, respectively. The current that flows in the secondary winding as a result of the induced

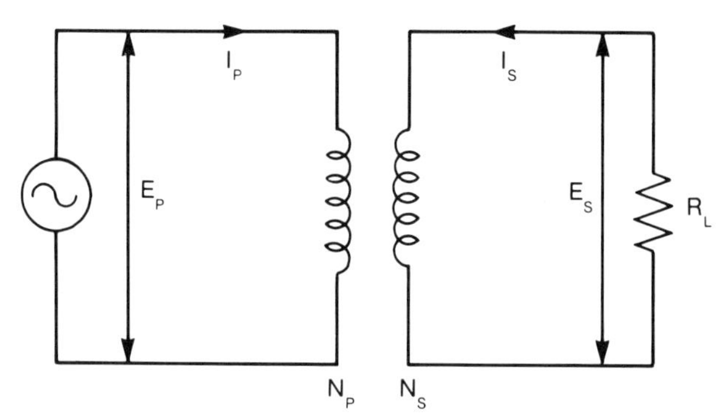

E_P is primary voltage (V); I_P is primary current (A); N_P is number of primary turns; I_S is secondary current (A); N_S is number of secondary turns; E_S is secondary voltage (V); R_L is load resistance (Ω).

Fig. 3.10. Electrical circuit illustrating the analogy between the transformer principle and induction heating

voltage must produce a flux which exactly equals the primary flux. Because the flux is proportional to the product of the number of turns and the currents in the windings, the relationship between the primary current I_P and the secondary current I_S is:

$$N_P I_P = N_S I_S$$

or

$$\frac{N_P}{N_S} = \frac{I_S}{I_P}$$

Hence, when the voltage is stepped up, the current is stepped down, and vice versa. It should be borne in mind, however, that these equations apply only to transformers with perfect coupling between the primary and secondary windings. In reality, coupling is not perfect, and the voltage and current ratios will each be slightly less than the turns ratio.

When a transformer is used for load-matching purposes, the impedance of the secondary is adjusted to match that of the electrical load, and the primary impedance matches that of the source. Usually, this involves reducing a high-voltage line supply to a lower-voltage, higher-current one. Because impedance is equal to E/I, the impedance ratio between the primary and secondary (Z_P/Z_S) can be calculated by means of the equation:

$$\frac{Z_P}{Z_S} = \frac{N_P^2}{N_S^2}$$

Thus, the impedance ratio is equal to the *square* of the turns ratio.

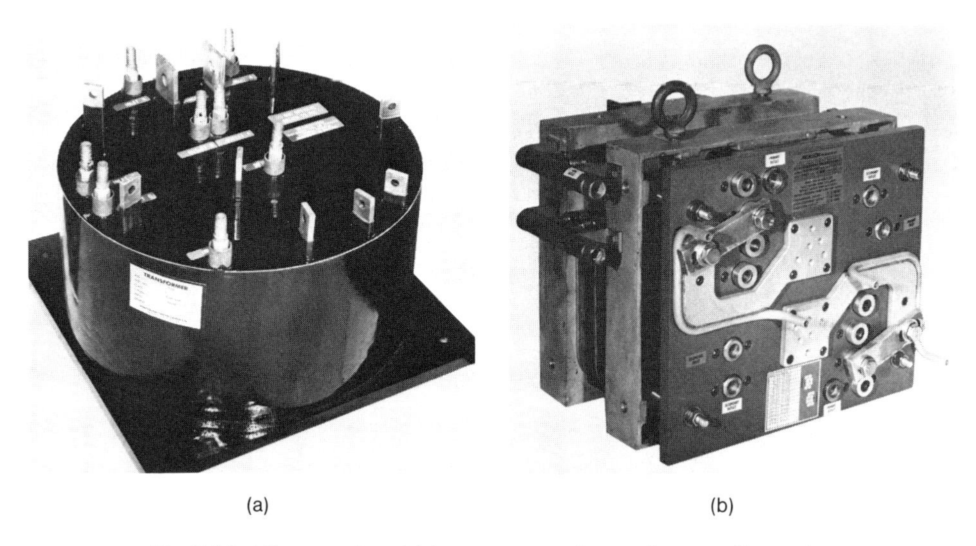

(a) (b)

Fig. 3.11. Photographs of (a) an autotransformer (source: Hunterdon Transformer Co.) and (b) rear view of an isolation transformer, showing adjustable taps (source: Jackson Products Co.)

When coil impedance is very high, a coil may operate directly across the output of the power supply (which is usually of high impedance). However, a matching transformer is frequently used. Matching transformers for low- and medium-frequency use generally have laminated iron cores. Where high-impedance work coils must be matched to the system and thus coil voltage is high, tapped autotransformers* (Fig. 3.11a) are utilized. These transformers have tapped secondaries so that they can be used to match coil impedance to supply impedance. Switches can be employed to change transformer primary taps to adjust the ratio where conditions dictate frequent adjustment (Fig. 3.12).

Where coil impedances are low, isolation transformers† are more commonly employed (Fig. 3.11b). In this case, ratio changes are generally made in the primary with bolt connections and an adjustable tap. The heavy current secondaries are water cooled to maintain low copper losses and increase efficiency.

---

*In an autotransformer, the coils are both magnetically and electrically connected. One section of the two windings is common to both the primary and secondary. Used when the desired output voltage is approximately 50% or more of the input voltage, this type of transformer offers low losses and less expensive construction. Thus, matching of high-impedance coils that require low transformer ratios most frequently relies on an auto-transformer.

†In an isolation transformer, the primary and secondary windings are coupled solely by the magnetic flux. The efficiency of energy transfer is related to the distance between the windings as determined by the voltage difference between the two (and hence the insulation required) and the efficiency of the transformer core material. Matching of a low-impedance induction coil, which requires a large transformer ratio, usually makes use of an isolation transformer.

Fig. 3.12. Tap-change switch used in adjusting voltage on the primary of a line-frequency induction heating system (source: American Induction Heating Corporation)

IMPEDANCE MATCHING AND TUNING FOR SPECIFIC TYPES OF POWER SUPPLIES

Fixed-Frequency Sources

Line-frequency power supplies, frequency multipliers, and motor-generators each produce output power at a single specific frequency. In order to obtain maximum power from the work coil, it is important to provide a tank circuit tuned to that specific frequency, whose impedance matches that of the power supply.

The schematic diagrams in Fig. 3.13 show the circuitry for both a high-impedance system using an autotransformer and a low-impedance tank circuit using an isolation transformer. Both systems have tapped transformer windings to match coil impedance to generator impedance and some means of connecting capacitance, as required, to tune the system to resonance at or near the operating frequency of the power supply. Because the capacitors are connected incrementally, it is not always possible to tune the circuit exactly to the output frequency of the generator, and thus, full output power is not always achieved.

Capacitor connections are generally made by connecting the terminals of individual capacitors to a common bus. This can be accomplished by use of bolted links, or large washers under the nut on the stud, where the stud is normally insulated from the bus. When changes must be made frequently, as in melting systems, a remote switching system is normally provided that operates individual contactors to connect the stud to the bus.

All fixed-frequency systems are set up with the workpiece in the coil,

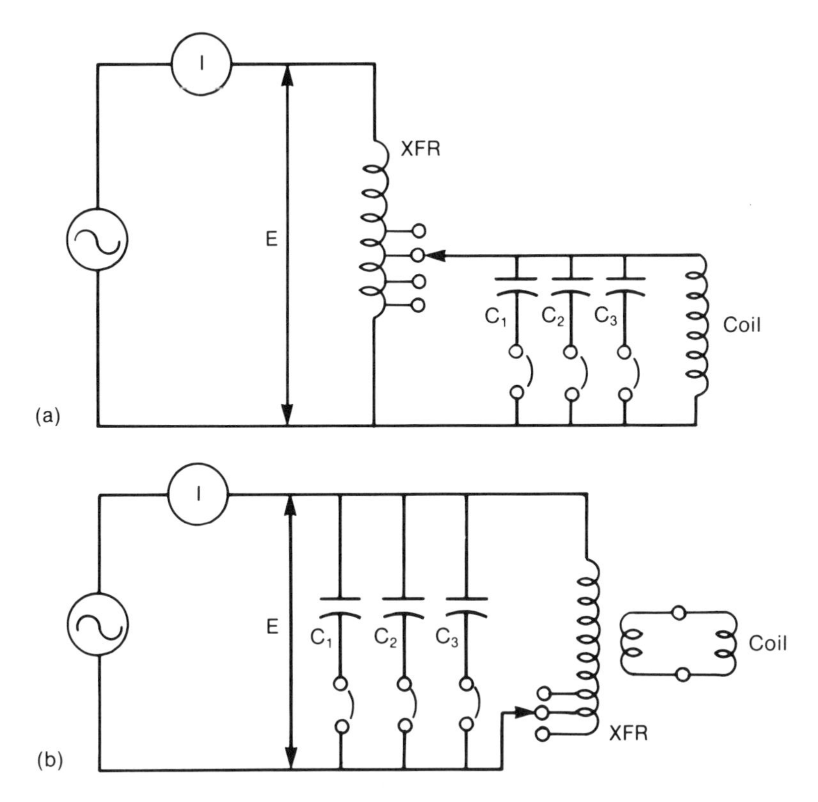

Fig. 3.13. Schematic circuit diagrams of low- to medium-frequency heat stations illustrating capacitor and transformer adjustments when using (a) an autotransformer (for high-impedance induction coils) and (b) an iosolation transformer (for low-impedance induction coils)

because the impedance would change dramatically if the part were not in place. Accordingly, balancing should be performed using short, low-power tests so that minimal heat is produced in the workpiece during setup and so that conditions for comparison therefore remain almost static.

Metering on fixed-frequency systems generally consists of a kilowatt meter, an ammeter, a voltmeter, and a kVAR or power-factor meter. These meters sometimes are read in percentages of their maximum ratings to simplify tuning. When the system is tuned properly, an increase in the output of the system will reflect increases in both current and voltage simultaneously. An increase in one and a decrease in the other are indicative of improper balance.

During the actual tuning operation, such as that for a motor-generator system, the field is slowly increased so that the meters begin to indicate. If the power factor is lagging, capacitance must be added to the tank circuit.* If the

*Note that this assumes a *parallel* resonant tank circuit. If the power factor is lagging, the lagging component of the current I_2 (in Fig. 3.5) is too great, and the leading component, or the current through the capacitor, must be increased. Because this leading component is equal to E/X_c, X_c must be *decreased*. Therefore, the capacitance must be increased.

power factor leads, capacitance must be removed. The power must be off while changes are made in the capacitor bank.

When the power factor has been corrected as close to unity as possible, power should be increased until either the voltage or the current reads the maximum permissible. Readings of all meters should then be taken. If, as machine power is increased, the voltage reaches a maximum before the current does, the primary-to-secondary ratio of the transformer is too high. The ratio must be adjusted, or turns must be removed from the work coil based on the equation:

Required turns = present turns × amperes obtained at rated voltage/rated amperage

Should rated current be achieved before rated voltage, the primary-to-secondary ratio is too low. The ratio must be adjusted, or turns must be added to the work coil as follows:

Required turns = present turns × rated voltage/voltage obtained at rated amperage

When the work coil is modified, the first operation must be a readjustment of the power factor (i.e., the circuit containing the coil must be retuned), because the inductive components of the tank have been changed. Then, once again, comparative readings should be taken.

If little or no indication occurs during a start-up procedure, it is indicative of an excessive amount of capacitance. All of the capacitor taps should be disconnected and slowly added again until a meaningful set of readings is achieved.

During setup of a motor-generator, the tank circuit is usually tuned to provide a slightly *leading* power factor. This helps to compensate for the reactance of the armature windings of such systems and enables rated voltage to be drawn from the power supply.* Operating with a slightly leading power factor also prevents exceeding the field-current rating when operating at rated voltage and power.

When magnetic materials go through the Curie temperature, power-factor changes may be considerable as the material loses its permeability (and coil inductance drops). The system may indicate a considerably lagging current at the completion of this change. It may then be necessary to connect additional capacitors to the tank at this point via contactors, possibly operated by a timer if the operation is highly repetitive.

In tuning of a fixed-frequency system during progressive heating operations, readings and adjustments should be made with the part moving through the coil at its normal speed.

---

*C. A. Tudbury, *Basics of Induction Heating*, Vol 2, John F. Rider, Inc., New York, 1960.

Variable-Frequency (Solid-State) Power Sources

Static (solid-state) power supplies generate frequencies which are determined by the electrical characteristics of the tank-circuit components. Accordingly, as the inductance of the coil changes due to heating of the part or loss of permeability above the Curie temperature, the supply frequency shifts accordingly. Tuning and load matching of such solid-state systems varies with the particular type of system, which is generally one of two kinds: (1) a constant-current or load-resonant inverter, or (2) a constant-voltage or swept-frequency inverter.

Constant-Current (Load-Resonant) Inverter. Load-resonant inverters are essentially variable-frequency power supplies. They have no resonant circuit, but, by means of a feedback system derived from the tank circuit, they key the power supply to operate at the tank-circuit frequency. This in effect allows them to vary in frequency as the resonant tank does when the coil/part impedance shifts during heating. Accordingly, once the tank circuit is tuned to within the allowable frequency range of the generator, the generator frequency shifts to match the tank frequency. It is therefore said to be "load resonant."

Parallel-tuned circuits, as used in load-resonant generators, are not self-starting inasmuch as they depend on an increase in load voltage to cut off the SCR's to start oscillation. Therefore, a piggyback or "pony" inverter is usually included in the generator design to initiate oscillation. To ensure that the system will start, it is standard procedure to connect all the available capacitance to the system when beginning, providing the lowest possible frequency, which is then within the range of the piggyback oscillation.

Power may be efficiently drawn from the system anywhere in the frequency range of the generator. The tank capacitors can be adjusted to a specific frequency in the equipment's range, which may be determined by the desired case depth in surface hardening applications, for example. It must be kept in mind that the capacitor combination utilized must have a high enough kVAR rating to match the maximum output at the delivered frequency. The frequency is generally displayed on the frequency meter of the generator or control station.

Because the frequency is dependent on the tank-resonant conditions, changes in the coil inductance will shift the frequency of operation. The system will always operate at the peak of the resonant curve, but no change in power output will occur as a magnetic material passes through the Curie temperature. The only change will be a shift in the operating frequency. If it is desired to select a specific frequency, depending on the application, capacitor taps can be adjusted to meet this requirement.

Because the system always operates at the resonant frequency of the tank circuit, its power factor is usually close to unity, and therefore impedance matching by means of a transformer-tap change is the only routine adjustment that need be made to secure optimum power output. This procedure is the same as that outlined for tuning of fixed-frequency motor-generator systems.

Constant-Voltage (Swept-Frequency) Inverter. The constant-voltage inverter also changes its frequency relative to the resonant frequency of the tank. However, it fires from a local oscillator within a fixed frequency range that is shifted by the resonant frequency of the tank circuit. Once the tank is tuned within the range of the local oscillator, it too will provide a constant output as the resonant tank frequency shifts, especially when the temperature passes through the Curie point during heating of magnetic steels. Nevertheless, it is necessary first to tune the system so that the tank frequency is within the frequency band of the local oscillator. This may be done through what is essentially the same procedure utilized for the load-resonant inverter. It must be assumed, using the frequency meter, that with all capacitance connected the total will bring the tank frequency within range. All remaining tuning steps would then be followed in turn.

There are conditions where the tank-resonant frequency falls outside the local oscillator range; under these conditions, tuning will be more difficult. It then must be determined, by measurement or calculation, at what frequency the tank operates, and capacitance must be added or removed to bring the frequency within the permissible limits of the oscillator.

Radio-Frequency (Vacuum-Tube) Power Supplies

High-frequency, vacuum-tube power supplies also make use of a tank circuit for the transfer of energy to the workpiece (Fig. 3.14). In this case, however, the circuit generally contains a tank coil in addition to the induction heating coil. As a workpiece is heated and its reflected resistance increases, the tank current I_T tends to decrease. However, the vacuum tube supplies an additional current I_P to make up for the losses and thus to keep the tank current constant. The plate ammeter of the radio-frequency (RF) supply reads this current I_P.

Under a "no-load" condition (no part in the work coil), the plate current I_{P_1} represents the minimum current supplied to the tank to replace the losses dissipated by the resistance R_P associated with the unloaded tank circuit. When the load is placed in the work coil, a "full-load" plate-current reading I_{P_2} indicates the new plate current required to supply not only the losses in R_P but also the additional losses of the load represented by R_L. Accordingly, the losses in the load are directly related to the difference between the two readings, $I_L = I_{P_2} - I_{P_1}$. Obviously, the difference between the full-load reading and the no-load reading represents the load current required by coupling of energy into the load or workpiece. Therefore, that combination of transformer or load-coil tap adjustments that produces the greatest differential between readings represents the best impedance match between the work coil/part and the generator tank.

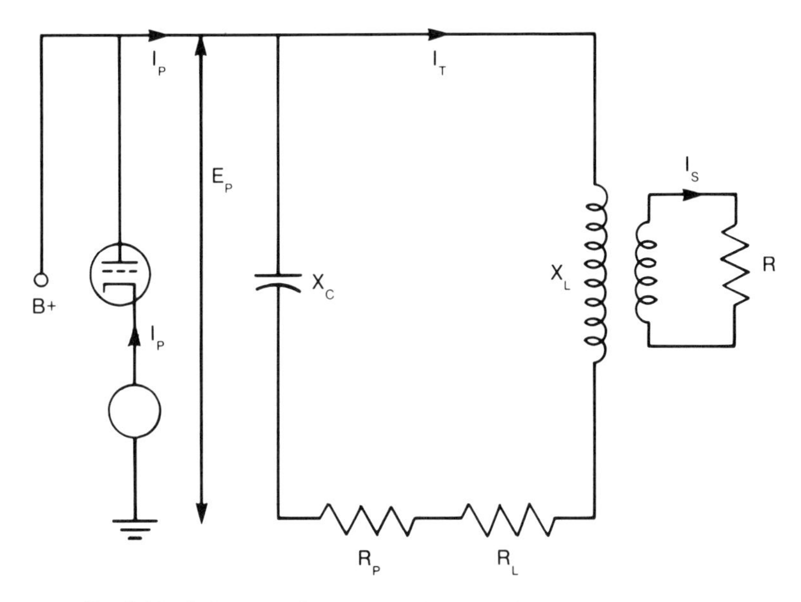

Fig. 3.14. Schematic illustration of a radio-frequency tank circuit

In practice, no-load and full-load readings are taken at each tank tap. The tap with the *greatest difference* between no load and full load is the most appropriate connection for the system. Once selected for a coil/part combination, it will always be the same. The same approach is also used in selection of the appropriate primary-to-secondary ratio if an RF power output transformer is used to connect the coil to the power supply.

Because the difference between no load and full load is proportional to the power into the part, it is important to try to maintain as low a no-load reading as possible. Generally the no-load reading should be at most 30 to 40% of the maximum allowable plate current. However, the physical constraints of the coil generally dictate this condition. When a high no-load reading occurs, the number of coil turns can be increased or the transformer turns ratio can be decreased.

In performing the tuning procedure it is best to use reduced power. However, once the power level is set, it should not be changed during this test. Further, the grid adjustment for these conditions should be maintained in accordance with machine specifications. Once tuning has been accomplished, the system should be run at full power to determine actual heating-rate and plate-current readings.

When certain loads such as small copper parts are being heated, it is often found that there is very little difference between the no-load and full-load readings. This is due to the fact that the resistivity of the material is low. It may also be physically impractical to add more turns to the coil. In these cases, it is sometimes practical to utilize a poor impedance match so as to increase the current in the work coil. Under these conditions, the no-load current is extremely high and will result in excess power being dissipated in both the tube and tank circuit. This loss, however, if within power-supply limitations, may be negligible in cost, compared with the increase in power to the workpiece and the resultant increase in heating rate.

For a specific part, if a small no-load/full-load differential occurs, and the part geometry permits, the number of coil turns may be increased to increase the heating rate. For example, if the maximum permissible plate current is 4 A and the no-load and full-load currents for a three-turn coil are 1 and 2 A, respectively, the part will draw 1 A. In this instance, approximately six turns can be added to the coil without exceeding the generator capacity.* It should be noted that if these additional turns are within the same surface area heated by the original coil, the power density will increase and the surface will heat at a higher rate, producing a greater temperature differential from the surface to the core of the part. If however, the additional turns are used to heat a larger area, the power density will stay the same but the throughput in terms of material per unit time will increase. In this same manner, the additional turns could be used in separate coils to heat multiples of the original part simultaneously, also increasing throughput. At some point, however, the impedance of the coil becomes so large that it cannot match the output of the power supply, and the heating rate decreases. Further, unless the system is matched by means of an output transformer, the voltage across the coil must be increased as the coil impedance increases. This can create a hazard to the operator and arcing from the coil to the part as well.

When the equipment is running at full power, a condition may develop in which the machine overloads or trips out due to excessive plate current. This indicates that the part is drawing power in excess of the power-supply rating. The coil-to-part coupling may then have to be decreased (greater distance from coil to part), or the number of coil turns may have to be decreased. Further, some modification may be made to the lead structure to correct this condition.

It is also important to note that when magnetic materials are heated above the Curie point, the power from radio-frequency supplies may decrease significantly because of radical changes in coil/part impedance. As with the motor-generator system, some technique may occasionally be necessary to retune the system so as to deliver higher power to the now nonmagnetic load. One such technique is construction of a series loading-coil arrangement (Fig. 3.15). The coil is closely coupled to the work so that it draws full power above the Curie temperature. This would ordinarily overload the system at temperatures below 760 °C (1400 °F). The loading coil, in series with the work coil, presents a high impedance to the generator, at the low tempera-

*This is based on the fact that each increment of three turns will increase the load current by 1 A.

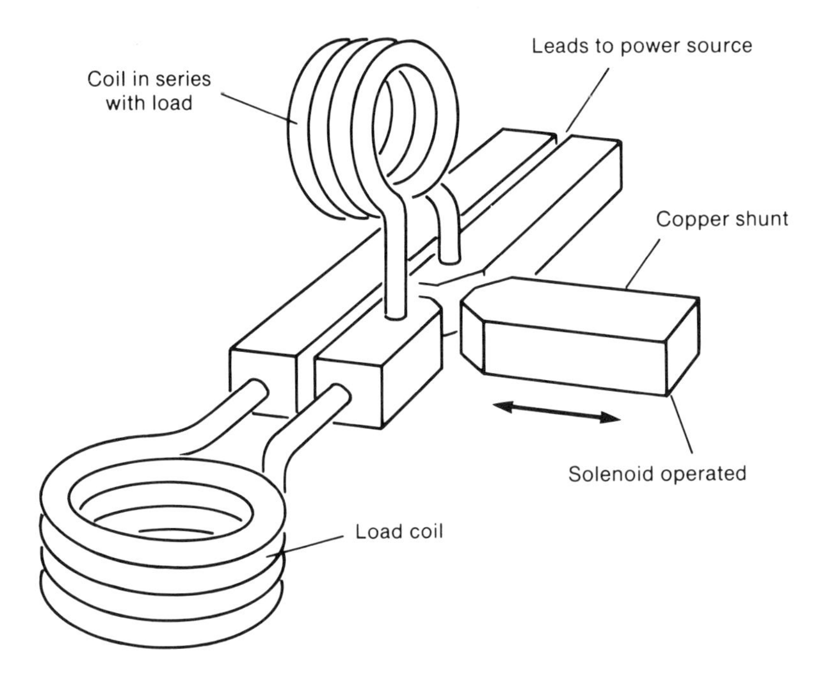

Fig. 3.15. Use of an auxiliary inductance ("series loading coil") to assist in tank-circuit tuning in heating of steel workpieces below, through, and above the Curie temperature (from F. W. Curtis, *High Frequency Induction Heating*, McGraw-Hill, New York, 1950)

ture, so that full power is supplied to the coil. Above the Curie temperature, the shorting bar eliminates the loading coil. It is important that the power be off during the brief period when the solenoid is activated to short the loading coil.

Chapter 4

Induction Heating Power Supplies

Besides the induction coil and workpiece, the induction generator (source of ac power) is probably the most important component of an over-all induction heating system. Such equipment is typically rated in terms of its frequency and maximum output power (in kilowatts). The discussion in this chapter addresses the selection of power supplies in terms of these two factors as well as the operational features of different types of sources. Auxiliary equipment, such as the cooling and control systems that are needed for all induction generators, is described in Chapters 5 and 7.

There are essentially six different types of power supplies for induction heating applications, each designed to supply ac power within a given frequency range. These six types are line-frequency supplies, frequency multipliers, motor-generators, solid-state (static) inverters, spark-gap converters, and radio-frequency (RF) power supplies. At the present time, the spark-gap oscillator is no longer in use, and the solid-state power supply has all but replaced the motor-generator. The ranges of frequencies and power ratings presently available are summarized in Fig. 4.1.

FREQUENCY AND POWER SELECTION CRITERIA

As discussed in Chapter 2, the frequency required for efficient induction heating is determined by the material properties (i.e., resistivity and relative magnetic permeability), the workpiece cross-sectional size and shape, and the need to maintain adequate skin effect. The interrelationship of these factors is described through the equation for reference depth and the definition of a "critical frequency" below which efficiency drops rapidly. In some cases, frequency requirements are also impacted by application requirements such as case depth in surface hardening of steels. In all situations, however, because equipment cost per kilowatt increases with frequency, a power supply of the

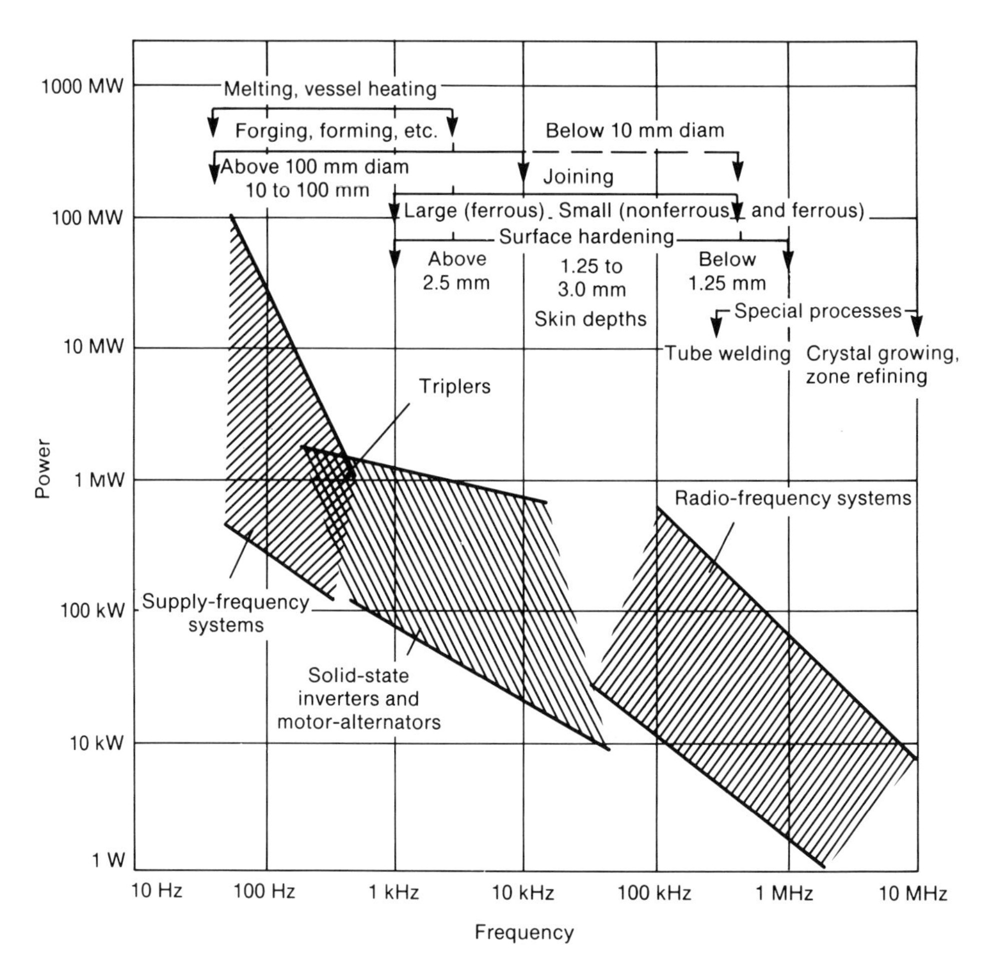

Fig. 4.1. Ranges of power ratings and frequencies for induction generators (from J. Davies and P. Simpson, Induction Heating Handbook, McGraw-Hill, Ltd., London, 1979)

lowest possible frequency that will accomplish the job is usually the best selection from an economic viewpoint. Typical frequency selections for a number of specific applications are summarized in Chapter 6.

Power requirements for induction heating are also dependent on the specific application. In through-heating applications, the power needed is generally based on the amount of material that is processed per unit time, the peak temperature, and the material's heat capacity at this temperature. Power specification for other operations, such as surface hardening of steel, is not as simple because of the effects of starting material condition and heat conduction to the unhardened core. Nevertheless, guidelines in such situations are available (Chapter 6).

When guidelines for power requirements, such as those in Chapters 2 and 6, are given, it should be emphasized that these often refer to the power actually transferred into the workpiece and not necessarily to the power drawn from the power supply. Power losses are of three forms:

- Coupling losses between the coil and the workpiece. As discussed in Chapter 2, the percentage of the power supplied to the coil that is transferred to the workpiece is a function of the resistivity and permeability of the workpiece material, the coil geometry, and the distance between the workpiece and the coil.
- Power losses between the output terminals of the power supply and the coil. These losses are associated with improper tuning of, and I^2R losses in, the appropriate tank circuit, imperfect impedance matching between the power supply and the workpiece/induction coil, and transmission losses between the power supply and the induction coil.
- Power losses within the power supply due to conversion of line frequency to higher-frequency ac.

It should also be emphasized that the method of rating power level for induction power supplies varies. For example, motor-generators are rated in kilowatts (rated current times rated voltage times power factor) at the *output terminals* of the generator. Because solid-state power supplies were initially direct replacements for motor-generators, manufacturers have continued this rating practice. Therefore, power output and efficiency ratings of solid-state inverters relate to power supplied at the output terminals. In terms of over-all efficiency and power delivered to the load, this neglects losses in transmission lines, losses in heat stations (due to improper tuning, for example), and the inefficiency of power transfer from the coil to the workpiece. Thus, the common practice of rating solid-state power-supply efficiencies at 90 to 95% refers solely to the power-supply conversion efficiency and not to the "line-to-load" energy transfer.

In radio-frequency systems, by contrast, IEEE Standard No. 54 defines the test used to determine power output. This standard describes the use of a calorimeter method for determining power into a load (Fig. 4.2). In practice, a water-cooled steel load is inserted into an induction work coil until the meters on the power supply indicate that it is operating at maximum rated voltage and current. The temperature of the water into and out of the load is monitored as well as the flow of water through the load. By this means, the power put into the load is readily calculated. The output power of the RF supply is then taken to be at this value, and the efficiency is computed as a "line-to-load" (over-all) efficiency. Figure 4.3 shows a typical data sheet used for rating RF power supplies. Because of the different manners in which power ratings are given for low-to-medium-frequency versus RF generators, equivalent ratings for the two types of systems are usually obtained by multiplying the quoted output power for low-to-medium-frequency units by a factor of approximately 0.65 to 0.70.

It is difficult to generalize on over-all efficiencies of different types of power

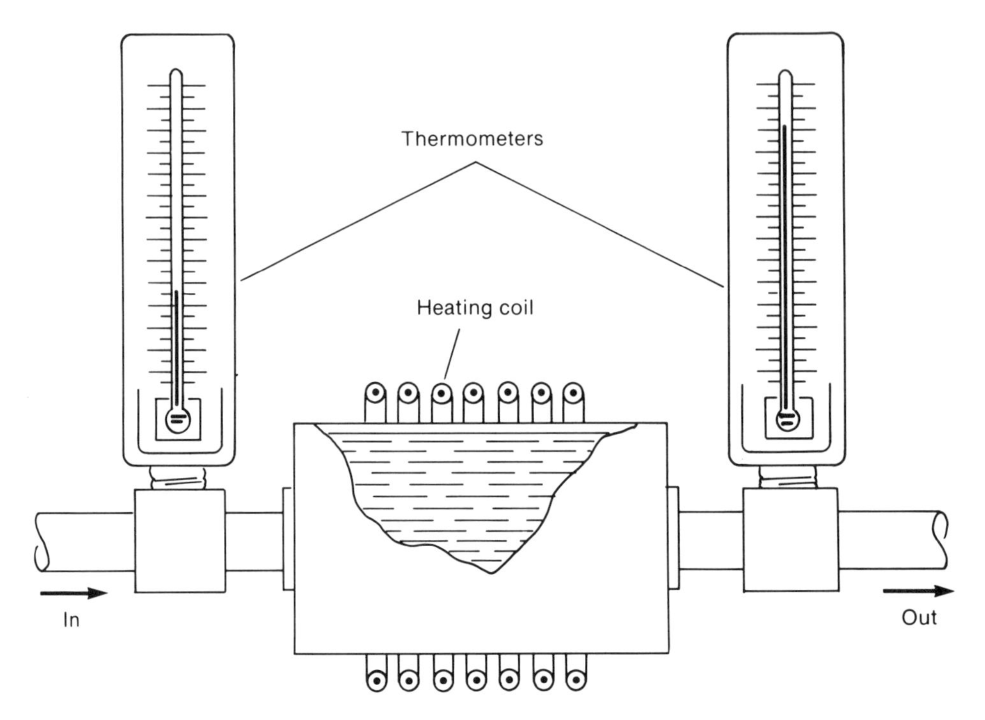

Fig. 4.2. Calorimeter used to determine the power output of a radio-frequency (RF) power supply by measuring water flow and temperature change (from F. W. Curtis, *High Frequency Induction Heating*, McGraw-Hill, New York, 1950)

supplies, let alone those of the same generic class, because many factors vary from part to part and from system to system. However, it is safe to assume that line-to-load efficiencies for low- and medium-frequency solid-state systems generally run between 60 and 65% and that those for RF (vacuum-tube) generators typically range from 50 to 55%.

With this as an introduction, the design and characteristics of each of the various types of power supplies are discussed next.

TYPES OF POWER SUPPLIES

Line-Frequency Induction Heating

When cross sections are large and through heating is desired, line-frequency systems often provide an efficient, low-cost method of utilizing an induction-based process. This is particularly true for large-tonnage applications.

Line-frequency induction heaters can be either single-phase or balanced three-phase systems, generally with low-voltage work coils that operate from the secondary of an isolation transformer. Depending on power requirements, systems of this type can be operated with standard line voltages of 220 or 440. When high powers are required, transformer primaries should be rated

GENERATOR HEAT RUN DATA

Function:	STATION 1 HR 1	HR 2	HR 3	HR 4		STATION 2 HR 1	HR 2	HR 3	HR 4	
Plate Voltage										KV DC
Plate Current										AMPS DC
Grid Current										AMPS DC
Filament Volts										V AC
Load Inlet Temperature										°F
Load Outlet Temperature										°F
Load Flow										GPM
KW Output										KW
Frequency										HZ
Enclosure Temperature										°F
Generator Inlet Flow										GPM
Generator Flow Temp In										°F
Generator Flow Temp Out										°F
Temperature Valve Setting										°F
"Load Reference Mark"										
Line Voltage										
Line Current										

GENERATOR "NO LOAD" DATA

Function										
Plate Voltage										KV DC
Plate Current										AMPS DC
Grid Current										AMPS DC
Frequency										HZ
Line Voltage										V AC
Line Current										AMPS AC

TESTED BY: ____________________ DATE: __________

Fig. 4.3. Test data sheet for recording performance of an RF power supply with a calorimeter (source: Ameritherm, Inc.)

accordingly. When coil impedances are high, the coil can be connected directly across the power lines.

The basic line-frequency system (Fig. 4.4) consists of a tapped transformer primary with the secondary wound to provide approximately the voltage required at the coil. The primary taps, operated by means of manual rotary switches (Fig. 4.5), permit adjustment of work-coil voltage to the specific load requirements. Power-factor-correction capacitors are placed across the primary windings of the transformer. Because currents in the system are large, it is desirable to keep the transformer and capacitor bank as close to the coil as possible. Accordingly, the heat station is generally mounted directly adjacent to the coil. When large-capacity systems are used, power-factor capacitors are sometimes mounted directly beneath the work area (Fig. 4.6) in order to reduce floor-space requirements.

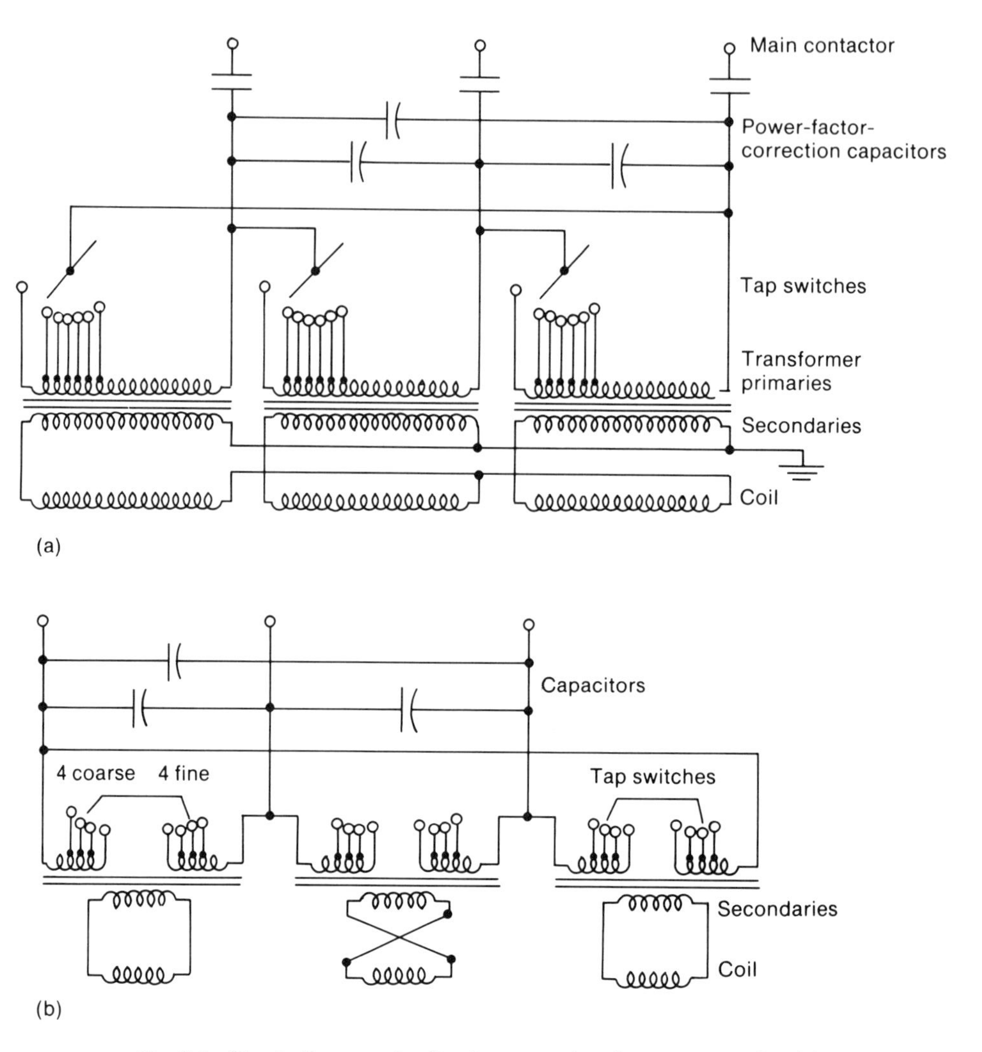

Fig. 4.4. Circuit diagrams for line-frequency heating systems using (a) single primary tap switching (high coil impedance) and (b) coarse/fine tap control (low coil impedance) (from J. Davies and P. Simpson, *Induction Heating Handbook*, McGraw-Hill, Ltd., London, 1979)

Temperature control of line-frequency systems is effected by turning the main power contactors on and off in response to a temperature signal derived from a thermocouple or infrared pyrometer. Since the heated mass is generally large and response time normally slow in these types of systems, this load-control technique is usually quite sufficient. When line input voltages exceed 3 kV, vacuum contactors can be used in the current primary (Fig. 4.7), or SCR thyristors can be used to provide stepless and/or on-off control.

Several examples of actual applications will serve to show the versatility of line-frequency induction heating techniques. A first example involves a small single-phase induction heater used for expanding bearings during a shrink-

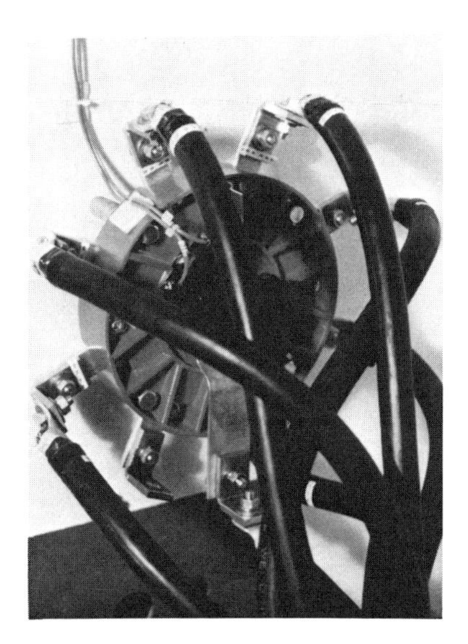

Fig. 4.5. Tap switch used in adjusting voltage on the primary of a line-frequency induction heating system (source: American Induction Heating Corp.)

fitting operation. Having a high-input impedance, and operating at 110 or 220 V, this application requires no power-factor correction. Essentially composed of a primary winding and a laminated core structure (Fig. 4.8), the system utilizes the bearing as a shorted secondary. Several laminated blocks are employed to control the flux path through the core and the center of the bearing (Fig. 4.9). In addition, a set of movable blocks is supplied with the equipment to enable it to fit a range of bearings. When the bearing size matches the transformer core opening, the bearing itself may be used to close the gap in the core.

On a somewhat larger power level, the same technique is used for heating ferrous or nonferrous rings, such as locomotive tires (Fig. 4.10). Typically a 75-kVA, 440-V, single-phase unit is used to heat titanium rings varying from 31 to 152 cm (12 to 60 in.) in diameter by up to 10 cm (4 in.) in cross section to 760 °C (1400 °F) in 1 to 5 min. for a stretch-forming application. Because of the requirement for opening and closing of the transformer core, this type of system is often referred to as a "split-yoke" design.

In another nonferrous application, line-frequency heaters are used to heat aluminum, copper, or brass billets prior to extrusion. The billets are carried on motor-driven rolls (Fig. 4.11) and are heated in three-phase, line-frequency coils. They are then passed into the extrusion press. On the system shown in Fig. 4.11, aluminum billets 28 to 48 cm (11 to 19 in.) in diameter and 102 to 229 cm (40 to 90 in.) long are extruded at temperatures ranging from 155 to 480 °C (310 to 900 °F). Three individual sets of coils are used to provide the necessary throughput, each heating one billet at a time and feeding the press

Fig. 4.6. Capacitor bank for a 3-MW line-frequency power supply; capacitors are located in a chamber beneath the induction coils (source: American Induction Heating Corp.)

on an alternate basis to provide the required throughput of approximately 38 tons per hour.

Line-frequency heaters, due to their efficiency and simplicity, have also been incorporated in systems for heating larger, ferrous forging billets and slabs for rolling-mill operations. The line-frequency system shown in Fig. 4.12 heats 15-by-15-by-114-cm (6-by-6-by-45-in.) round-cornered square preforms for forging diesel-engine crankshafts and is rated at 18 tons per hour at 6000 kW. Line-frequency systems are also utilized in dual-frequency heating lines for forging. The depth of penetration for steel below the Curie temperature is considerably less than that above it. Therefore, the minimum diameter of steel billet that can be heated efficiently at 60 Hz is approximately 6 cm (2.4 in.) at temperatures below 705 °C (1300 °F).

Probably the largest induction heating installation, line-frequency or otherwise, is the slab-heating facility illustrated in Fig. 4.13. It is rated at a total of 210 MW. Six heating lines each with three heating stations are used to preheat 600 tons of steel per hour transported from a storage yard or directly from a continuous caster. Slabs as large as 30 cm (12 in.) thick by 152 cm (60 in.) wide by 793 cm (312 in.) long are heated to 1260 °C (2300 °F) for rolling. The slabs, which may weigh as much as 30 tons each, are heated in coils which encompass their longest dimension (Fig. 4.14) rather than being wound across their width. Efficiency is optimized in this manner, because flux-field cancellation across the long dimension is less than if the coil were wound conventionally.

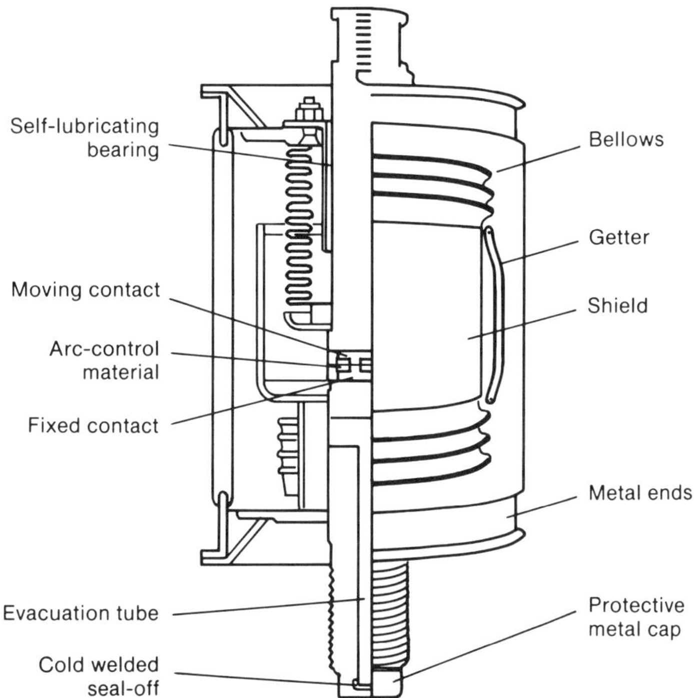

Fig. 4.7. Vacuum and air break contactors used for line-frequency induction heating (from J. Davies and P. Simpson, *Induction Heating Handbook*, McGraw-Hill, Ltd., London, 1979)

Frequency Multipliers

Frequency multipliers are used to obtain multiples of the line supply, most often 180 and 540 Hz. Like line-supply installations, they are used primarily for large heating and melting applications.

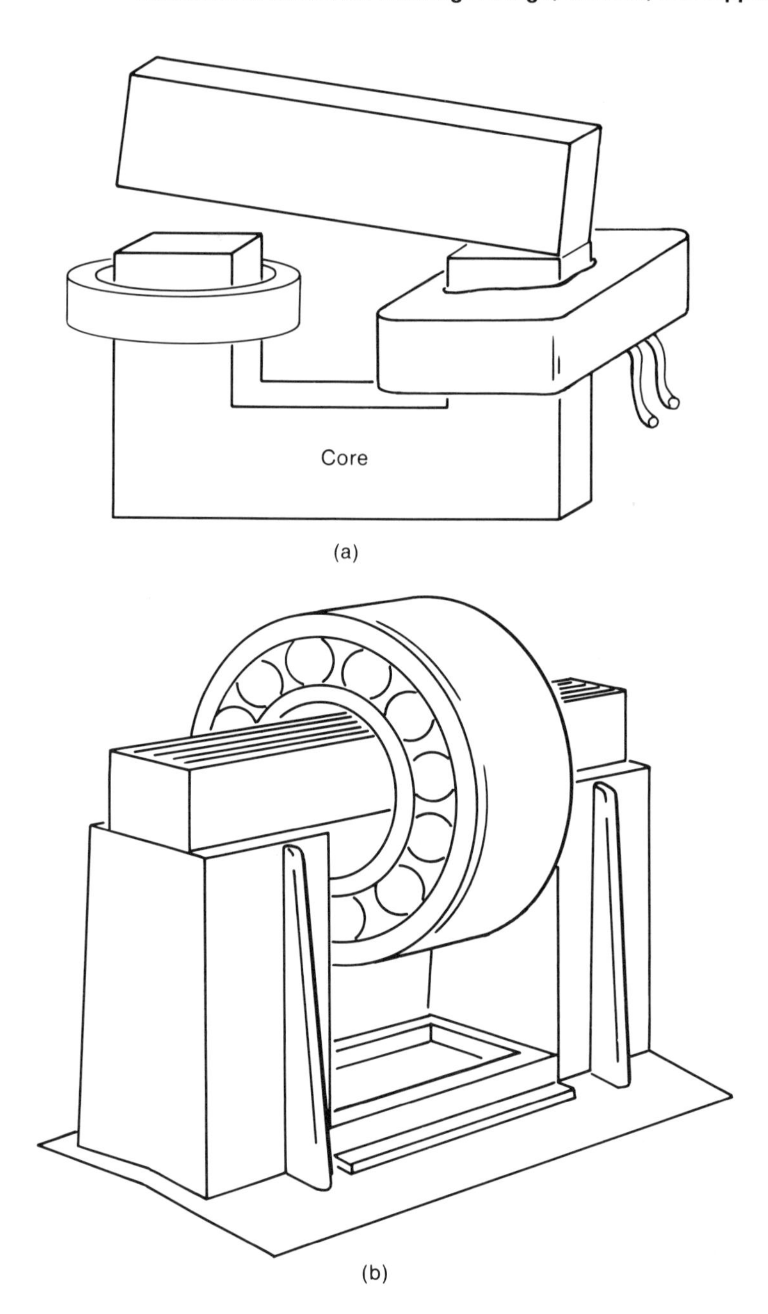

Fig. 4.8. Laminated core and primary winding for an induction unit used to expand bearings: (a) "split-yoke" system and (b) production bearing heater (from C. A. Tudbury, *Basics of Induction Heating*, Vol 2, John F. Rider, Inc., New York, 1960; Read Electric Sales and Supply)

The 180-Hz supply, also known as a tripler because it is derived directly from 60-Hz line current, is composed of three saturable reactors, each of which is connected between one leg of a Y-connected secondary and a single-phase load. The voltages from the secondaries are 120° apart. Passing them

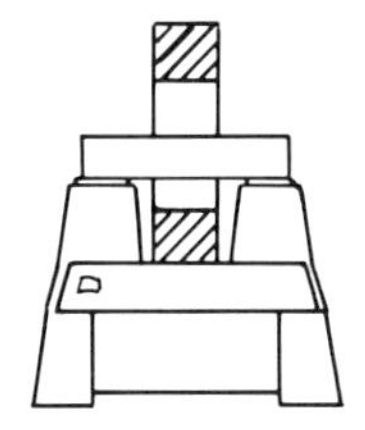

Large bearings are placed on the angled support surface, with the largest possible yoke through the bearing bore and resting on the contact pillars. Maximum bearing width, 110 mm.

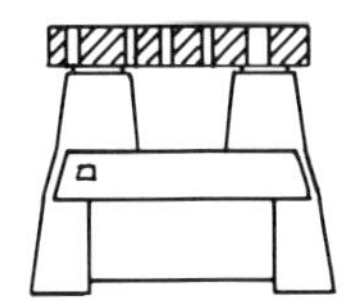

Workpieces equal in size to, or broader than, the distance between the contact pillars may be placed directly on the pillars with no yoke required.

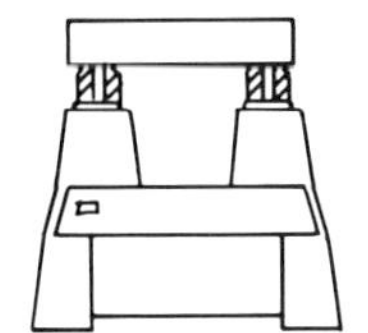

Small workpieces (with or without bore) can also be placed on top of the pillars. If only one workpiece has to be heated, place a yoke of near equal size on the other pillar and connect to largest yoke.

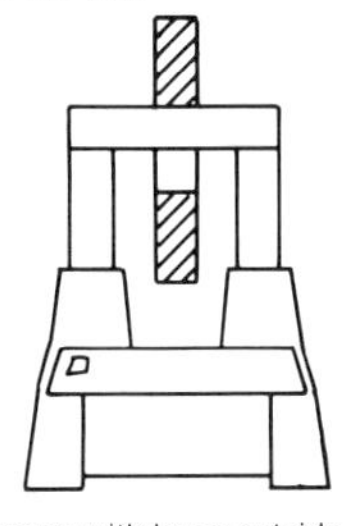

For flanges with large outside diameters, the contact pillars can be raised using a pair of yokes.

Should the workpiece have a very small bore or no bore at all (e.g., pulleys, plates, etc.) adaptor yokes can be used.

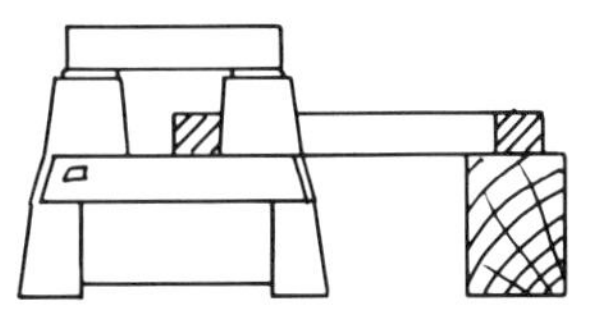

Large rings can also be heated in a horizontal position using a rest with the largest yoke in position.

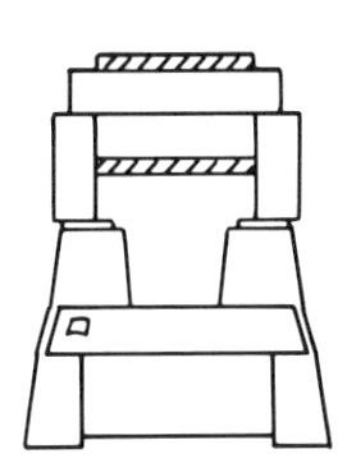

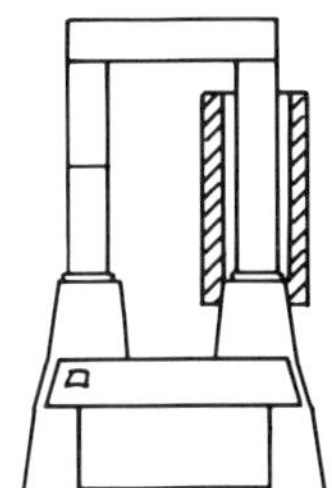

For long pieces, the working space can be increased by using a pair of support yokes, or the space can be raised by combining suitable yokes in a vertical position.

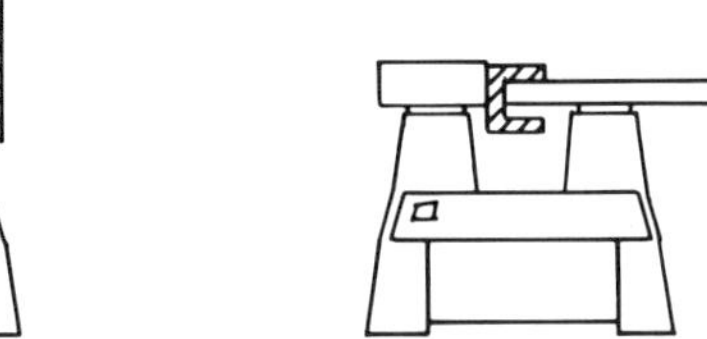

Combinations of various yokes may be used for workpieces with stepped bores, such as couplings, pulleys, etc.

Fig. 4.9. Adjustment of the laminated core structure for heating bearings of different sizes; note that the flux path must form a complete loop (source: SKF Industries, Inc.)

through a single-phase output, therefore, produces three pulses in the same time period as that in which a normal 60-Hz system would produce one waveform. Accordingly, the frequency is essentially tripled. The 180-Hz system is similar to the 60-Hz heater in that the circuit containing the load must be tuned to the frequency of the power supply—namely, 180 Hz.

Systems may be cascaded to produce other multiples of the line frequency, such as 540 Hz.

Motor-Generators

A motor-generator is a rotary-driven system composed of a motor coupled to a generator. The motor and the generator can be individual units with a

Fig. 4.10. "Split-yoke" system used to expand locomotive tires for shrink fitting (source: Cheltenham Induction Heating, Ltd.)

mechanical coupling, or they can be constructed on a single shaft. Both horizontal (Fig. 4.15) and vertical (Fig. 4.16) types have been used. A cooling system is used to remove the heat dissipated by the current in the windings and that developed in the structure and motor laminations. Early motor-generator systems utilized air cooling, but water-cooled heat exchangers built within the unit shell have since become standard.

Frequencies generated using motor-generators are determined by rotational speed and the number of poles in the generator. Although nominally rated at 1, 3, and 10 kHz, standard frequencies are actually 960, 2880, and 9600 Hz.

The generator consists of a toothed rotor rotating inside a stator. In the

Fig. 4.11. Induction installation rated at 6.6 MW that is capable of heating 38 tons of aluminum billets per hour to a temperature of 480 °C (900 °F) (source: American Induction Heating Corp.)

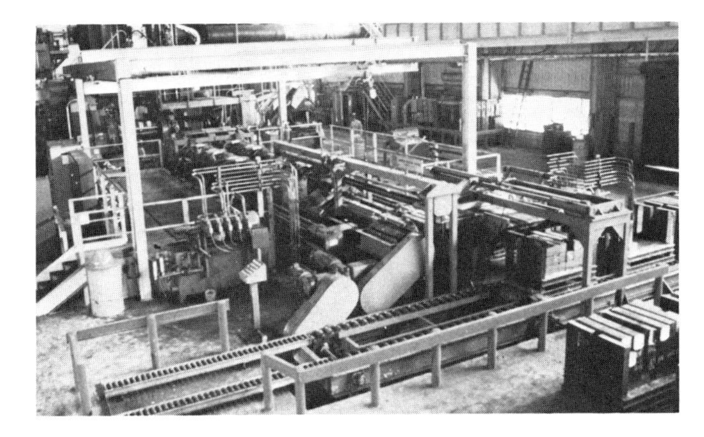

Fig. 4.12. Line-frequency induction system used to heat 15-by-15-by-115-cm (6-by-6-by-45-in.) round-cornered square steel billets for forging of diesel-engine crankshafts (source: American Induction Heating Corp.)

Fig. 4.13. One stage of the McLouth steel induction slab heating facility, rated at a total of 210 MW, used to preheat slabs as large as 30 cm by 152 cm by 8 m (12 in. by 60 in. by 26 ft) to the rolling temperature of 1260 °C (2300 °F) (source: Ajax Magnethermic Corp.)

slots of the stator, there are two windings.* One, called the field winding, is connected to an external source of direct current and forms a magnetic field around the rotor. When the rotor rotates, alternating currents are induced in the second, or output, winding. These currents result from changes in the magnetic flux in the slots (where the stator windings are located) as the rotor sweeps past them.

---

*Formally, when both sets of windings (field and output) are on the stator, the equipment is called an inductor-alternator. Some motor-generators have the output windings on the rotor, rather than on the stator.

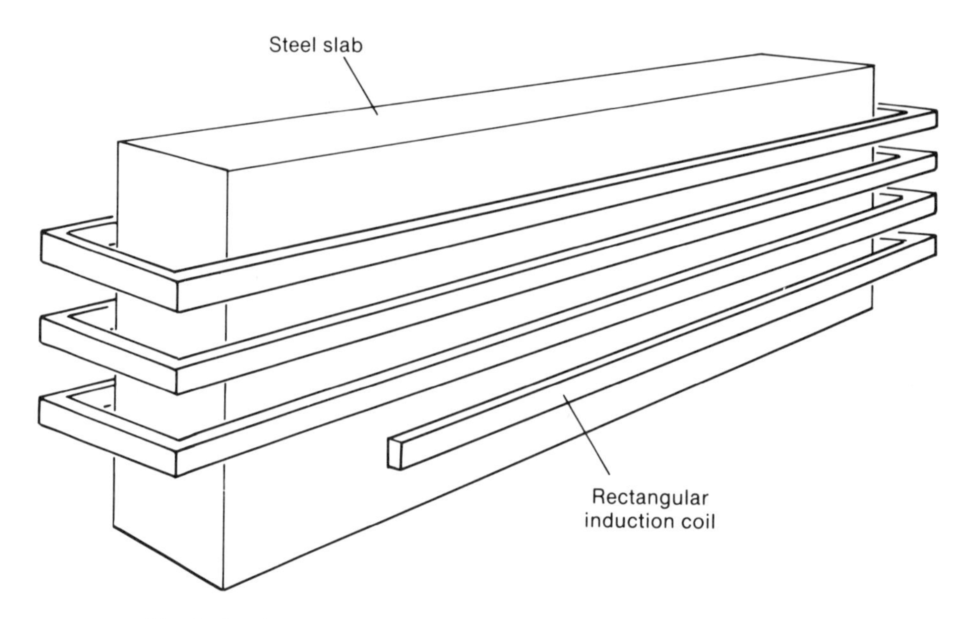

Fig. 4.14. "Ross" coil used in the heating of steel slabs (source: Ajax Magnethermic Corp.)

The frequency of the generator, f (in hertz), is determined by the number of pairs of poles (P_2) on the rotor and the speed n of the rotor in revolutions per minute, through the equation $f = P_2n/60$. A maximum frequency of 10 kHz is considered to be the practical limit for rotary equipment of this sort.

The output voltage from a motor-generator is a single-phase voltage generally of 200, 400, or 800 V. This output voltage is adjusted by controlling the voltage to the field winding. Because the output voltage has a tendency to decrease as the load increases, automatic voltage regulators are often

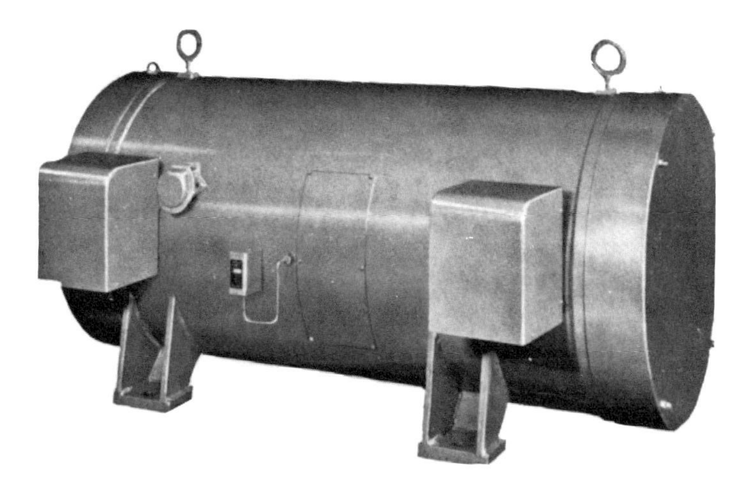

Fig. 4.15. Fully enclosed, horizontal motor-generator set that is water cooled (source: Westinghouse Electric Corp.)

Fig. 4.16. Vertical motor-generator set (source: TOCCO, Inc.)

employed to control the field. The regulators change the field voltage as the output voltage varies, to maintain a stable output.

Motor-generators have high inertia and thus require considerable power to attain operating speeds. Accordingly, when starting, excessive motor inrush currents required are brought to lower levels through the use of a reduced-voltage starter. Moreover, the power needed to spin the heavy rotor is the same, whether at light load or full power. For this reason, system efficiency is low when the generator is loaded lightly (i.e., when only a fraction of its rated power is drawn).

The motor-generator is a fixed-frequency supply. Hence, the tank circuit containing the load must be tuned. Because they operate at a fixed frequency, motor-generators of the same frequency can be connected in parallel to increase system capacity. Generally, to do this, the units must be exactly the same in terms of voltage, poles, etc., to prevent one generator from "hogging" the load or possibly even feeding another.

Since the motor-generator is a fixed-frequency device, it also operates similarly to the motor-generator that supplies voltage to a standard power line. Thus, several separate and independent induction heating work stations can be connected to a single motor-generator and operated on a simultaneous basis. They can be connected or disconnected from the supply at will. However, the stations must be tuned as closely as possible to the generator frequency, and provision must be made to prevent the total connected load at any time from exceeding the generator rating.

A typical heat station operated from a motor-generator (Fig. 4.17) usually comprises power-factor-correction capacitors and an autotransformer or isolation transformer for impedance matching (Fig. 4.18). When an isolation transformer is used, the capacitors are connected in parallel with the primary. The transformer primary is tapped to provide a range of voltage adjustments. When an autotransformer is used (with high-impedance coils) the capacitors are generally connected in parallel with the work coil. With very-high-impedance coils, it is sometimes possible to match the generator output impedance without the use of an intermediate transformer. In either case, provision is made to add or subtract capacitance for proper frequency matching. It should be noted, though, that since power-factor correction makes use of discrete values of capacitance, it is not generally possible to tune the system *exactly* to the resonant frequency of the motor-generator. Because of this, it is generally advantageous to adjust the power factor initially so that it is in

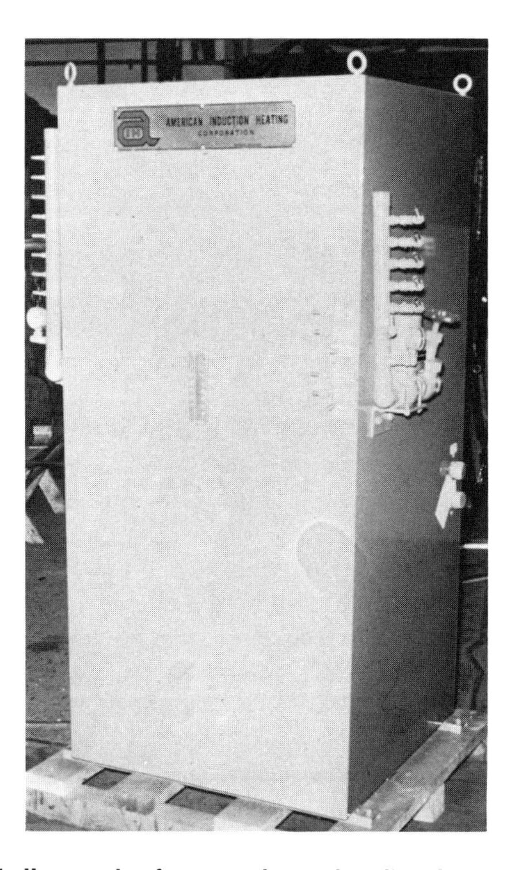

Fig. 4.17. Heat station for use at low and medium frequencies with an impedance-matching transformer (source: American Induction Heating Corp.)

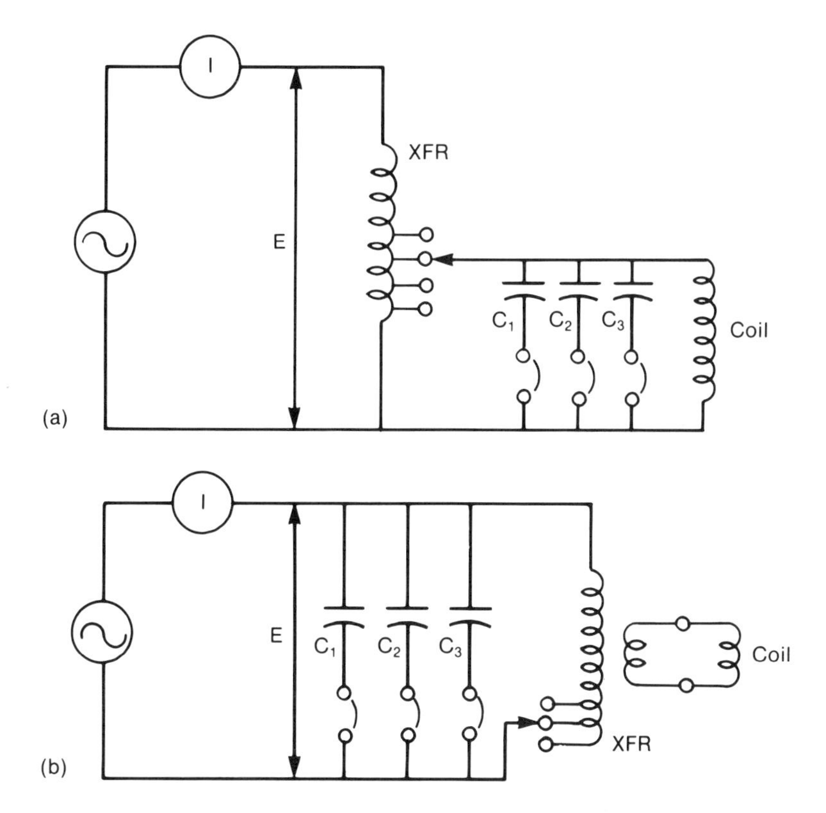

Fig. 4.18. Schematic circuit diagrams of a low-to-medium-frequency heat station illustrating capacitor and impedance-matching adjustments when using (a) an autotransformer for high-impedance induction coils and (b) an isolation transformer for low-impedance induction coils

a lagging or leading condition. Such a condition can partly or wholly compensate for decreases (more common) or increases (less common) in tank inductance during heating. However, operating for any period in the lagging condition can cause excessive field current with associated field relay tripping.

Solid-State Inverters

With the development of high-power silicon-controlled rectifiers (SCR's) having rapid turn on/turn off times, it became possible to provide an equivalent to the motor-generator system. While eliminating a number of basic disadvantages, not the least of which is inefficient operation, the solid-state inverter provides some new and unique operational capabilities.

The first development in this field was that of the swept- (or variable-) frequency inverter. With this system (Fig. 4.19a), line voltage is converted to direct current and then is applied by means of a capacitive voltage divider to an SCR (dc-to-ac) inverter circuit. The system utilizes internal commutation to turn off the SCR's in the inverter circuit. A feedback circuit from the resonant tank, together with power- and voltage-control signals, is fed to a local

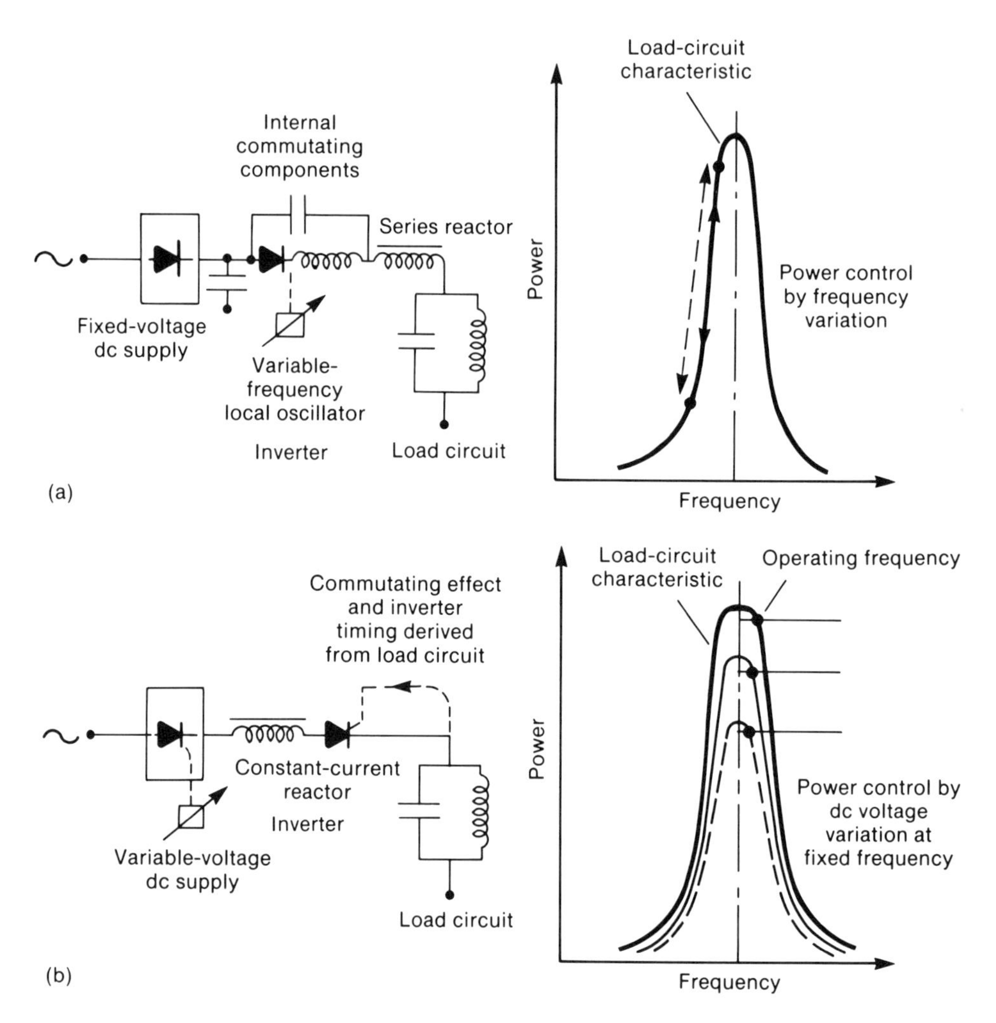

Fig. 4.19. Circuit schematics for (a) a swept-frequency inverter and (b) a load-resonant or constant-current solid-state inverter (from J. Davies and P. Simpson, *Induction Heating Handbook*, McGraw-Hill, Ltd., London, 1979)

oscillator whose output signal then controls the inverter firing rate. The system then operates at a frequency along the straight portion of the tank's resonant-frequency curve, but not at the resonant frequency. By offsetting the firing rate of the oscillator from the resonant frequency of the tank, the unit can be made to operate at any point along this curve. It can be seen on this curve that a slight shift in frequency will cause a large change in power output. Therefore, controlling the local oscillator firing rate within these limits provides a wide range of power output control. It should be noted, however, that as the Q value of the circuit changes, the shape of the resonant-frequency curve changes. For high-Q circuits, the slope of the curve is sharp, and a small change in frequency provides a wide range of power control. For circuits having low Q values, the necessary frequency shift to achieve a comparable power adjustment becomes wider and therefore sometimes impossible to achieve.

The other common solid-state inverter, known as the load-resonant or current-fed inverter (Fig. 4.19b), also utilizes a dc power source. However, this inverter is provided with an SCR rectifier for voltage control. The SCR inverter uses no local oscillator but derives its commutating effect directly from the resonant tank circuit. Thus, it is load-resonant. In actuality, for reasons based on the operational characteristics of the SCR's, the system is usually operated with a slightly leading power factor and thus an operating frequency slightly higher than the load-resonant frequency. Because power output control of the load-resonant inverter is done by means of SCR's in the dc supply, the Q value of the load is not a factor in this respect. The system continues to operate at a frequency slightly above that of the tank circuit.

In both types of solid-state inverters, the system frequency in essence is tuned to the load, whereas in motor-generators the load is tuned to the fixed frequency of the generator. It should be remembered that with motor-generators, as the work-coil inductance changes during heating, especially for ferrous loads passing through the Curie temperature, the tank circuit must be returned to optimize power transfer. On the other hand, with a load-resonant solid-state system, the resonant frequency of the tank changes, even when going through the Curie point, as determined by the load. The swept-frequency inverter, if set for a constant power output, also will shift its frequency as the load is heated. Thus, either solid-state system compensates automatically for changes in the load when passing through the Curie temperature, unlike the motor-generator.

Both types of solid-state systems can operate with either a series or parallel tank circuit. The parallel circuit is most commonly used because it permits remote mounting of the tank circuit (heat station) with minimum losses in the transmission lines. The transmission voltage used is limited to the inverter voltage unless a transformer is applied at the power source. With a series circuit, the coil voltage is a function of the Q value of the circuit and can be quite high, necessitating the use of high-voltage capacitors for tuning. Further, since the inductance of the transmission line becomes part of the tank inductance, the distance from the power supply to the work coil is limited for reasons of efficiency.

Figure 4.20 compares the relative over-all efficiencies of motor-generators, swept-frequency inverters, and load-resonant inverters. The low efficiency of the motor-generator is due to the high inertia of the rotor, windage losses, and similar problems produced by the mechanical system. These must be overcome, even for the lightest loads, since the generator must always operate at rated speed regardless of output power. The low efficiency of the swept-frequency inverter at light loads is due to the fact that it is operating at a low point on the resonant-frequency curve.

Because solid-state inverters were initially designed to replace motor-gen-

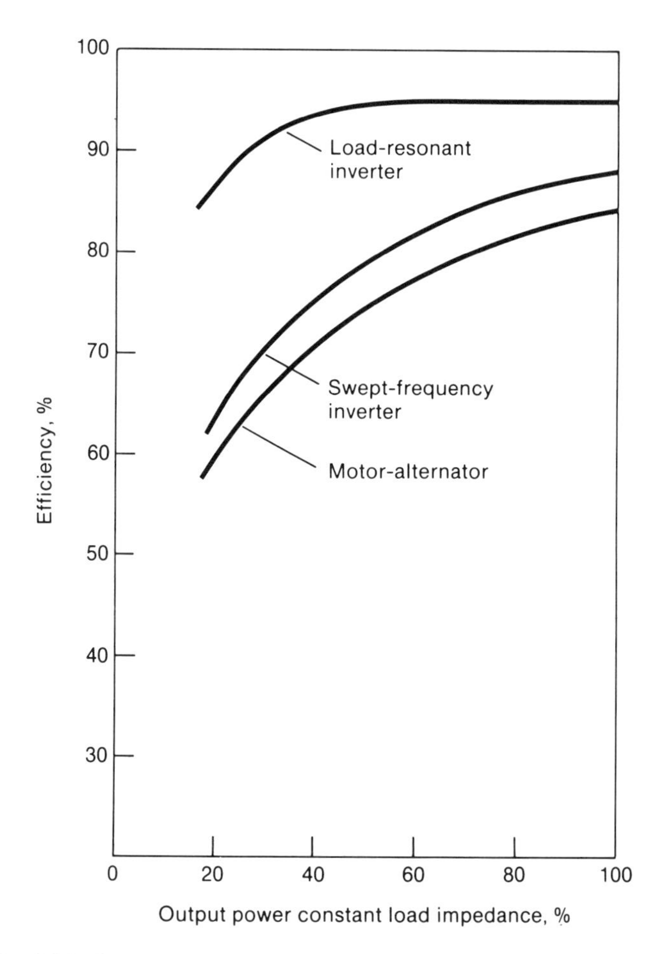

Fig. 4.20. Comparison of the over-all (system) efficiency of a typical motor-generator, a swept-frequency inverter, and a load-resonant (constant-current) inverter (from J. Davies and P. Simpson, *Induction Heating Handbook*, McGraw-Hill, Ltd., London, 1979)

erators, the power ratings of these systems were based on standard motor-generator rating techniques—that is, power is rated at the terminals of the generator, neglecting the losses in the heat station, coil, etc. For this same reason, power-supply frequencies were also chosen to match those of available motor-generator sets to reduce confusion. However, solid-state power-supply frequencies are really not limited because the generators are either load-resonant or controlled by a local oscillator. Thus, a frequency can be selected which is optimum for the application. In effect, all solid-state generators are actually variable-frequency inverters. For a specific piece of equipment, apparent power at the terminals will vary with frequency. Therefore, power supplies generally are designed to operate within a specific frequency *range*. Further, heat-station capacitors must have sufficient kilovar capacity to handle the frequency range used, and magnetic components (transformers and reactors) must be rated for operation at these frequencies as well.

Spark-Gap Converters

Although no longer produced commercially, the spark-gap converter was one of the earliest means of providing high-frequency ac power in the radio-frequency region. It was used primarily in metal melting, in particular for precious metals.

In its simplest form (Fig. 4.21), the spark-gap converter includes a step-up transformer (T_1) which increases the line voltage to a higher level, which is subsequently impressed across the spark gap. The spark gap is in parallel with a capacitor (C_1) and the primary of transformer T_2. This is an output transformer; its secondary is connected to the work coil. In operation, as the primary voltage rises in each half-swing of the 60-Hz line current, the voltage across the gap also rises until it exceeds the breakdown level. This voltage is also equal at that instant to the voltage impressed across the series circuit formed by the capacitor and the output transformer primary. The capacitor charges to this peak voltage. When the voltage exceeds the breakdown level of the gap, the low-resistance arc that has formed discharges the capacitor through the gap, passing an oscillating current through the combination of the capacitor and the transformer primary. As the capacitor discharges, the voltage across the gap decreases until the arc can no longer be sustained. On the reverse half-cycle of the 60-Hz line current, the cycle is repeated with the current in the circuit going in the reverse direction.

In later developments of the system, vacuum tubes were used to increase stability, and the gaps were contained in a hydrogen atmosphere to reduce the need for constant maintenance.

Frequencies generated by this system are generally in the range of 80 to 200 kHz. However, many frequency components are produced simultaneously so that the output is not purely sinusoidal. Efficiency of these systems is rather low—on the order 15 to 50% line-to-load.

Radio-Frequency (Vacuum-Tube) Power Supplies

Operational Characteristics. The operation of a radio-frequency (RF) power supply is best understood with reference to the parallel resonant tank circuit con-

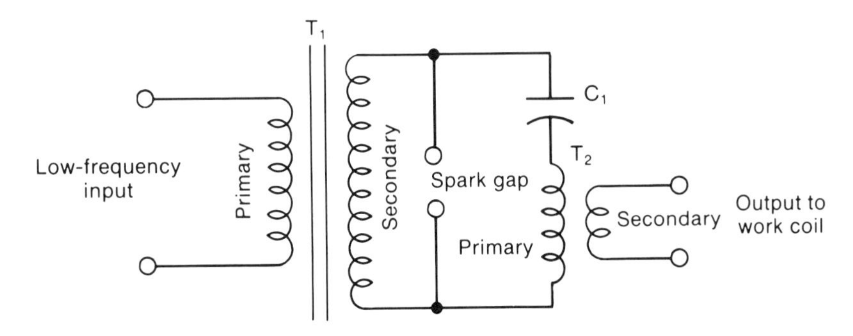

Fig. 4.21. Simplified circuit diagram for a spark-gap induction generator (from *Induction Heating*, American Society for Metals, Metals Park, OH, 1946)

taining a capacitor C, inductor L, and resistor R (Fig. 4.22). When switch S is closed momentarily, the capacitor is charged to the voltage E_{DC} of the supply (neglecting inductance and lead losses). When the switch is opened, the capacitor discharges through the circuit composed of L, C, and R. After one cycle, the capacitor is again charged with its original polarity, but at a somewhat lower voltage dependent on the losses in the circuit. If the switch were again closed momentarily, the capacitor would again be charged to the supply voltage E_{DC}. In this manner, the losses in the resonant circuit are replaced every cycle by the operation of the switch, and thus an oscillation of constant amplitude is maintained in the tank circuit. If there were no resistive losses in the tank circuit, it would continue to "ring" indefinitely. However, because the load is resistive, some means must be provided to replace these losses.

In an RF power supply, the switch and source of electromotive force (emf), E_{DC}, is replaced by a vacuum tube having three components—i.e., a triode. The grid of the tube, by varying its voltage with respect to the cathode/filament, controls the flow of electrons to the plate. The tube operates in a class "C" condition (Fig. 4.23) in which the plate current flows through the tube in short pulses. This results in high power and high efficiency.

The power to the system is generally obtained by rectifying ac current from the power line which has been raised to high voltage by a transformer. The dc current thus obtained has its negative potential connected to the filament and its positive to the plate. It is then necessary to operate the tube in synchronization with the oscillating voltage in the tank so that it acts as the switch. The tank-circuit oscillation is sustained by providing makeup power from the dc supply, through the tube at the resonant frequency of the tank. A small amount of the total tank voltage, 180° out of phase with it, must be supplied to the grid of the tube to maintain this oscillation. This "feedback voltage" is generally derived from the tank circuit.

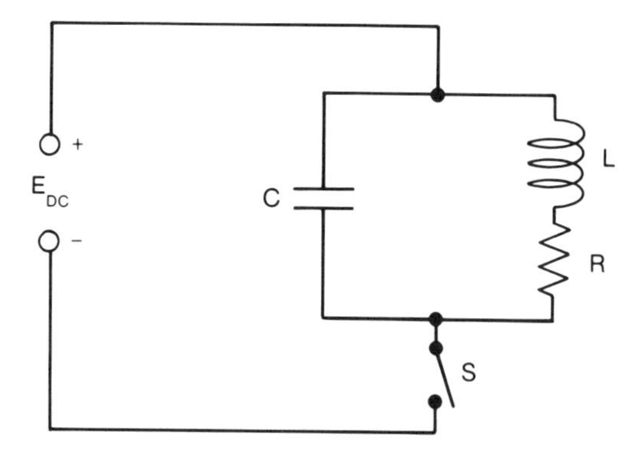

Fig. 4.22. Simplified electrical tank circuit for a radio-frequency oscillator (from E. May, *Industrial High Frequency Electric Power*, Wiley, New York, 1950)

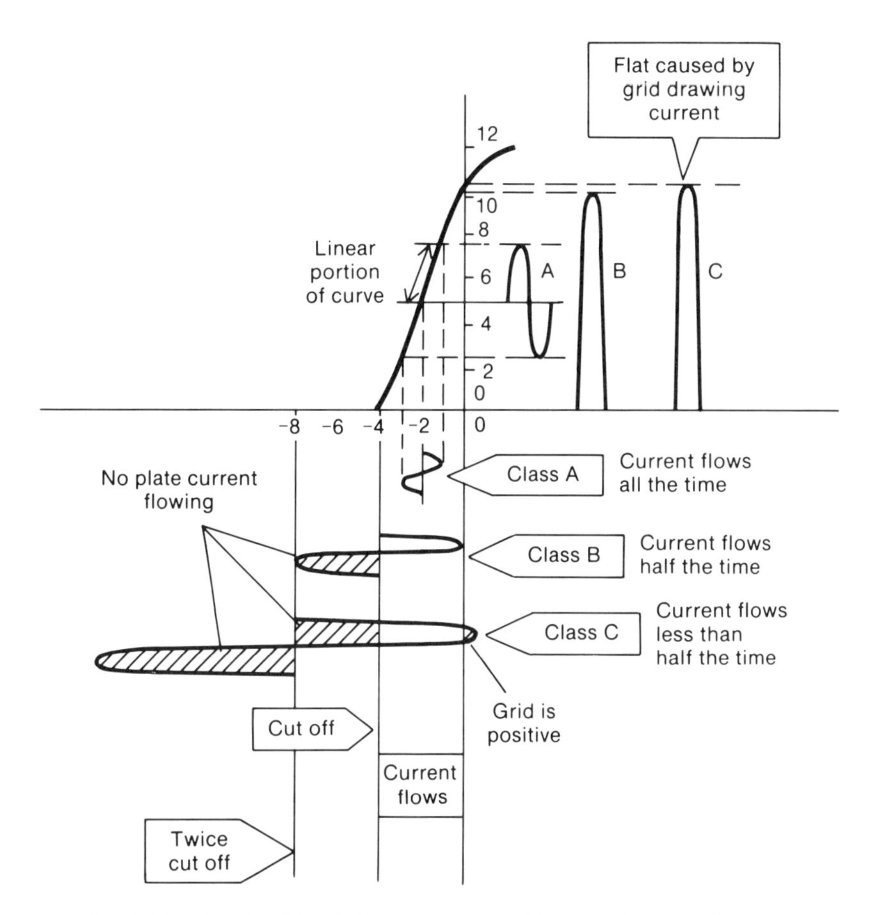

Fig. 4.23. Relationship of plate current to grid current in an oscillator operated under "class C" conditions (from V. Valkenburgh, *Basic Electronics*, Vol 2, Hayden Book Company, Inc., Rochelle Park, NJ, 1955)

In an actual oscillator circuit (Fig. 4.24), the dc power supply is connected to the tube circuit with the positive lead in the tube plate circuit via choke L_A and the negative lead connected to the filament. The tank circuit is in parallel with the tube; blocking capacitor C_A passes the RF current, but blocks the dc from the tank. The tank inductance is composed of the tank coil L and the work coil L_W. A magnetically coupled tickler coil L_G couples energy from the tank coil to the grid and filament of the tube via the grid capacitor C_G and a grid resistor R_G. This resistor keeps the average potential of the grid in proper relation to the filament for class C operation.

Methods of developing the grid feedback signal vary from oscillator to oscillator. In induction heating power supplies, standard oscillators are the inductively coupled tickler-coil system (Fig. 4.25a), the Hartley oscillator (Fig. 4.25b), and the Colpitts oscillator (Fig. 4.25c). With the latter two systems, the feedback voltage is automatically controlled as a percentage of the tank voltage, either by the transformer tap ratio (Hartley oscillator) or by the

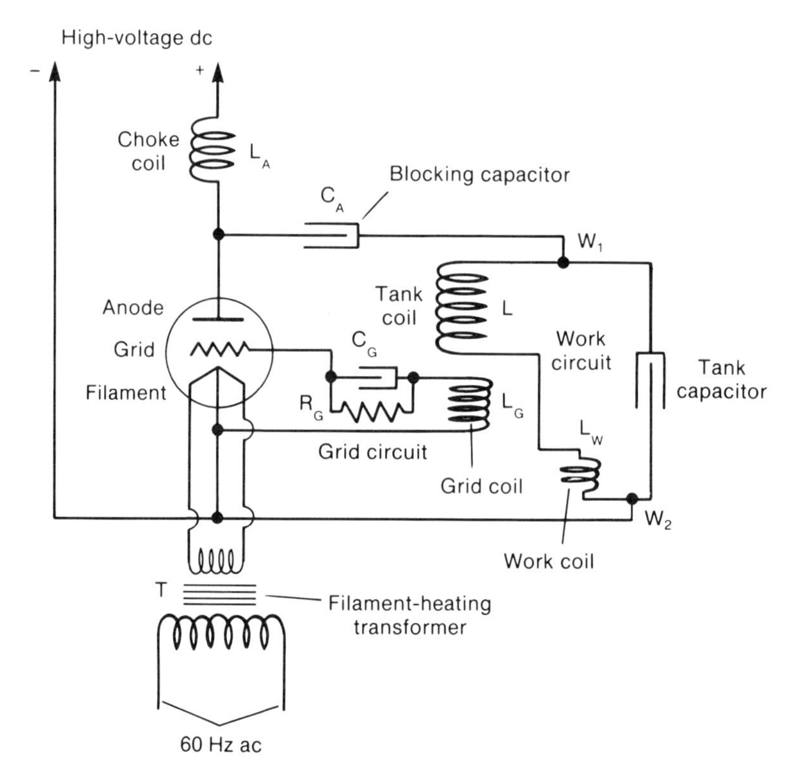

Fig. 4.24. Circuit diagram for a simple vacuum-tube-type high-frequency generator (from *Induction Heating*, American Society for Metals, Metals Park, OH, 1946)

ratio of grid tank capacitance to plate tank capacitance (Colpitts oscillator). With the tickler-coil technique, the coil is mechanically adjusted in relation to the tank coil and thus can be regulated to provide optimum drive for most conditions.

Radio-frequency tank circuits most frequently found in commercial equipment utilize either a work coil in series with the tank coil (Fig. 4.26a) or a work coil operating from an RF output transformer (Fig. 4.26b). The series-connected system, usually utilizing a tapped tank coil, is often referred to as a DTL (or "Direct Tank Loaded") technique, because the work coil is directly connected as part of the tank-circuit inductance. On the other hand, in transformer coupled output circuits, the work coil is isolated from both the transformer primary and the tank circuit. The work coil is usually center or midpoint grounded to minimize voltage potential and, therefore, arcing from coil to ground.

Design Considerations. Most common RF generators oscillate in the range of 200 to 450 kHz. It is apparent that as the coil inductance changes, the resonant frequency of the tank will vary as well. Thus, unless crystal controlled

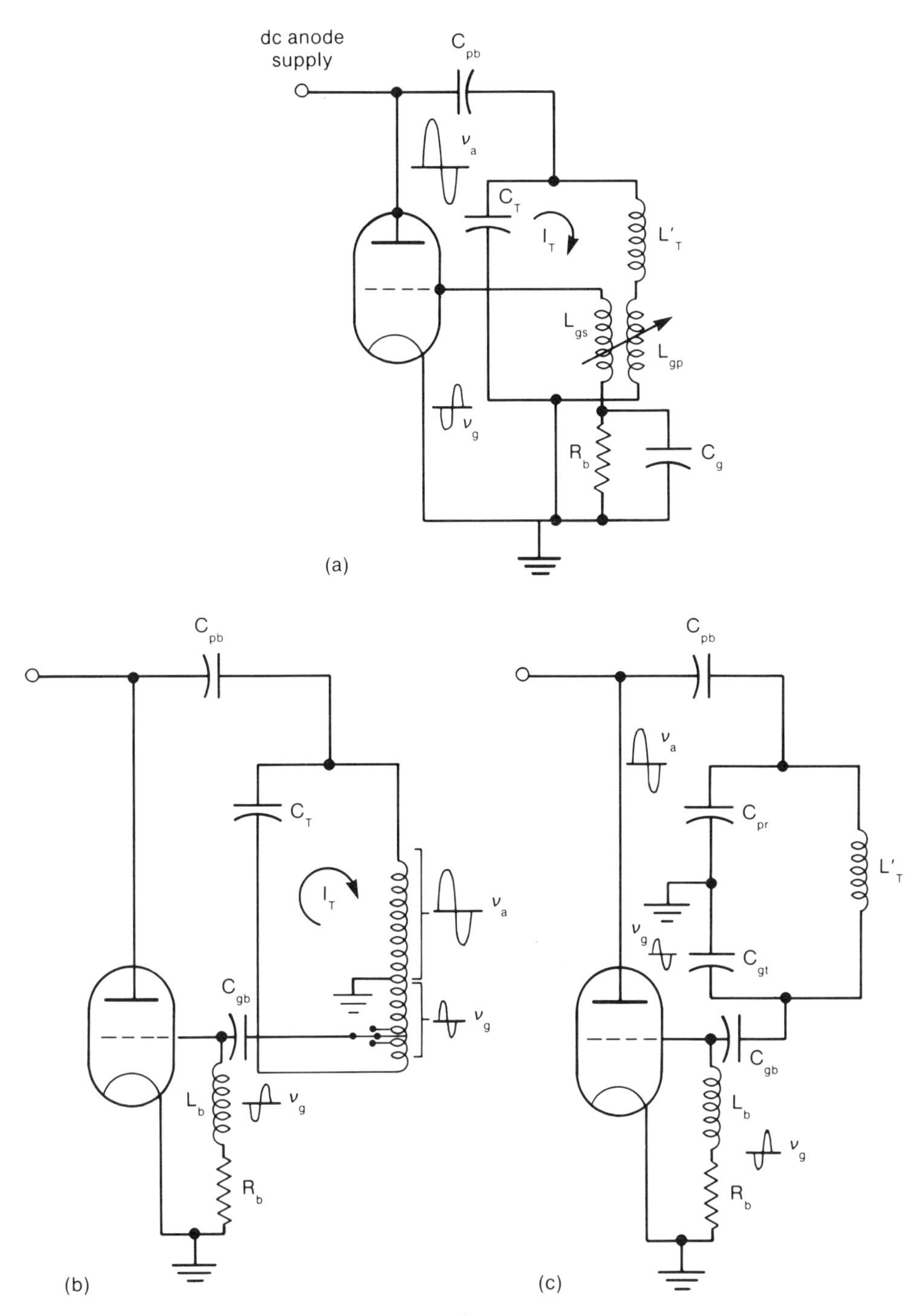

Fig. 4.25. Oscillator circuits used in high-frequency induction generators: (a) grid coupled, (b) Hartley, and (c) Colpitts (from J. Davies and P. Simpson, *Induction Heating Handbook*, McGraw-Hill, Ltd., London, 1979)

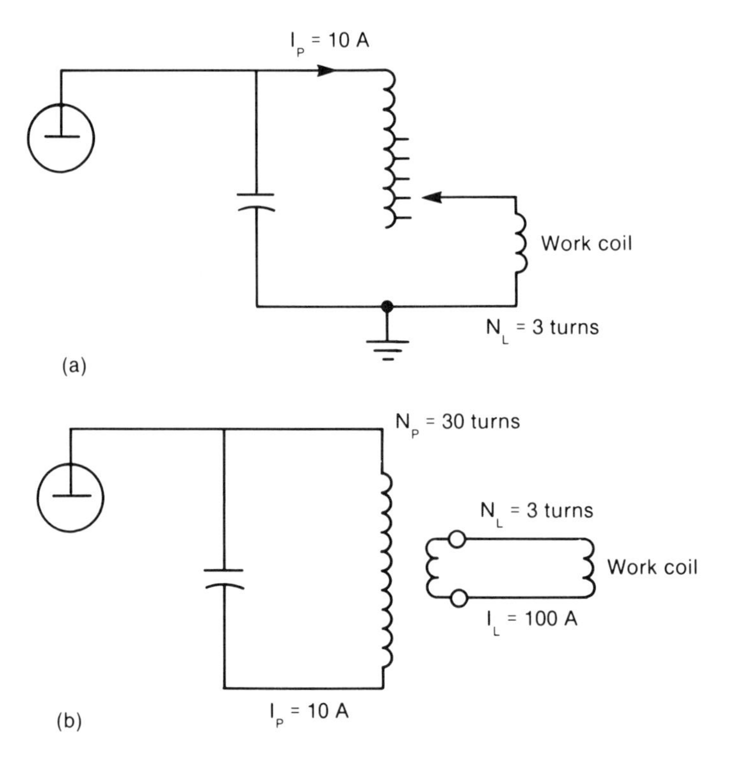

Fig. 4.26. Output tank circuits of common high-frequency induction generators: (a) direct tank loaded and (b) transformer coupled

(as in sputtering systems), the frequency will shift within the above limits during operation. Unlike the motor-generator, which tends to hold the voltage constant, the RF generator tries to maintain a constant circulating current in the tank circuit.

Although all RF power supply ratings are in terms of "power into a calorimetric load," they will not necessarily all respond similarly to a particular coil/part combination. A power rating refers to the capability of an individual system to supply the rated power to this load. It does not refer to the ease with which it can deliver that power. Typically, if the supply can deliver 10 kW to the test load with a 2.5-cm (1-in.) coupling distance, as compared to a power supply that requires the coil to be within 1.3 cm (0.5 in.) of the load to deliver the same power, its "loadability" is superior.

Loadability refers directly to the current in the tank circuit and thus the current in the work coil. Tank-circuit current, I_T, can be calculated from the formula $I_T \cong 0.75\ E_P/X_C$, where E_P is the dc voltage applied to the tank and X_C is the capacitive reactance of the tank capacitor at resonant frequency. The ratio of tank kVA to the kW rating of the generator is defined as the loadability and may be used to compare competitive power supplies.

Transmission losses between the work coil and the tank circuit can be considerable. If an RF output transformer is used and is outside the power-supply enclosure, or if a low-impedance heating coil is used at a distance from a DTL-connected generator, significant power losses will occur. In general, a coil or transformer remote from an RF generator should have an impedance at least ten times that of the impedance of the transmission line between the two. Moreover, in the case of a DTL system, high voltages can occur on these lines with resultant operator hazard. Radio-frequency interference can occur as well. Proper care must therefore be taken to duct these leads in an aluminum or other low-conductivity conduit.

It is also possible to locate the entire tank circuit remotely in a way similar to heat stations running off low- or medium-frequency power supplies. In this case, the remote tank (either DTL or transformer outputted) is connected to the supply by means of semiflexible coaxial cable. This also simplifies moving the tank circuit itself (Fig. 4.27) rather than the coil, for greater efficiency, when movement is required. Grid feedback in these systems is generally derived from a tapped feedback transformer (Fig. 4.28) that is in the power-

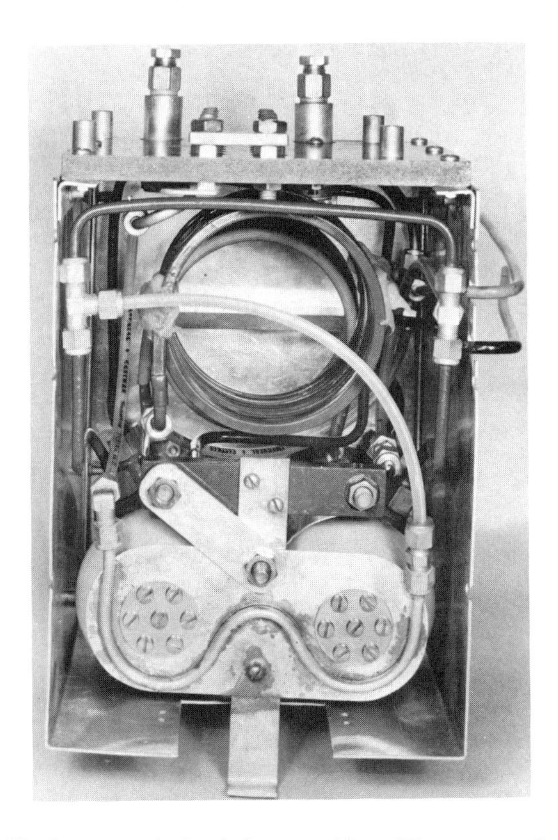

Fig. 4.27. Remote tank circuit for use with an RF power supply, including work coil, output transformer, and tank capacitors (source: Lindberg Cycle-Dyne, Inc.)

Fig. 4.28. Multitapped feedback transformer that provides adjustable grid drive when the tank circuit is located remote from the power supply (source: Lindberg Cycle-Dyne, Inc.)

supply enclosure. The primary is connected between the plate of the tube and the tube filament to derive a frequency signal. A secondary winding provides a voltage which is 180° out of phase to drive the grid.

Radio-Frequency (Solid-State) Equipment

Solid-state equipment capable of operating in the radio-frequency range is also now available (Fig. 4.29). Systems in use operate at 50 to 450 kHz and

Fig. 4.29. Interior of a 15-kW solid-state RF generator which operates at 50 to 250 kHz, showing microprocessor and associated controls at left, and plug-in power boards at right (source: Ameritherm, Inc.)

a wide range of power outputs. Unlike the inverters used at 50 kHz and below, these units do not have SCR's, but are instead powered by MOSFET (Metal Oxide Semiconductor Field Effect Transistor) output devices.

A solid-state RF supply is similar to an SCR inverter in that line voltage enters the system through an isolation transformer which then produces a dc voltage level that is adjustable through an SCR voltage control. This filtered dc output is the input for the solid-state RF supply's power boards. The MOSFET's on each board and the boards themselves are connected in parallel. A sensing circuit in the cable leading to the tank circuit feeds the tank resonant-frequency signal to the microprocessor. The MOSFET's are driven at this frequency by the microprocessor. The software will also adjust the system operating frequency in response to the change in tank resonant frequency that results when the temperature of a magnetic load rises above the Curie point. As does the load-resonant inverter, the solid-state RF system operates at the peak of the resonant curve, and its efficiency is therefore similar.

The software for these microprocessor-driven systems can be modified to provide constant current, constant power, or constant voltage, depending on the conditions required for the particular application. The software also permits interrogation of the system through a computer-keyboard interface.

Chapter 5

Auxiliary Equipment for Induction Heating

Besides the power supply, other auxiliary pieces of equipment are required in most induction heating installations. These include systems for cooling the power supply and induction coil, power-timing devices, temperature-control devices, and materials-handling systems. The first two types of equipment are discussed in this chapter. Temperature monitoring and control are addressed in Chapter 7, and materials-handling concepts are described in Chapter 10.

EQUIPMENT COOLING SYSTEMS

All induction heating systems require coolants (primarily water), because current-carrying components dissipate waste heat through I^2R losses. Obviously, the greater the resistance of the current path and/or the current, the greater the loss in the system, and the more waste heat generated. In particular, the induction heating coil, capacitors, and output transformer constitute a major area of power loss. In tube-type radio-frequency (RF) systems, the plate of the tube may dissipate as much as 50% of the input power.

The following questions should be considered when choosing a cooling system for induction heating:

1. What is the maximum temperature that any component can tolerate?
2. How much water flow is necessary to remove the heat from the system?
3. How much water pressure must be developed to provide the required water flow through the equipment?

Each of these questions is answered by inspecting the specifications of the equipment manufacturer. However, these are essentially mechanical considerations. There are, in addition, specifications regarding the chemistry of the cooling water that greatly affect its use as a coolant and which must be considered in planning and operation. Furthermore, many of the components that are water cooled are at various voltages above ground. In an RF system, for instance, the plate of the oscillator tube is generally several thousand volts above ground, yet must be cooled with water. To accomplish this, water is carried to the tube through a length of nonconductive hose, passed through the tube cooling channel, and then returned to ground through a similar hose column. In essence, this hose column is a high-resistivity path with water as the conductor. At a 10-kV level, approximately 5 m (16 ft) of hose must be used on each side of the column to provide a sufficient length of water path to prevent the water from carrying current and thus dissipating additional heat. This presupposes that the resistivity of the water is sufficiently high.

Another consideration in a water-cooling system regards the regulation of the water temperature to a level above the dew point in the area or cabinet where the system operates. A temperature below this point will result in condensation on the cooled components. Condensation, particularly in high-voltage areas, can cause arcing between components of differing voltages and the resultant failure of these components. In addition, any technique that reduces water consumption will provide an economic means of cutting water and sewage costs.

In perhaps the simplest type of cooling system (Fig. 5.1), water is passed from a tank, through a pump, to the induction heater, with the return path going back to the pump. Raw water from the tap is passed through a temperature-control valve whose sensor is in the output waterline of the generator. Because additional tap water is added to the system, an overflow pipe passes the excess to a drain.

In operation, cooling water is added to the system from the tap as the exit water temperature exceeds its limits. Thus, tap water is added only on temperature demand, reducing input water requirements. This system also main-

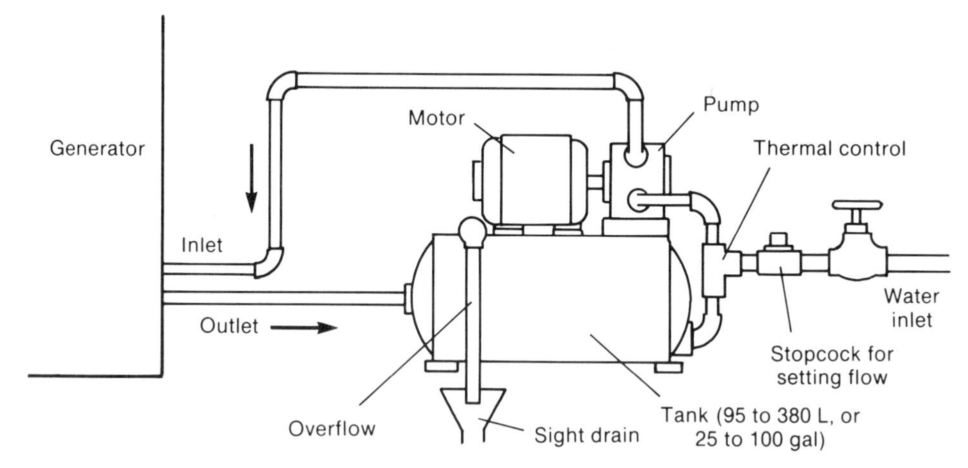

Fig. 5.1. Simple water-cooling system designed to control temperature with minimum water throughput (from F. W. Curtis, *High Frequency Induction Heating*, McGraw-Hill, New York, 1950)

tains the ambient temperature of the generator above the dew point. However, it does not protect the system from a water-supply problem of low resistivity or excess solids content.

Water-Cooling Systems

Distilled water, tap water, and deionized water have all been used in cooling systems for induction heating equipment. Deionized water is "hungry" and will attack the zinc in brass fittings, which after much use may become spongy. In contrast, distilled water will not attack brass. Tap water, on the other hand, if high in resistivity and reasonably clean, will, in effect, become equivalent to distilled water after several passes through the equipment.

If solids are contained in tap water, they have a tendency to plate out on the walls of the component being cooled. This results in the buildup of an insulating layer along the coolant path, similar to boiler scale, which gradually reduces the heat-exchange capability of the system. Thus, although pressure, flow, and input water temperature are maintained, the system could overheat. Table 5.1 covers a typical specification for water quality for an induction heating generator. Solids-content limits in parts per million are generally considered for cases in which constant flow brings contaminants into the system on a continuous basis. However, once the system is sealed, additional solids are not brought into the cooler, and the tap water becomes similar to distilled water.

When the cooling system is used in a plant where temperatures may go below freezing for any period of time, the coolant should be composed of a mixture of water and ethylene glycol to prevent freeze-up. A 40% solution of ethylene glycol is satisfactory in most localities. It should be noted, however, that most antifreeze solutions, although they are ethylene glycol based, should not be used. Unlike pure ethylene glycol, the additives in antifreeze reduce the resistivity of the coolant and thus cause problems in hose columns.

Table 5.1. Typical cooling-water specification for an induction heating power supply (source: American Induction Heating Corp.)

- Minimum pressure differential, 207 kPa (30 psi)
- Maximum inlet water temperature, 35 °C (95 °F)
- pH between 7.0 and 9.0 (i.e., slightly alkaline)
- Chloride content < 20 ppm
  Nitrate content < 10 ppm
  Sulfate content < 100 ppm
  $CaCO_3$ content < 250 ppm
- Total dissolved solids content < 250 ppm;
  no solids to precipitate at T ≤ 57 °C (135 °F)
- Resistivity ≥ 2500 Ω·cm at 25 °C (77 °F)
- Must contain a magnetite eliminator and corrosion inhibitor
- Antifreeze (50% maximum), "uninhibited" ethylene glycol

Many water-base cooling systems make use of a heat exchanger, especially those which use distilled water as the coolant. In distilled-water systems, tap water is passed through the tubes of a "water-to-water" exchanger, cooling the distilled water but not mixing with it. The temperature-demand valve remains connected to the tap waterline. With small induction heaters (7.5 kW or less), water-to-air heat exchangers, similar in construction to an automotive radiator with a fan, are sometimes used.

Vapor-Coolant Systems

When requirements for coolant exceed 45 L/min (12 gal/min), many installations now make use of vapor-cooled systems. These are (in the smaller sizes) air-to-water heat-exchange systems. Packaged in a single unit, or with the exchange unit separate from the pumping station, they offer a method of reducing water consumption.

Typically, vapor-cooled systems (Fig. 5.2) utilize a mist of water sprayed onto the fins of an air-to-water heat exchanger. This water is lost as a vapor or mist of steam, but because of the high heat input per unit weight of water, the system requires only about 1% of the normal water usage of a conven-

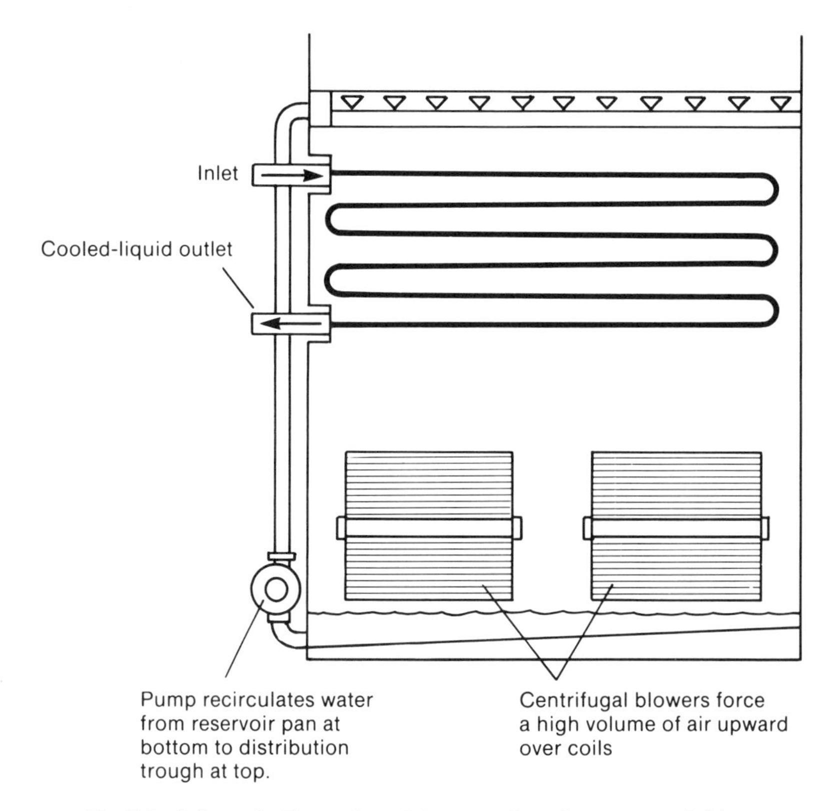

Fig. 5.2. Schematic illustration of the operation of a vapor-cooled heat-exchanger system (source: Water Saver Systems, Inc.)

tional water-cooling system. The system still isolates the equipment coolant from the water used for the mist spray. However, during constant use, solids are deposited on the fins from the vapor, reducing their energy-transfer capabilities. Therefore, provision must be made to wash down the system occasionally to remove these solids. Further, because the mist contains no ethylene glycol, the sump for the excess spray must be heated in the winter months.

Figure 5.3 shows a schematic representation of a typical system. Here, the heat exchanger is mounted outside the building, and the pumping station is adjacent to the induction heaters. This system, dependent on proper sizing, can operate more than one installation at a time. In many cases, the system will also act as a cooler for other processes, furnaces, etc. in the same area.

The primary advantage of a vapor-cooled system as compared with a water-base system results from the fact that the former is a closed system. Water towers can be "open" in that the cooling of the water is performed by exposing it to air. In addition to impurities absorbed during this exposure, chemicals such as algicides must usually be added to the coolant. Because these can

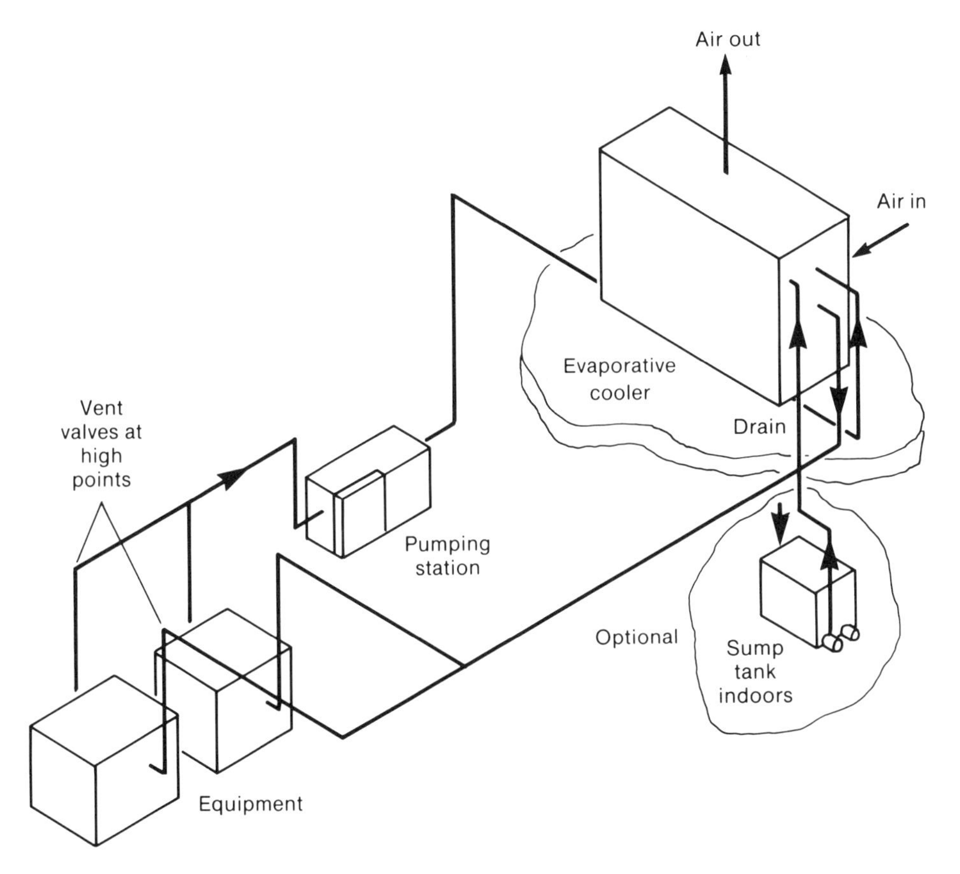

Fig. 5.3. Vapor-cooled heat-exchanger system with a pumping station adjacent to the power-supply equipment (source: Water Saver Systems, Inc.)

be detrimental to the system, a bleed-off method is usually employed to minimize buildup of water impurities over a period of operating time.

Figure 5.4 shows a further refinement of the system depicted in Fig. 5.3. Because cooling-capacity requirements are generally greatest during the summer months, the vapor-cooled system is placed outside the plant during this period. The heat lost from the exchanger is passed to the atmosphere by the vapor. In the winter, however, it is desirable to recover this lost heat and use it efficiently. This is done by adding an auxiliary air-to-water heat exchanger to the system. The valves to the outside vapor cooler are closed, and the valves to the air-to-water system are opened. The air-to-water exchanger is located in the plant air system and may be modulated by tempered air drawn back from the air ducts. Thus, the heat normally lost is recovered and used to warm the plant air. Approximately 95% of the waste heat can be recovered in this manner.

TIMERS

Control of induction heating systems can be on an open-loop or a closed-loop basis. Closed-loop systems make use of a control signal (most often temperature) in a feedback circuit to regulate the operation of the induction

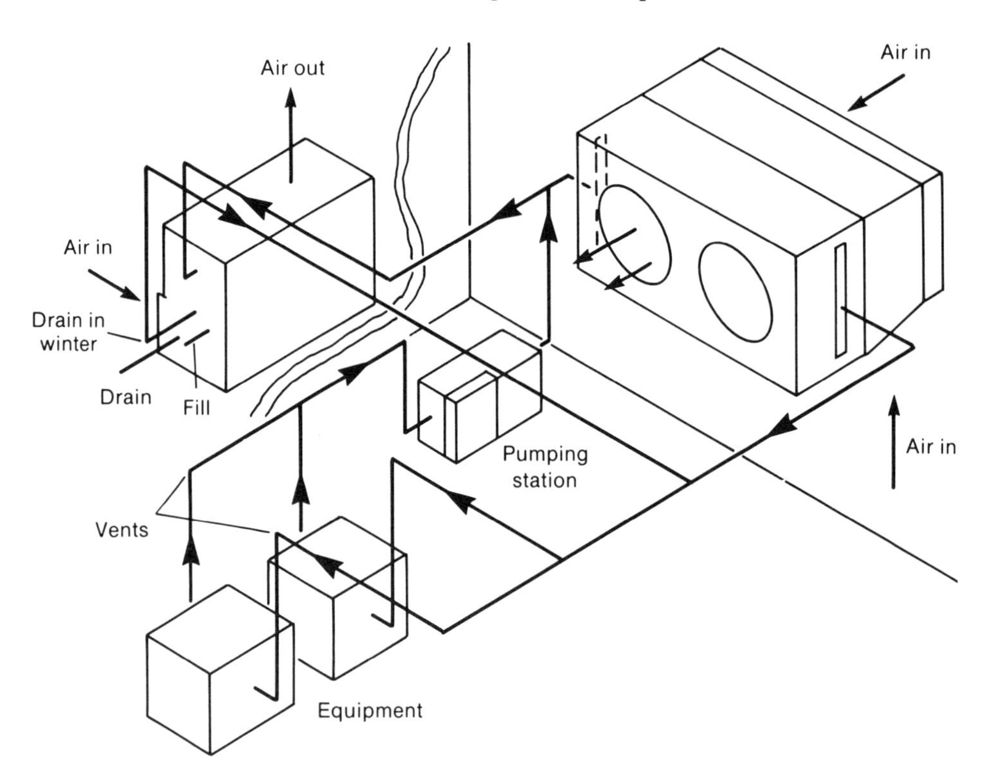

Fig. 5.4. Vapor-cooled heat-exchanger system equipped with a recuperator that returns waste heat to the plant air-circulation system (source: Water Saver Systems, Inc.)

power supply. The control signal is compared with a prescribed, or setpoint, value, and the difference is fed into the feedback circuit. Systems of this sort are described in Chapter 7. By contrast, open-loop systems are not nearly as sophisticated, relying primarily on measurements of heating time to ensure reproducible and accurate results. For applications in which process parameters are not extremely critical, methods utilizing the control of heating time have been found quite satisfactory.

The basic technique requires timer control of the main circuit contactor for a specific time duration. This method is used on some induction heating systems from line frequency through radio frequency. There are, however, some important factors that should be noted. First, the heavier the load, the larger the power supply, and thus the greater the physical size of the contactor. Due to the large mass of the contactor armature and the magnetic circuit that makes it operate, several cycles of the line frequency can be required to pull the contactor into its fully closed position. In small contactors, the portion of the line cycle in which the solenoid is energized can also vary the time for full closure. In large systems, where long-duration heating times are often used, the resultant variation is not normally a problem. However, when short-duration timing is employed, as in RF heating of small workpieces, the final temperature may vary considerably. This is particularly true when timing cycles are under 3 s.

Timers for most induction heating systems can be either electromechanical or electronic. When electromechanical timing is used, normal procedure is to keep the timer motor running and energize or de-energize the clutch for timing functions. For timing cycles of 10 s or less, the use of an electronic timer is recommended. In power supplies driven by silicon-controlled rectifier (SCR) power controllers, the circuit breaker or main contactor (or both) is generally used as a primary disconnect and a backup means of disconnecting the load from the power source. In such a mode, this means is generally used to isolate all power in the equipment for maintenance problems as well as to interrupt all power under certain fault or emergency shutdown conditions. Contact closure in the gate-trigger circuit of the SCR package can then be used to turn the power-supply output on or off. This is a high-speed technique that eliminates the delay in contactor closing and, more importantly, increases the life of the contactor by eliminating repetitive mechanical operation.

As timing periods become shorter, the magnetic-hysteresis effect in saturable reactors prevents their being used for high-speed on/off operation. The delay between reactor energization and the point at which the reactor achieves its preset power level can be considerable when compared with heating time. Although considerably faster than in the reactor, this same effect is also apparent in the firing cycle of the SCR controller when extremely accurate, short-duration heating times (2 to 3 s or less) are desired. These operational times are generally associated with RF heating systems.

When short heating cycles are necessary, the SCR or reactor is generally

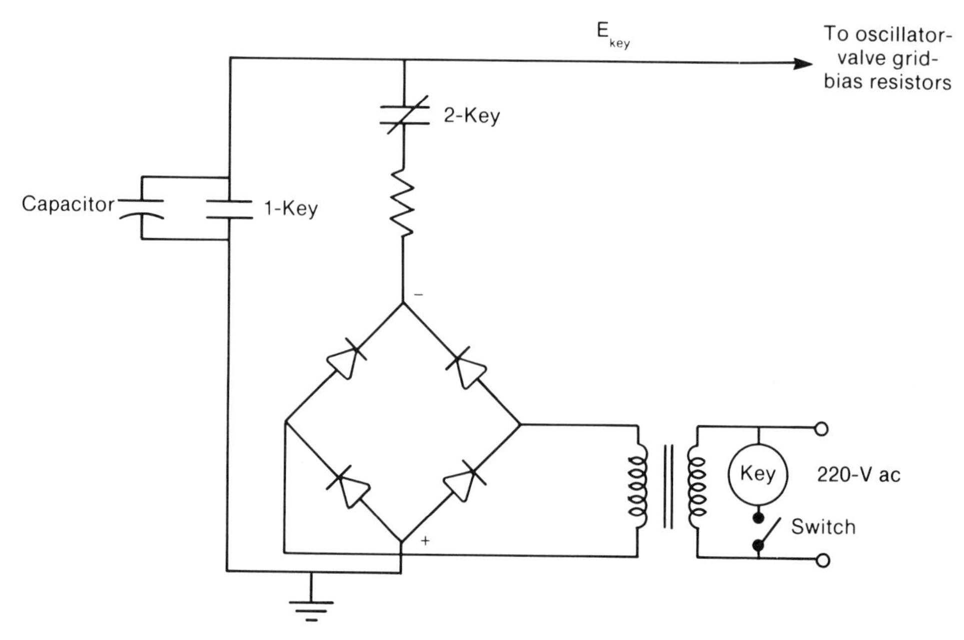

Fig. 5.5. Circuit schematic for a grid-trigger system used for high-speed operation of an RF power supply (from J. Davies and P. Simpson, *Induction Heating Handbook*, McGraw-Hill, Ltd., London, 1979)

kept energized. System timing is then produced by a "grid-trigger" device (Fig. 5.5) that applies a negative bias to the grid of the tube, driving it to cutoff. This makes the grid of the tube highly negative and thus prevents current flow from cathode to anode. A very light relay-contact closure is all that is necessary to short the bias voltage through a shunt resistor, restoring the normal grid bias. The speed of control of the system is limited only by the operational speed of the relay, because the tube is operative at the resonant frequency of the tank.

Chapter 6

Process Design for Specific Applications

Prior to the construction and implementation of any induction heating process, the details of the actual heating requirements must be considered. Such considerations include type of heating, throughput and heating time, workpiece material, peak temperature, and so forth. The major types of induction applications include through heating, surface heating (for surface heat treatment), metal melting, welding, and brazing and soldering. Selection of equipment (frequency, power rating) and related design considerations are summarized in this chapter. Coil design for the various applications is described in Chapter 8, and materials-handling details are outlined in Chapter 10.

DESIGN PROCEDURES FOR THROUGH HEATING

On the basis of total kilowatt hours of electricity expended each year, through heating (without melting) is probably the largest single use of induction heating in the metals industry. The principal applications of induction through heating include heating prior to hot working and through heating for purposes of heat treatment. The important hot working operations in this regard are forging, extrusion, and rolling. Typical geometries for the first two processes are simple round or round-cornered square billets. Induction heating also finds application in the rolling of slabs, sheets, blooms, and bars. In the area of through heat treating, induction is used for processes such as hardening, tempering, normalizing, and annealing. As in hot working, the part geometries in these instances tend not to be too complex, typical ones being bar stock, tubular products, and sheet metal.

The important features in the design of induction through-heating processes are related to the characteristics of the power supply—namely, its frequency and power rating.

Selection of Frequency for Through Heating

The selection of frequency for through heating is usually based solely on calculations of reference depth using the critical frequency as a guideline. From Chapter 2, the reference depth is given by d (in.) = 3160 $\sqrt{\rho/\mu f}$ when the resistivity ρ is in units of $\Omega \cdot$in., or d (cm) = 5000 $\sqrt{\rho/\mu f}$ when ρ is in $\Omega \cdot$cm. For round bars, the critical frequency is that frequency at which bar diameter is approximately four times reference depth; for slabs and sheets, the critical frequency is that at which thickness is about 2.25 times reference depth. For these simple geometries, the critical or minimum frequency f_c is therefore given by

$$\left.\begin{aligned} f_c &= 1.6 \times 10^8\ \rho/\mu a^2 \quad &\text{(a in in.)} \\ f_c &= 4 \times 10^8\ \rho/\mu a^2 \quad &\text{(a in cm)} \end{aligned}\right\} \text{Round bar}$$

or

$$\left.\begin{aligned} f_c &= 5.06 \times 10^7\ \rho/\mu a^2 \quad &\text{(a in in.)} \\ f_c &= 1.27 \times 10^8\ \rho/\mu a^2 \quad &\text{(a in cm)} \end{aligned}\right\} \text{Slab or sheet}$$

where a is bar diameter or slab/sheet thickness.

The critical-frequency expressions are most readily evaluated for nonmagnetic materials, or those for which $\mu = 1$. In these cases, f_c is simply a function of geometry and the material's resistivity. Values of resistivity ρ for various metals are given in Table 6.1. It can be seen that ρ is not constant, but varies with temperature. For this reason, estimates of f_c are usually based on the value of ρ corresponding to the *peak* temperature to which the metal is to be induction heated.

To illustrate, critical frequency will be calculated for several situations. Consider first the heating of a 1.91-cm- (0.75-in.-) diam round bar of aluminum to 510 °C (950 °F). From Table 6.1, the resistivity at this temperature is approximately $10.44 \times 10^{-6}\ \Omega \cdot$cm ($4.1 \times 10^{-6}\ \Omega \cdot$in.). Because $\mu = 1$, the equation for critical frequency yields f_c = 1170 Hz. If the bar were made of an austenitic stainless steel, on the other hand, the resistivity at the same temperature would be approximately $101.6 \times 10^{-6}\ \Omega \cdot$cm ($40 \times 10^{-6}\ \Omega \cdot$in.), and the critical frequency would be approximately ten times as great, namely 11,380 Hz.

For the case of magnetic steels, the above equations for f_c apply as well. However, μ is a function of temperature. Below the Curie temperature, $\mu \approx$ 100 for typical power densities. Above this temperature, $\mu = 1$. If the round bar in the above example were of a carbon steel that was to be tempered, it would have a resistivity of about $63.5 \times 10^{-6}\ \Omega \cdot$cm ($25 \times 10^{-6}\ \Omega \cdot$in.). Assuming $\mu = 100$, the critical frequency would thus be 71 Hz. If this bar were heated to a higher temperature, say 980 °C (1800 °F), at which $\mu = 1$ and $\rho =$

Table 6.1. Approximate electrical resistivities of various metals (from C. A. Tudbury, *Basics of Induction Heating*, John F. Rider, Inc., New York, 1960)

Material	Approximate electrical resistivity, μΩ·cm (μΩ·in.), at temperature, °C (°F), of: 20 (68)	95 (200)	205 (400)	315 (600)	540 (1000)	760 (1400)	980 (1800)	1205 (2200)
Aluminum	2.8 (1.12)	...	...	6.9 (2.7)	10.4 (4.1)	...	...	...
Antimony	39.4 (15.5)	...	...	...	...	...	...	...
Beryllium	6.1 (2.47)	...	...	...	11.4 (4.5)	...	...	...
Brass (70Cu-30Zn)	6.3 (2.4)	...	...	...	...	...	...	...
Carbon	3353 (1320.0)	...	...	...	1828.8 (720.0)	...	...	...
Chromium	12.7 (5.0)	...	...	...	...	...	...	...
Copper	1.7 (0.68)	...	...	3.8 (1.5)	5.5 (2.15)	...	9.4 (3.7)	...
Gold	2.4 (0.95)	...	...	...	...	...	12.2 (4.8)	...
Iron	10.2 (4.0)	14.0 (5.5)	...	...	63.5 (25.0)	106.7 (42.0)	123.2 (48.5)	...
Lead	20.8 (8.2)	27.4 (10.8)	...	49.8 (19.6)	...	...	...	...
Magnesium	4.5 (1.76)	...	...	...	...	...	...	...
Manganese	185 (73.0)	...	...	...	...	...	...	...
Mercury	9.7 (3.8)	...	...	...	...	...	...	...
Molybdenum	5.3 (2.1)	...	...	...	...	...	...	33.0 (13.0)
Monel	44.2 (17.4)	...	...	...	...	...	...	...
Nichrome	108.0 (42.5)	...	...	114.3 (45.0)	...	114.3 (45.0)	...	...
Nickel	6.9 (2.7)	...	...	29.2 (11.5)	40.4 (15.9)	...	54.4 (21.4)	...
Platinum	9.9 (3.9)	...	...	...	...	...	...	...
Silver	1.59 (0.626)	...	...	...	...	6.7 (2.65)	...	...
Stainless steel, nonmagnetic	73.7 (29.0)	...	...	99.1 (39.0)	...	...	130.8 (51.5)	...
Stainless steel 410	62.2 (24.5)	...	...	...	101.6 (40.0)	...	127 (50.0)	...
Steel, low carbon	12.7 (5.0)	16.5 (6.5)	...	...	59.7 (23.5)	102 (40.0)	115.6 (45.5)	121.9 (48.0)
Steel, 1.0% C	18.8 (7.4)	22.9 (9.0)	...	...	69.9 (27.5)	108 (42.5)	121.9 (48.0)	127.0 (50.0)
Tin	11.4 (4.5)	...	20.3 (8.0)	...	...	...	...	...
Titanium	53.3 (21.0)	...	...	...	...	...	...	165.1 (65.0)
Tungsten	5.6 (2.2)	...	...	...	...	...	...	38.6 (15.2)
Uranium	32.0 (12.6)	...	...	...	...	...	...	...
Zirconium	40.6 (16.0)	...	...	...	...	...	...	...

117×10^{-6} Ω·cm (46×10^{-6} Ω·in.), the critical frequency would increase substantially to approximately 13,000 Hz.

Variable-frequency power supplies typically offer a twofold or threefold range of frequency. Thus, for nonmagnetic materials, the same generator can be used to heat parts of a given size over a range of temperatures. In contrast, for magnetic materials the power-supply frequency must be selected on the basis of the maximum temperature to which the parts will ever be heated.

To avoid calculation of critical frequencies, tables or graphs which show frequency requirements as a function of bar size are usually employed. Sample compilations are given in Table 6.2 and Fig. 2.9 and 6.1.

When a frequency higher or lower than the critical frequency is used, the heating efficiency increases or decreases, respectively, as was discussed in Chapter 2. The increase in efficiency above f_c is rather small for large frequency increases. Conversely, the efficiency drops rapidly when frequencies below f_c are utilized. This effect was illustrated in Fig. 2.10. It is also shown in Fig. 6.2 for heating of round steel bars below or above the Curie temperature. These plots give efficiencies for heating bars of varying sizes with a given frequency. Because a given frequency can be used effectively to heat material of a greater diameter than that for which this frequency equals f_c, the curves in Fig. 6.2 all show plateaus above a critical bar size and a rapid drop-off below this size.

Figure 6.2 also shows the effect of coupling (distance between the coil inner diameter and workpiece outer diameter). As this distance increases (i.e., as the coupling diminishes), the efficiency is lowered. This trend is related to the fact that fewer lines of magnetic flux pass through the workpiece as the coupling is decreased. Therefore, according to Faraday's Law, the induced voltage and thus the induced current are lower. Because the current through the induction coil and its resistance remain fixed, the coil dissipates a given amount of power. However, the amount of energy dissipated in the workpiece decreases with poorer coupling. Therefore, the percentage of the total energy which is used to heat the workpiece decreases, and efficiency goes down. A similar argument can be used to explain the low efficiency associated with the heating of a tube using a solenoid coil located within the tube. In this case, the field strength is small *outside* the coil turns, and relatively small amounts of current can be induced in the workpiece. The coupling in these situations is low even when the air gap between the coil and workpiece is small. Thus, this type of heating arrangement should be avoided unless space limitations or other considerations demand it.

Charts of the efficiency of heating of tube and pipe products using solenoid coils located around the outer diameter provide useful information on the influence of workpiece geometry. Important factors here are the outer diameter and wall thickness of the tube or pipe. Several useful plots for medium-carbon steel products heated either below the Curie temperature (such as may be required for through tempering) or above it (such as in nor-

Table 6.2. Approximate smallest diameters that can be heated efficiently by the equipment and to the temperatures indicated (from C. A. Tudbury, *Basics of Induction Heating*, Vol 1, John F. Rider, Inc., New York, 1960)

		Approximate smallest diameter, in.(a), for efficient heating by:						
		Motor-generators/solid-state supplies					Vacuum-tube oscillators	
Material	Final temperature, °C (°F)	60 Hz	1 kHz	3 kHz	10 kHz	50 kHz	450 kHz	2000 kHz
Aluminum:								
Extrusion	510 (950)	3.2	0.78	0.45	0.25	0.11	0.037	0.018
Aluminum soldering	650 (1200)	3.7	0.88	0.52	0.28	0.12	0.042	0.020
Brass:								
Soft soldering	260 (500)	3.0	0.73	0.42	0.23	0.10	0.035	0.016
Silver soldering	650 (1200)	3.6	0.88	0.51	0.28	0.12	0.041	0.020
Melting	955 (1750)	4.1	0.99	0.58	0.32	0.14	0.047	0.022
Copper:								
Brazing	980 (1800)	3.1	0.76	0.44	0.24	0.11	0.036	0.017
Silver soldering	650 (1200)	2.6	0.63	0.37	0.20	0.09	0.029	0.014
Melting	1095 (2000)	3.3	0.81	0.47	0.26	1.2	0.039	0.018
Lead:								
Soft soldering	260 (500)	6.7	1.6	0.99	0.52	0.23	0.078	0.037
Melting	325 (620)	7.3	1.8	1.0	0.56	0.25	0.084	0.040
Steel:								
Stress relieving	540 (1000)	1.4	0.35	0.20	0.11	0.05	0.017	0.008
Hardening	870 (1600)	11.0	2.7	1.5	0.85	0.38	0.125	0.060
Forging	1205 (2200)	11.3	2.8	1.6	0.88	0.39	0.130	0.062
Melting	1510 (2750)	11.5	2.8	1.7	0.90	0.40	0.133	0.063
Zinc:								
Soft soldering	260 (500)	3.6	0.87	0.50	0.28	0.12	0.041	0.020

(a) 1 in. = 25.4 mm.

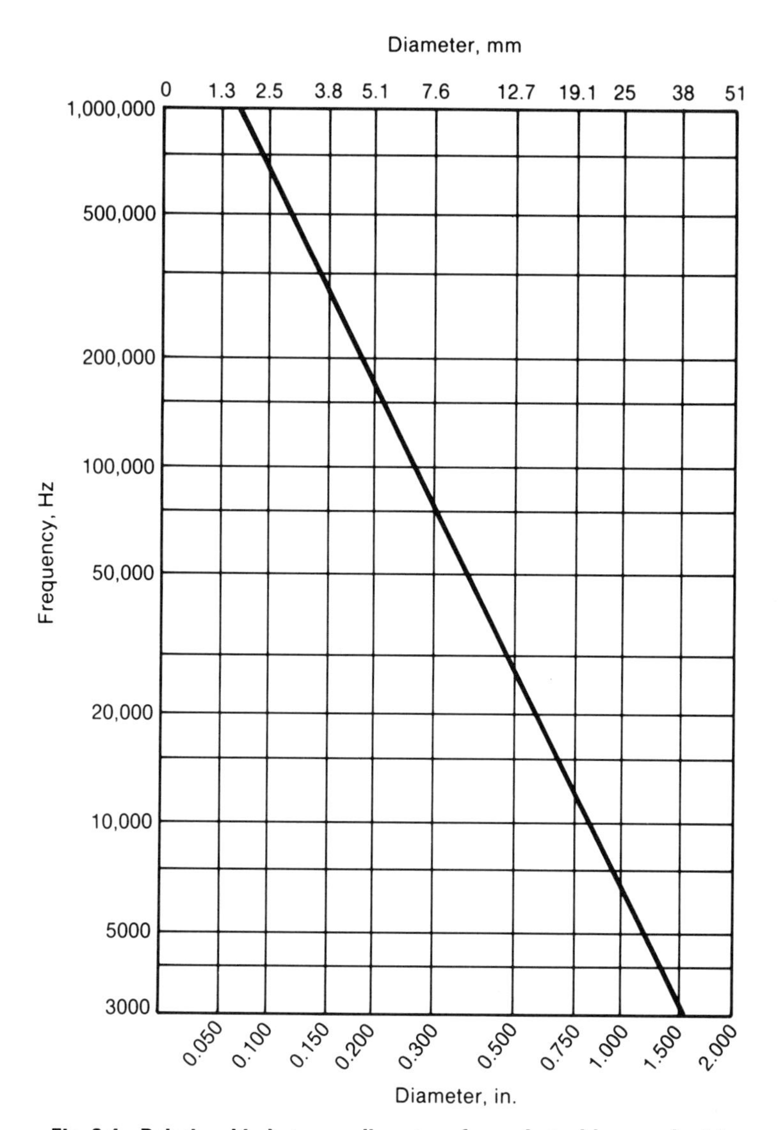

Fig. 6.1. Relationship between diameter of round steel bars and minimum generator frequency for efficient austenitizing using induction heating (from F. W. Curtis, *High Frequency Induction Heating*, McGraw-Hill, New York, 1950)

malizing or hardening operations) are depicted in Fig. 6.3. From these curves it is apparent that, for a given power-supply frequency and tube geometry (OD, ID), the efficiency is higher for heating below the Curie temperature. This is a result of the smaller reference depth below the Curie temperature. Similarly, the increase in efficiency with increased frequency for a given geometry and temperature range are associated with higher OD-to-d ratios as for the induction heating of solid bars. Two other trends are not so obvious—those related to the effects of outer diameter and wall thickness for a given

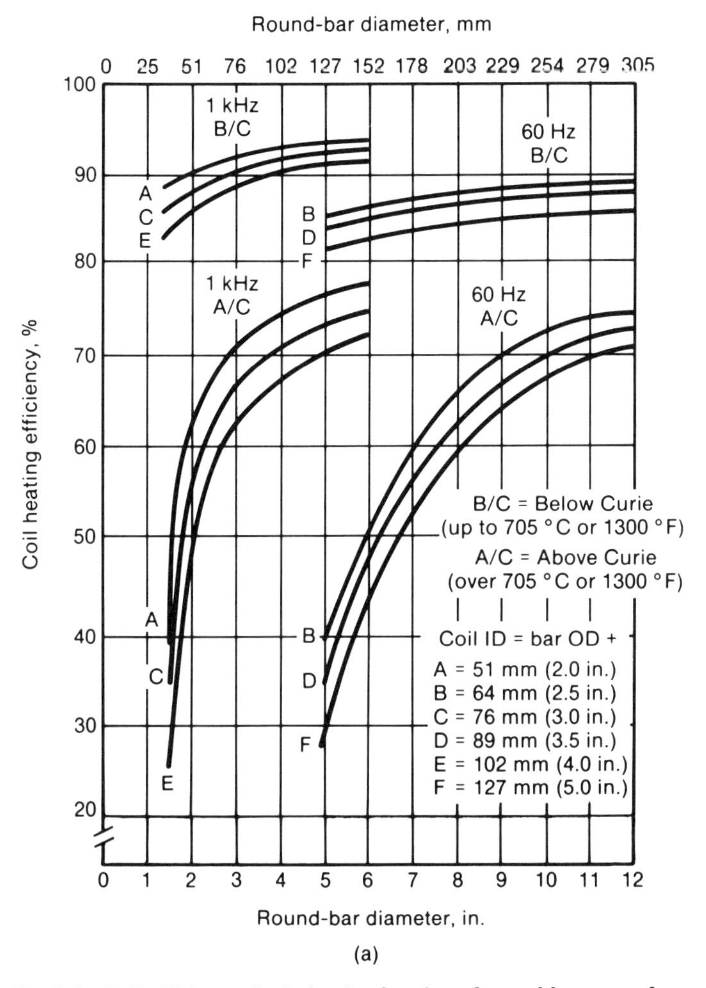

Fig. 6.2. Coil efficiency for induction heating of round bars as a function of bar diameter, size of air gap between coil and bar, and induction power-supply frequency: (a) 60 Hz and 1 kHz; (b) 180 Hz and 3 kHz; (c) 10 kHz (from G. F. Bobart, "Innovative Induction Systems for the Steel Industry," ***Proc. Energy Seminar/Workshop on New Concepts in Energy Conservation and Productivity Improvement for Industrial Heat Processing Equipment*****, Chicago, March, 1982)**

frequency and temperature range. With regard to the effect of outer diameter at a fixed wall thickness, it should be noted that the reference depth is the same irrespective of OD but that the equivalent resistance scales linearly with OD. Thus, a greater fraction of the power goes into the workpiece (when reflected through the coil circuit) as the OD increases. Such behavior is entirely comparable to the data in Fig. 6.2 for solid bars. The effect of wall thickness at fixed OD and frequency is more difficult to understand. As the wall becomes thinner, the over-all workpiece resistance increases. However, its equivalent resistance as reflected in the coil circuit tends to decrease, as can

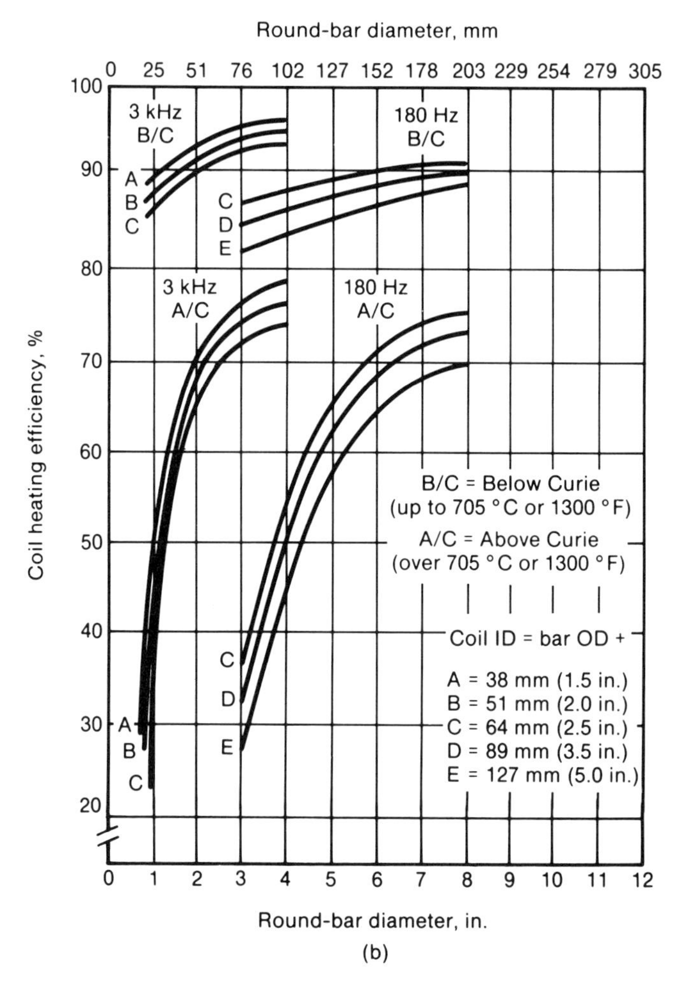

Fig. 6.2. continued

be inferred from the data in Fig. 2.12. Hence, efficiency goes down with decreasing wall thickness. For the case of an infinitely thin-wall tube, there is actually no workpiece and no heat dissipated in the workpiece, thereby leading to zero efficiency.

Critical frequencies for through heating of slabs and sheets depend on coil geometry. The relationships given earlier in this section pertain solely to flat workpieces heated by solenoid (i.e., encircling-type) coils. A plot of the critical frequencies of a number of metals for such induction heating arrangements is shown in Fig. 6.4. The similarity between this plot and that for round bars (Fig. 2.9) is evident. Note, however, the slightly lower values of f_c for the sheet geometry. For example, the value of f_c for a 2.54-mm- (0.1-in.-) diam round steel bar heated to 815 °C (1500 °F), which is above the Curie temperature, is approximately 10^6 Hz; but the f_c value for a 2.54-cm- (1.0-in.-) thick sheet heated to the same temperature is only 250,000 Hz.

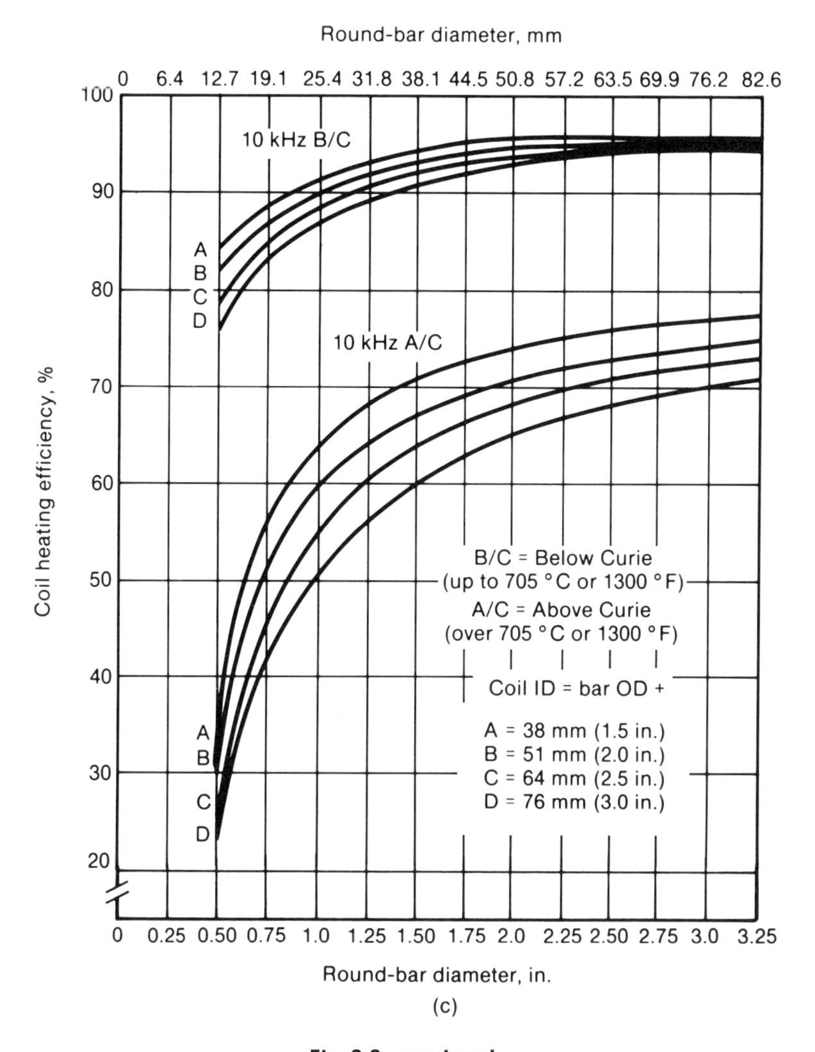

(c)

Fig. 6.2. continued

The critical frequency for a sheet material is lower when it is heated from one side only. This can be done using inductors such as pancake coils which set up fields of magnetic induction whose flux is *perpendicular* to the sheet surface, rather than parallel to it as with a solenoid coil. Such an arrangement is often referred to as transverse-flux induction heating. Figure 6.5 gives the required frequency for efficient heating utilizing a special inductor of this type designed to provide maximum temperature uniformity across the strip width. Heating of steel sheet 2.54 mm (0.1 in.) thick to 815 °C (1500 °F) using a transverse flux inductor requires a frequency of approximately 2000 Hz, which is two orders of magnitude less than that required if a solenoid coil is employed. The ability to use a much lower frequency impacts heating costs in two ways. First, the higher-frequency (RF) power supply that would be

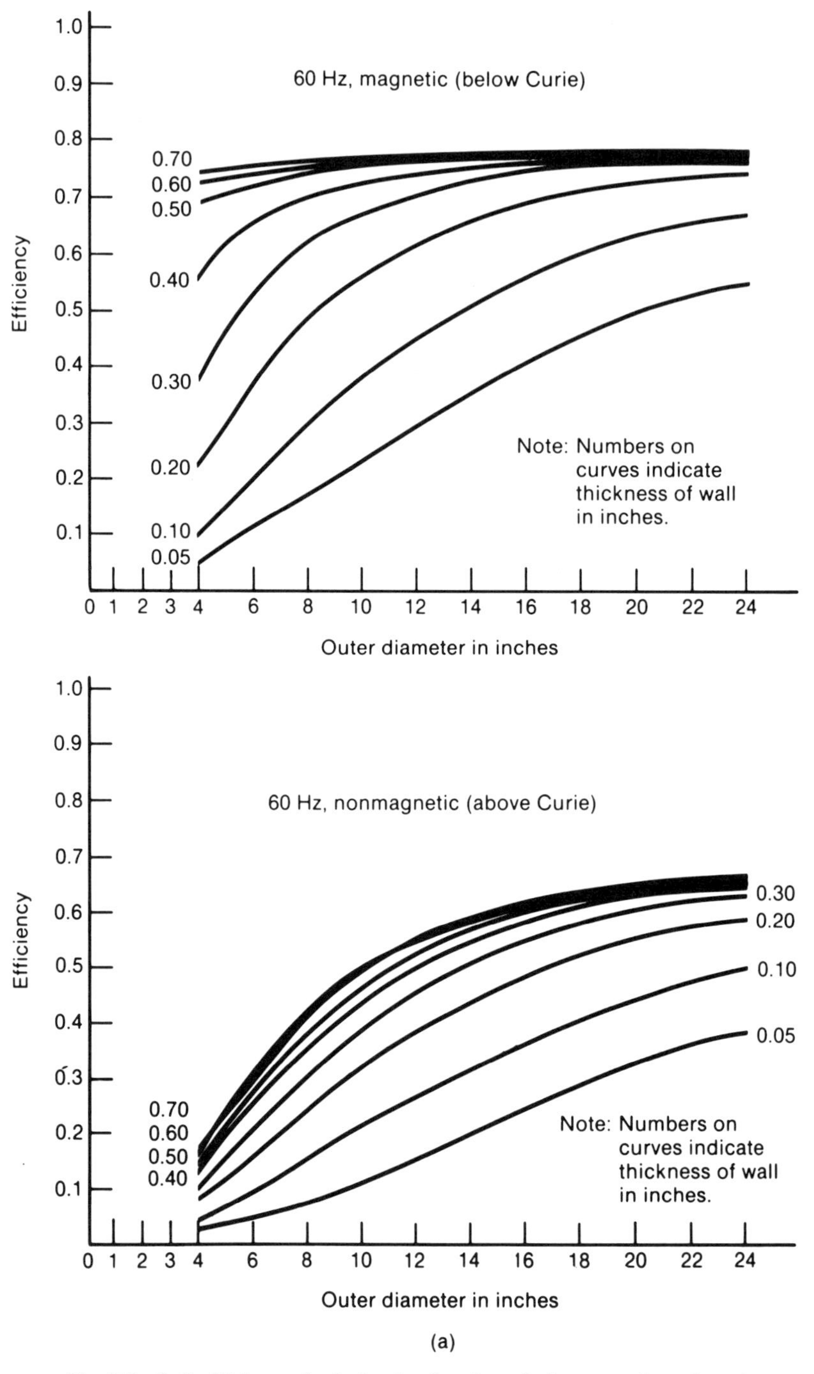

Fig. 6.3. Coil efficiency for induction heating of pipes as a function of wall thickness and outer diameter using power-supply frequencies of (a) 60, (b) 180, and (c) 960 Hz (from Brochure SA9906, Westinghouse Electric Corp., Baltimore)

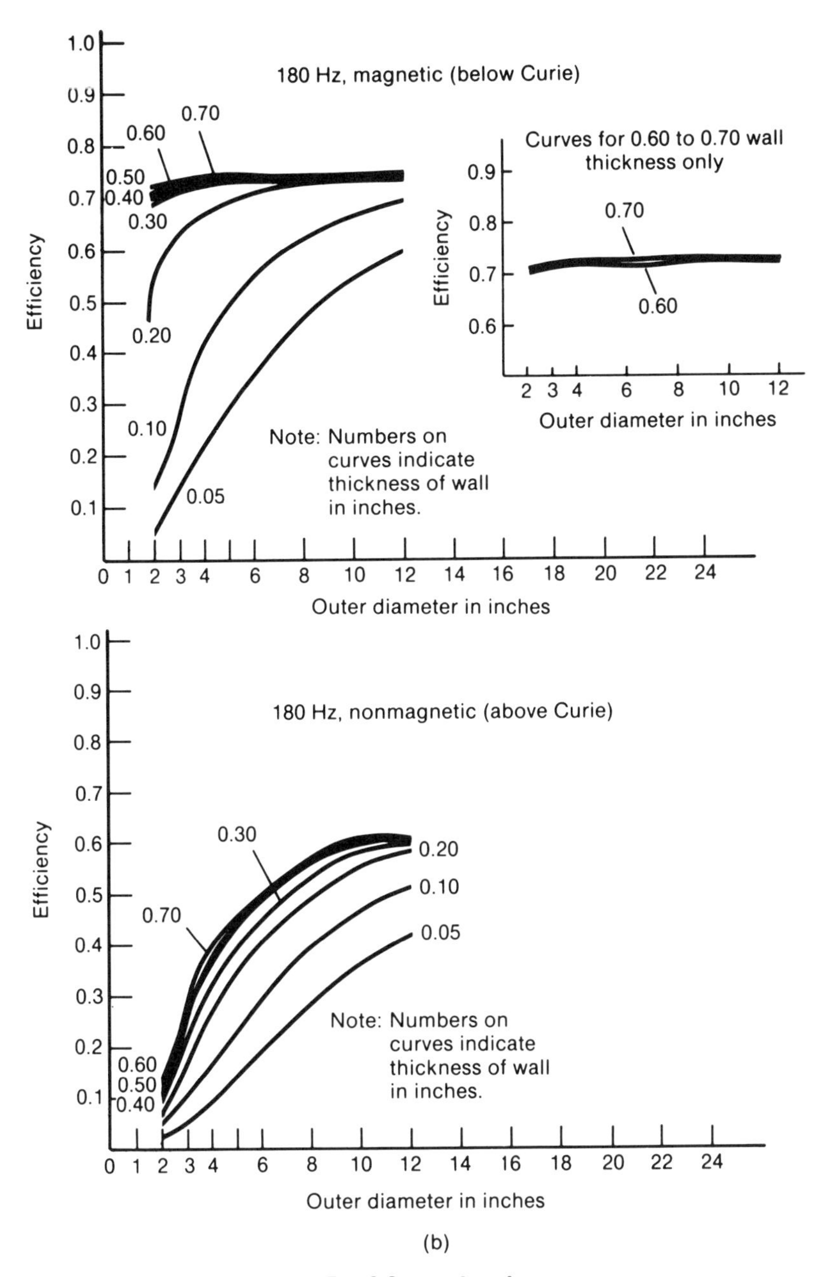

Fig. 6.3. continued

needed to provide 250,000-Hz power is considerably more expensive on a per-kilowatt basis than the medium-frequency unit that would supply 2000 Hz. Secondly, solid-state units are considerably more efficient in converting line-frequency power to the rated induction heating frequency than are vacuum-tube RF generators. Thus, in selection of frequencies for heating of sheet and

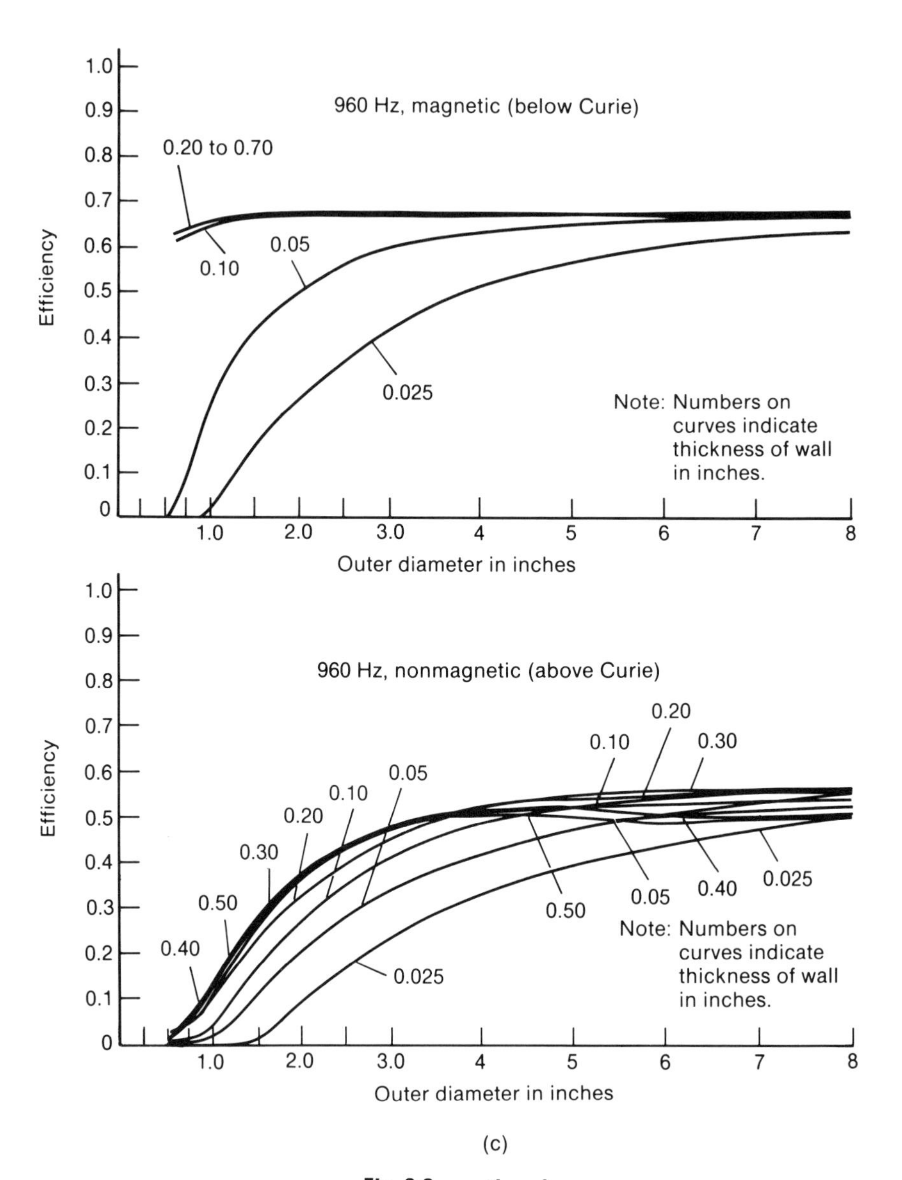

(c)

Fig. 6.3. continued

slab materials, coil design should be considered as well as efficiency and economics.

Selection of Power Rating for Through Heating

During selection of power rating for induction through heating, two factors—material throughput and required temperature uniformity—should be evaluated. The first of these, material throughput, determines the over-all power rating because:

Required power = (kilograms/hour throughput) × (kilowatt hours/kilogram) × (efficiency factor)

The factor "kilowatt hours/kilogram" is known as the heat content; it is the amount of thermal energy required to heat a given material usually from room temperature to a higher, specified temperature. As might be expected, the value of heat content increases with the temperature to which the metal is heated. Heat contents for a number of materials are summarized in Fig. 6.6. Tables 6.3 and 6.4 also provide related information—i.e., typical processing temperatures for several metals and the heat contents of these metals when raised to these temperatures.

The efficiency factor in the simple equation used to calculate power requirements is determined by the coupling efficiency of the coil to the workpiece (that is to say, the percentage of the energy supplied to the coil that is trans-

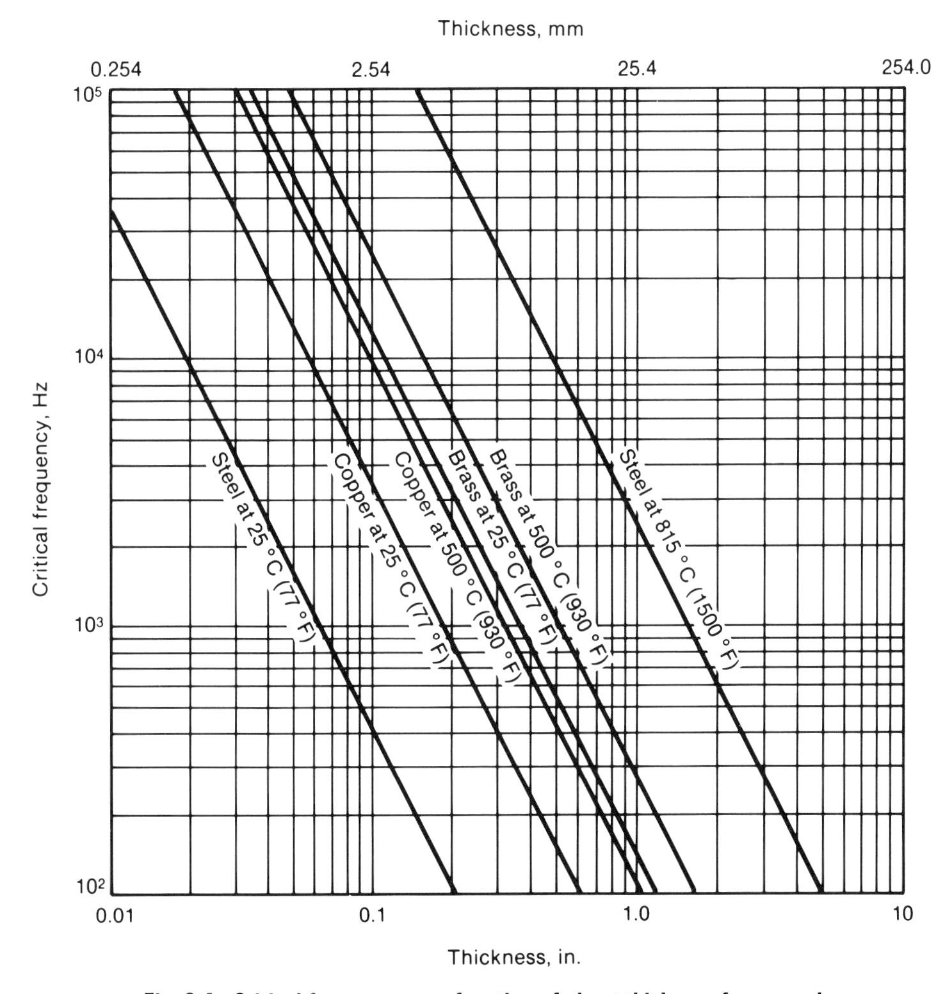

Fig. 6.4. Critical frequency as a function of sheet thickness for several different metals induction heated using a solenoid coil (from G. H. Brown, C. N. Hoyler, and R. A. Bierwirth, *Theory and Application of Radio Frequency Heating*, Van Nostrand, New York, 1947)

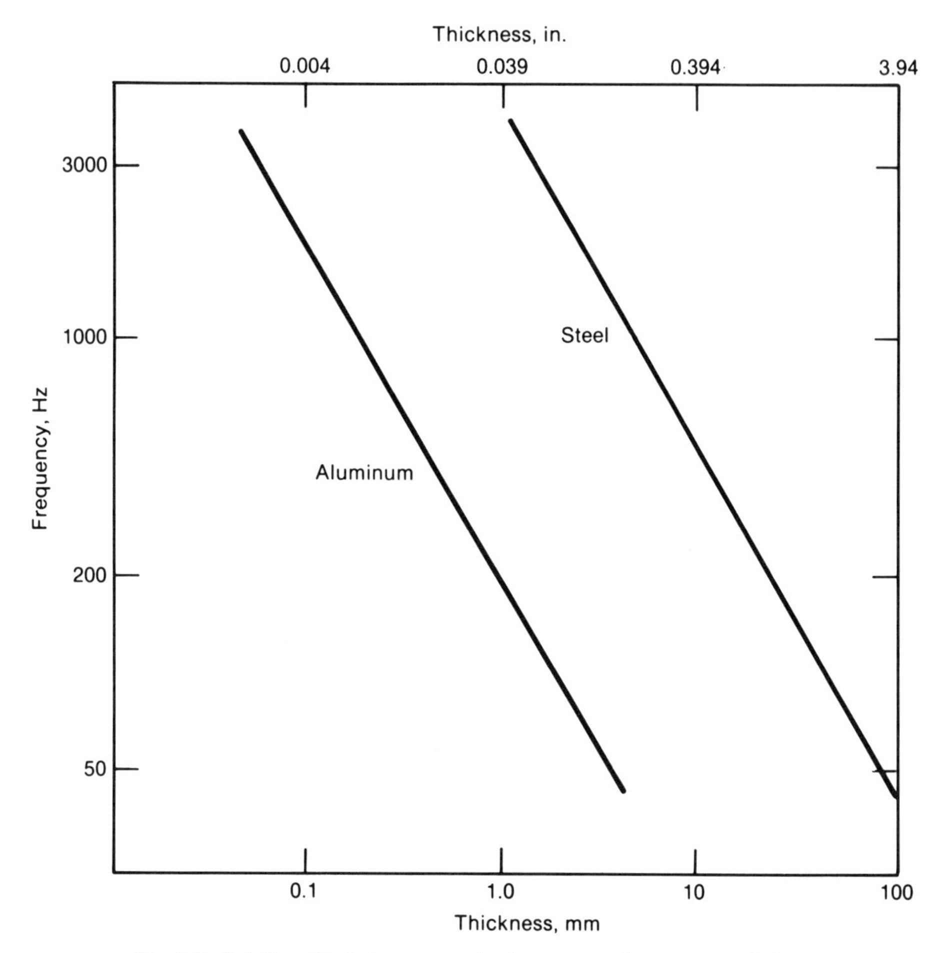

Fig. 6.5. Relationship between required generator frequency and sheet thickness for transverse-flux induction heating of aluminum and ferritic steel alloys (from R. Waggott, *et al.*, *Metals Technology*, Vol 9, December, 1982, p 493)

ferred to the workpiece) and by thermal losses due to radiation and convection. These factors were discussed in Chapter 2.

Requirements on temperature uniformity must also be considered in selection of power rating. Temperature uniformity during through heating is influenced, however, by both power-supply frequency and power density. Low frequencies enhance uniformity by providing large reference depths, or depths of current penetration. For this reason, frequencies as close as possible to the critical frequency should be selected. Lower frequencies lead to low heating efficiencies. Use of high frequencies is more expensive from an equipment standpoint and mandates the use of soaking time after the heating cycle to allow for heat conduction from the shallow-heated zone to enable temperature equalization.

Power density, or power per unit of surface area, has a similar effect on temperature uniformity. Excessively high power densities may lead to overheating (or even melting) of the workpiece surface. Moderate power densities allow moderate heating of the workpiece surface and sufficient time for the heat to be conducted to the center of the workpiece. The variations of the surface and center temperatures of an induction heated round bar are depicted schematically in Fig. 6.7. At the beginning of heating, the surface temperature increases much more rapidly than the temperature at the center. After a while, the rates of increase of the surface and center temperatures become comparable due to conduction. However, a fixed temperature differential persists during heating. The allowable temperature differential permits the generator power rating to be selected. The basic steps in selection of power rating for heating of round bar are as follows:

- Select the frequency and calculate the ratio of bar diameter to reference depth (a/d). As mentioned above, the frequency for through heating should

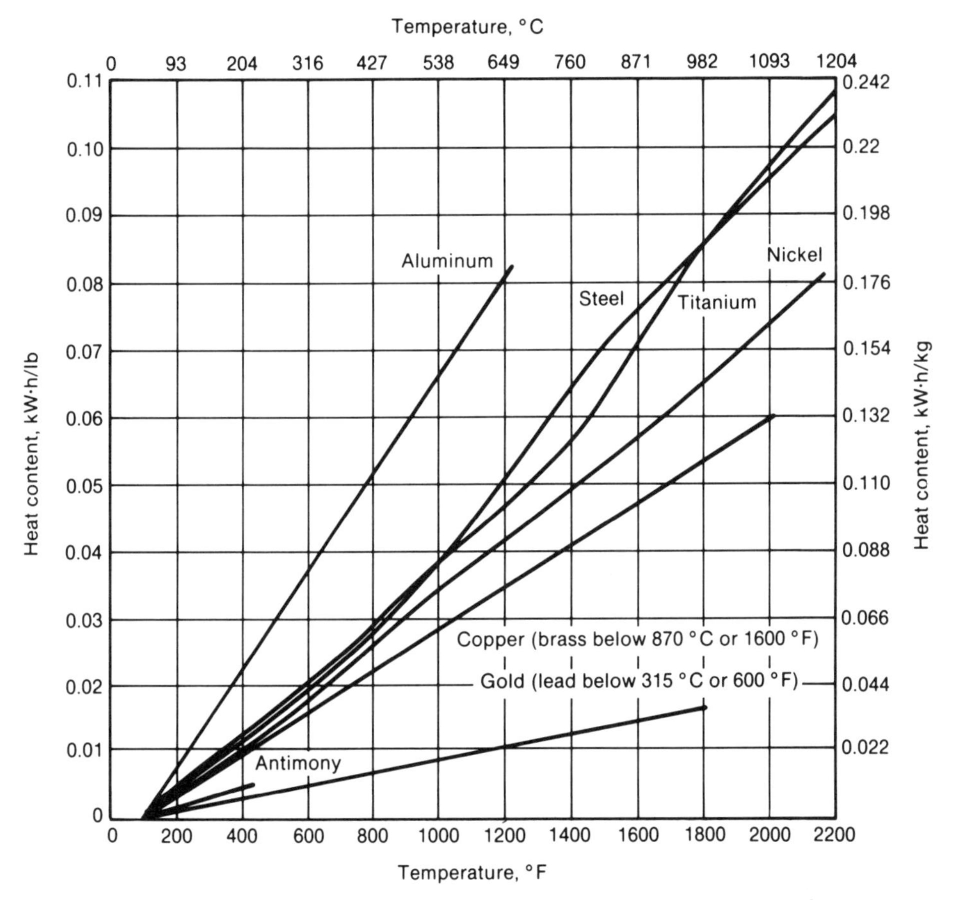

Fig. 6.6. Heat content (i.e., heat capacity) above 20 °C (70 °F) of several different metals as a function of temperature (from C. A. Tudbury, *Basics of Induction Heating*, Vol 1, John F. Rider, Inc., New York, 1960)

Table 6.3. Temperatures required for typical metalworking processes (from G. F. Bobart, "Innovative Induction Systems for the Steel Industry," *Proc. Energy Seminar/Workshop on New Concepts in Energy Conservation and Productivity Improvement for Industrial Heat Processing Equipment*, Chicago, March, 1982)

	Required temperature, °C (°F), for processing of:							
		Stainless steel						
Process	Carbon steel	Magnetic	Nonmagnetic	Nickel	Titanium	Copper	Brass	Aluminum
Hot forging	1230 (2250)	1095 (2000)	1150 (2100)	1095 (2000)	955 (1750)	900 (1650)	815 (1500)	540 (1000)
Hardening	925 (1700)	980 (1800)	...	760 (1400)	900 (1650)	815 (1500)	650 (1200)	480 (900)
Annealing/normalizing	870 (1600)	815 (1500)	1040 (1900)	925 (1700)	815 (1500)	540 (1000)	540 (1000)	370 (700)
Warm forging	760 (1400)	...	650 (1200)	650 (1200)	...	...	...	...
Stress relieving	595 (1100)	595 (1100)	595 (1100)	595 (1100)	595 (1100)	280 (500)	290 (550)	370 (700)
Tempering	315 (600)	315 (600)	315 (600)	315 (600)	315 (600)	...	...	...
Curing of coatings	230 (450)	230 (450)	230 (450)	230 (450)	230 (450)	230 (450)	230 (450)	230 (450)

Table 6.4. Average energy requirements for induction heating in typical metalworking processes (from G. F. Bobart, "Innovative Induction Systems for the Steel Industry," *Proc. Energy Seminar/Workshop on New Concepts in Energy Conservation and Productivity Improvement for Industrial Heat Processing Equipment*, Chicago, March, 1982)

	Required energy(a), kW·h/ton, for processing of:							
		Stainless steel						
Process	Carbon steel	Magnetic	Nonmagnetic	Nickel	Titanium	Copper	Brass	Aluminum
Hot forging	400	375	430	450	375	700	400	300
Hardening/aging	250	260	...	300	325	600	325	275
Annealing/normalizing	225	210	375	400	300	425	375	210
Warm forming	175	...	250	240	...	...	...	...
Stress relieving	150	150	200	250	225	200	200	210
Tempering	70	70	100	120	110	...	...	...
Curing of coatings	50	50	75	90	80	175	110	125

(a)Based on in-line continuous process.

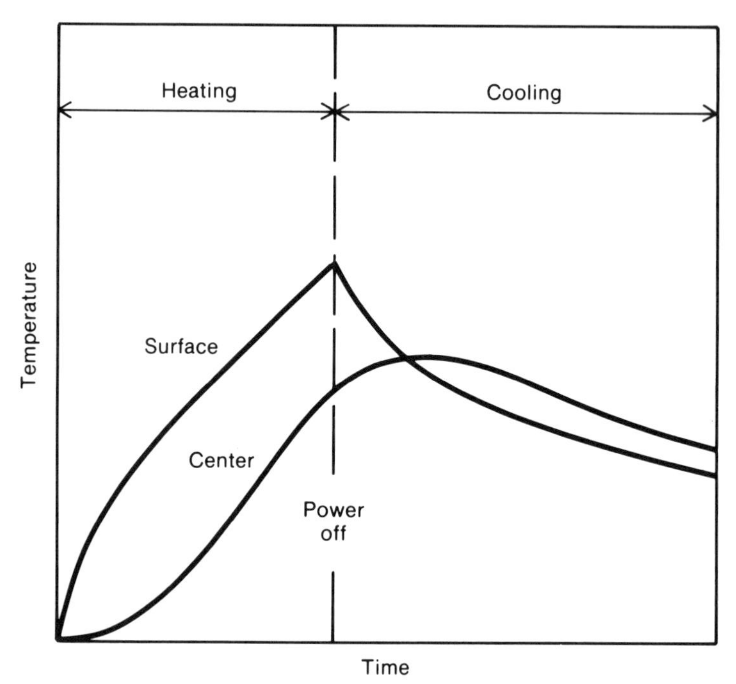

Note that, following an initial transient, the surface-to-center temperature differential is constant during the heating cycle. After heating, however, the surface cools more rapidly, leading to a temperature crossover.

Fig. 6.7. Schematic illustration of the surface and center temperature histories of a bar heated by induction (from C. A. Tudbury, *Basics of Induction Heating*, Vol 1, John F. Rider, Inc., New York, 1960)

be close to the critical frequency, and the value of a/d thus will be approximately 4 to 1.

- Using the value of thermal conductivity for the metal being heated (Fig. 6.8) and a/d, estimate the induction thermal factor K_T (Fig. 6.9).
- The power per unit length of the bar is calculated as the product of K_T and the allowable temperature differential between the surface and center ($T_s - T_c$). Multiplying this by the length of the bar yields the required net power (in kilowatts), to which the losses due to imperfect coupling and radiation must be added.

In Fig. 6.9, it can be seen that the K_T curves drop rapidly above a/d ratios of four to six. This trend demonstrates that lower power levels are needed at high a/d ratios to maintain the same temperature uniformity as that at the a/d ratio corresponding to the critical frequency.

To illustrate the above procedure, consider the through heating of a steel bar 152 mm (6 in.) long by 102 mm (4 in.) in diameter to a surface temperature of 980 °C (1800 °F) and a center temperature of 870 °C (1600 °F). Selecting a 540-Hz generator, the a/d ratio is found to be approximately four.

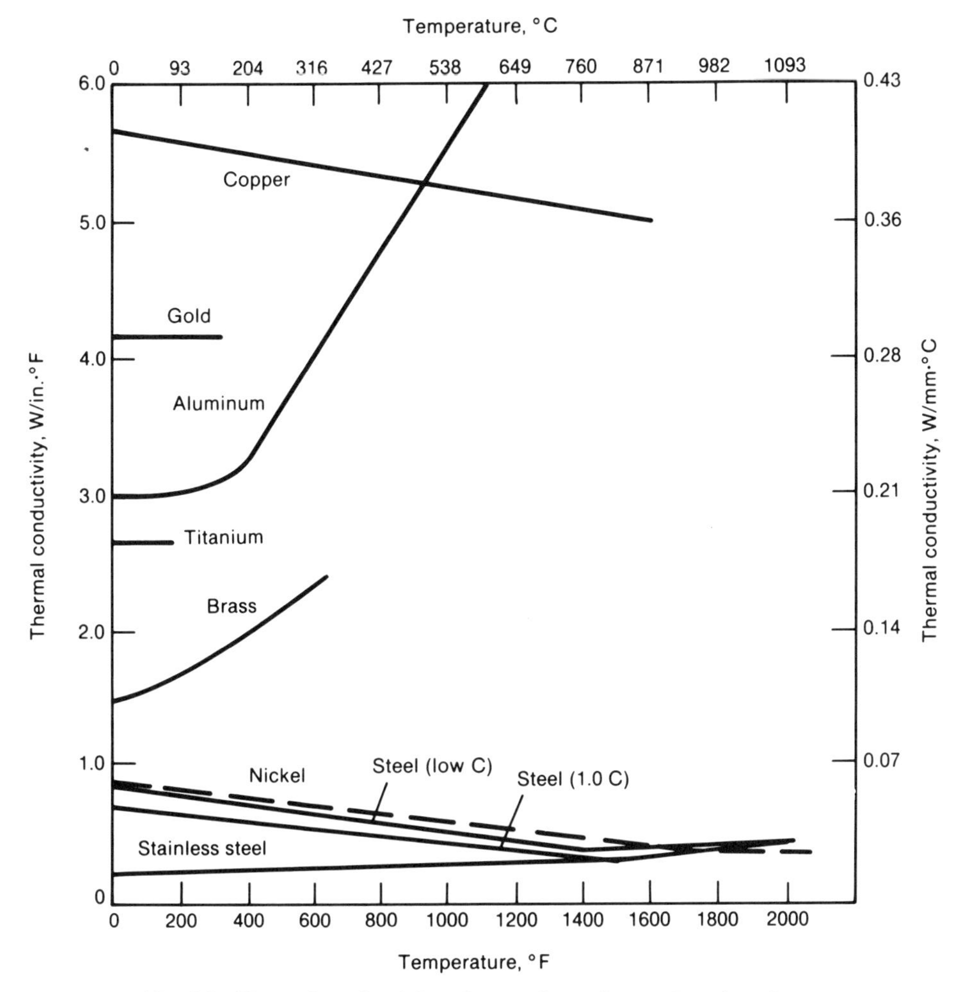

Fig. 6.8. Thermal conductivity of several metals as a function of temperature (from C. A. Tudbury, *Basics of Induction Heating*, Vol 1, John F. Rider, Inc., New York, 1960)

The average value of the thermal conductivity for carbon steels when heated to hardening temperature is about 0.050 W/mm·°C (0.70 W/in.·°F). From Fig. 6.9, K_T is found to be equal to 0.00113 W/mm·°C (0.016 W/in.·°F.) Thus, the power required is (0.00113)(110)(152) = 19 kW. This corresponds to a power density (power per unit of surface area) of 19/(152) (102)π = 0.00039 kW/mm^2 (0.25 kW/in.^{2}). If the temperature differential were to be cut by a factor of two, the power rating and the power density would be similarly decreased. There is, however, a lower limit to which the power can be decreased. This is determined by the heating time itself. The heating time is equal to the amount of energy needed divided by the rate of energy input, or power. For the above example, the workpiece weighs approximately 9.66 kg (21.3 lb), which, using data from Fig. 6.6, requires (21.3)(0.082) = 1.75 kilowatt hours to heat to 925 °C (1700 °F). Therefore the heating time will be either 328 s (for a surface-to-center temperature differential of 110 °C, or 200 °F) or 656 s (for a temperature differential of 55 °C, or 100 °F) when the power rating is halved.

In actual induction through-heating operations, power densities somewhat higher than those suggested by calculations such as the ones above are often used when a soak or dwell period precedes forming, quench hardening, or another operation. As suggested by the schematic illustration in Fig. 6.7, tem-

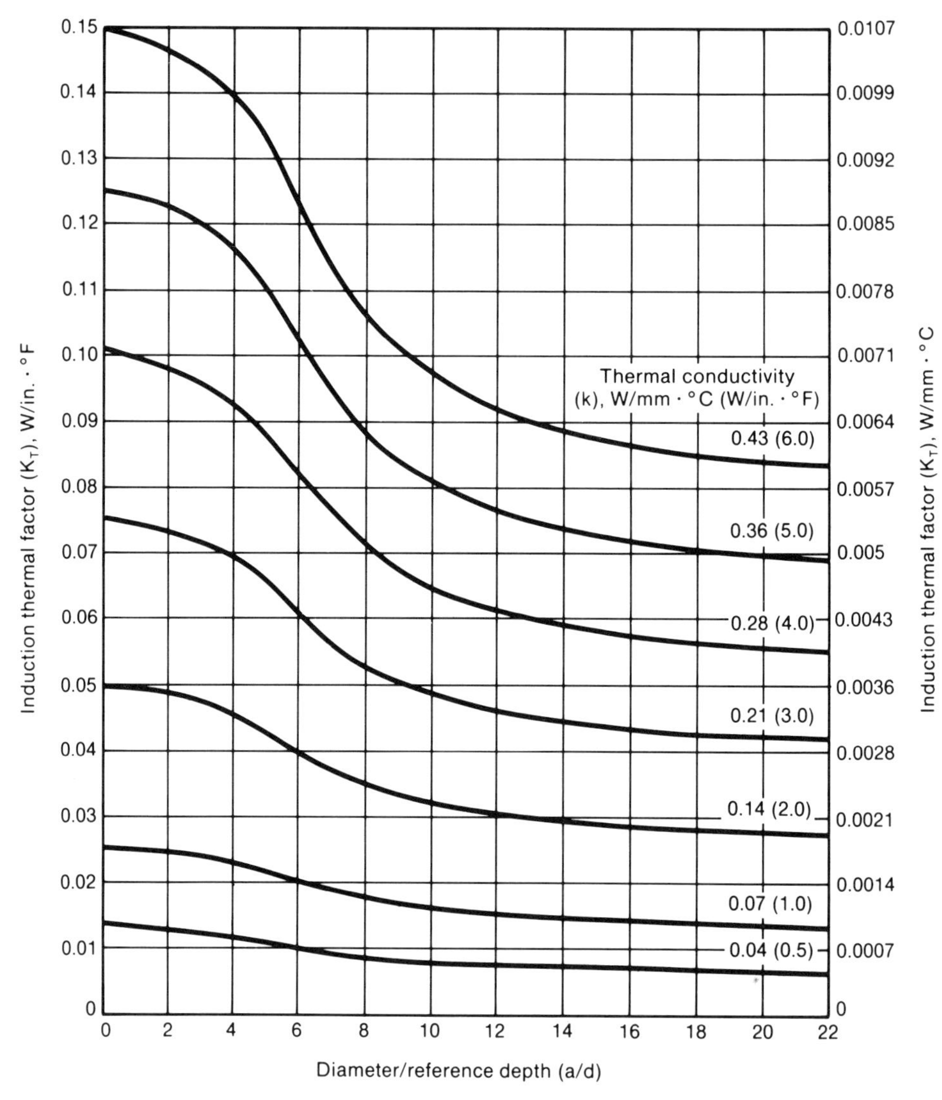

Fig. 6.9. Induction thermal factor for round bars as a function of ratio of bar diameter to reference depth (a/d) and thermal conductivity (from C. A. Tudbury, *Basics of Induction Heating*, Vol 1, John F. Rider, Inc., New York, 1960)

Table 6.5. Approximate power densities required for through heating of steel(a) (from T. H. Spencer, *et al.*, *Induction Hardening and Tempering*, ASM, Metals Park, OH, 1964)

Frequency Hz(b)	Power density(c), kW/in.^{2}(d), for through heating to temperatures of: 150 to 425 °C (300 to 800 °F)	425 to 760 °C (800 to 1400 °F)	760 to 980 °C (1400 to 1800 °F)	980 to 1095 °C (1800 to 2000 °F)	1095 to 1205 °C (2000 to 2200 °F)
60	0.06	0.15	(e)	(e)	(e)
180	0.05	0.14	(e)	(e)	(e)
1,000	0.04	0.12	0.5	1.0	1.4
3,000	0.03	0.10	0.4	0.55	0.7
10,000	0.02	0.08	0.3	0.45	0.55

(a) For hardening, tempering, and forming operations. (b) Power-density values in this table are based on use of proper frequency and normal over-all operating efficiency of equipment. (c) In general, these power densities are for section sizes of 1.27 to 5.08 cm (1/2 to 2 in.). Higher inputs can be used for smaller section sizes, and lower inputs may be required for larger section sizes. (d) 1 kW/in.^{2} = 0.155 kW/cm^2. (e) Not recommended for these temperatures.

perature uniformity is enhanced during this period as a result of heat conduction and radiation effects. After the power is turned off, the hotter surface cools as it loses heat to the center (by conduction) and to the surroundings (by radiation). In contrast, the center temperature continues to rise because of heat conduction from the surface.

Calculations of temperature uniformity as presented above can be avoided by reference to tables of power densities ordinarily used for through heating of various metals. One such listing, for through heating of magnetic steels, is shown in Table 6.5. The values of power in Table 6.5 are based on typical efficiencies and proper selection of frequency (which lead to a/d ratios usually in the range of four to six). Large-diameter bars, which can be heated efficiently with low-cost, low-frequency power supplies, typically employ low power densities. This is because of the longer times required for conduction of heat to the centers of these larger pieces. (It should also be kept in mind that the reference depth increases above the Curie temperature, at which the relative magnetic permeability drops to unity, sometimes necessitating higher-frequency units for heating to high temperatures. An exception to this practice is the use of 60-Hz sources for induction heating of very large parts such as slabs in steel mills.)

DESIGN PROCEDURES FOR HEAT TREATING

Several types of heat treatments can be conducted using induction heating. These include through and surface hardening, normalizing, and tempering of steels, through annealing of steels and nonferrous metals, and annealing of pipe welds. Design of induction through heat treating operations is very similar to those procedures employed for through heating prior to hot working, which were discussed in the previous section. In this section, only guidelines for surface and localized hardening and annealing processes are discussed. A more in-depth discussion of induction heat treatment is presented in the book by Semiatin and Stutz.*

Surface Hardening

Surface hardening of a steel part consists of raising a surface layer of the part to a temperature (denoted by Ac_3 or Ac_{cm}, depending on carbon content) at which it will be transformed to austenite and rapidly cooling the part to produce a hard martensitic structure in this region. Design of surface hardening treatments demands consideration of the workpiece material and its starting condition, the effect of rapid heating on Ac_3 or Ac_{cm} temperature, property requirements, and equipment selection.

Workpiece Material. Some typical induction surface hardened steels are:

- Medium-carbon steels, such as 1030 and 1045, used for automotive shafts, gears, etc.
- High-carbon steels, such as 1070, used for hand tools
- Alloy and stainless steels used for bearings, automotive valves, and machine-tool components
- Tool steels, such as M2 and D2, used for cutting tools and metalworking dies.

Cast irons (e.g., iron-carbon alloys which contain more than 2 wt % carbon) can also be induction hardened and tempered. Despite the inability to obtain a uniform austenite phase during the hardening operation, induction is still readily applied to many cast iron parts, such as cast crankshafts, to obtain hard, wear-resistant surfaces (55 to 60 HRC) and to avoid distortion. Frequency and power density are selected by use of guidelines similar to those for steels (discussed below), based on the size of the part and the case depth needed. Shallow cases require high frequencies and power densities, and deep cases require lower frequencies and power densities. The principal difference between hardening of steels and hardening of cast irons lies in the need to control temperature more closely for the latter materials. For steels, the austenitizing temperature can be varied over a fairly wide range without measurably affecting the hardening characteristics, provided that long soaking times and austenite grain growth are avoided. In contrast, temperature control is very

*S. L. Semiatin and D. E. Stutz, *Induction Heat Treatment of Steel*, ASM, Metals Park, OH, 1986.

important for cast irons because the carbon content of the austenite phase for a given alloy increases with temperature.

Starting Material Condition. The starting material condition can have a marked effect on response to induction hardening. The Ac_3 or Ac_{cm} temperature increases with coarseness of microstructure as well as with heating rate, as shown in Fig. 6.10 for 1042 steel. In particular, at a given heating rate, higher austenitizing temperatures are required for coarser microstructures. Such a behavior has a large effect on surface hardening response. For example, if the same induction heating parameters (e.g., power density, frequency, and heating time) were used for surface hardening of a given steel in three different starting conditions, the final case depths would vary even though the thermal histories of the three bars would be similar. It is apparent that such differences are caused by the different austenitizing temperatures for the various conditions. An example of this effect is shown in Fig. 6.11 for 1070 steel bars that were heated to a set temperature and quenched immediately. The results illustrate the fact that steels with a starting quenched-and-tempered microstructure tend to develop much deeper hardened cases than identical alloys induction heat treated under identical conditions, but containing an annealed microstructure. For the particular experiments used in deriving the data in Fig. 6.11, bars 2.54 cm (1 in.) in diameter were heated to 925 °C (1700 °F)

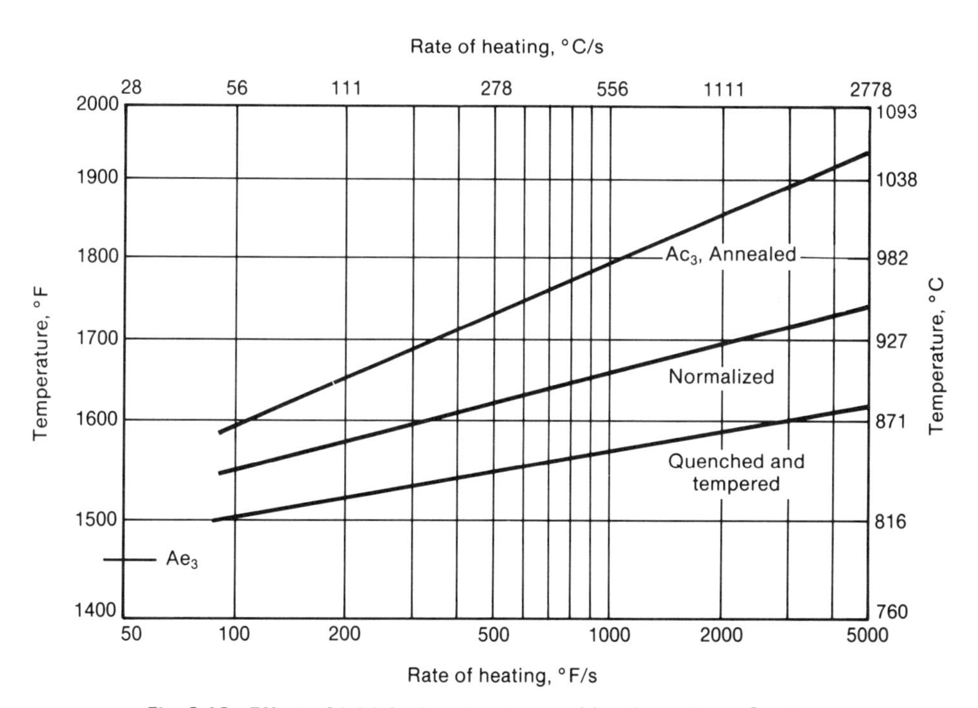

Fig. 6.10. Effect of initial microstructure and heating rate on Ac_3 temperature for 1042 steel (from W. J. Feuerstein and W. K. Smith, *Trans. ASM*, Vol 46, 1954, p 1270)

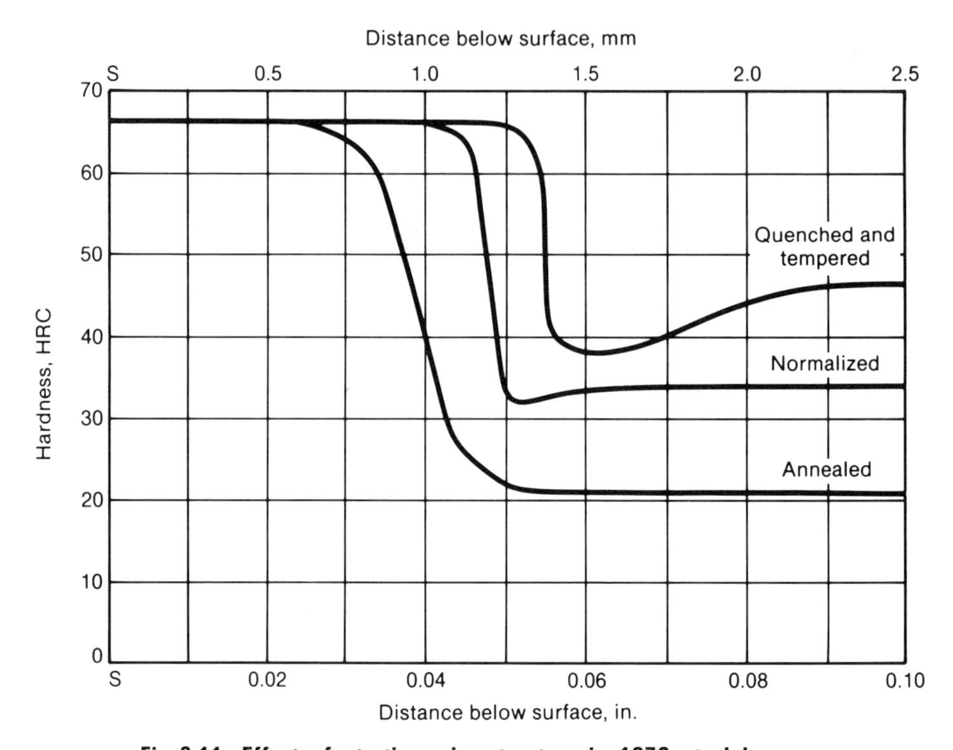

Fig. 6.11. Effect of starting microstructure in 1070 steel bars on surface-hardening response using a 450-kHz induction generator operated at a power density of 2.5 kW/cm^2 (15.9 kW/in.^{2}) (from T. H. Spencer, *et al.*, *Induction Hardening and Tempering*, ASM, Metals Park, OH, 1964).

in 1 s and water quenched. At this heating rate, the Ac_3 temperatures were found to be approximately 910 °C (1670 °F), 855 °C (1570 °F), and 815 °C (1500 °F) for the annealed, normalized, and quenched-and-tempered microstructures, respectively. It is thus apparent that the quenched-and-tempered bars were austenitized to a greater depth than the bars with the other microstructures and were, therefore, hardened to a greater depth.

The starting microstructure is also an important consideration in induction hardening of cast irons. It is important that carbon in the alloy will readily dissolve in the iron at the austenitizing temperature because of the short times inherent in induction processes. For example, the combined carbon in pearlite is readily soluble, and high percentages of pearlite (usually around 50 to 70% in gray iron and nodular iron) in the matrix are desirable; for this reason, cast irons are often normalized prior to hardening to obtain a pearlitic structure and to improve austenitizing response. Sufficient pearlite to give a combined carbon content of 0.40 to 0.50% is usually adequate to provide the hardnesses needed in cast iron parts. The combined carbon behaves essentially the same as carbon in steel of similar carbon content. On the other hand, graphitic carbon in the form of large flakes (gray cast iron) or large nodules (nodular cast iron, also often known as ductile iron) remains essentially unchanged during heat treatment. As another example, in nodular cast irons which have been previously quenched and tempered, the carbon in the matrix is in the form of many fine nodules. When this microstructure is induction heated to hardening temperatures, the carbon dissolves and diffuses rapidly, enabling quick austenitization and excellent hardening response.

Austenitizing Temperature. The time required to form a totally austenitic microstructure in a steel depends on the austenitizing temperature selected and the starting microstructure. In all cases, the speed with which austenite is formed is controlled by the diffusion of carbon, a process which can be accelerated a great deal by increasing temperature. For example, the time for complete austenitization in a plain carbon steel of eutectoid composition (0.8% C) with an initial microstructure of pearlite can be decreased from approximately 400 s (at an austenitizing temperature of 730 °C, or 1345 °F) to about 30 s (at an austenitizing temperature of 750 °C, or 1380 °F), as shown in Fig. 6.12. Each of these times can be considerably increased if the starting microstructure is a spheroidized one with large carbide particles. This is because the diffusion distance for carbon, which must be transported from the carbon-rich carbide phase, is considerably larger than in pearlite with thin lamellae of ferrite and carbide. Conversely, the finer bainitic and martensitic microstructures tend to be reaustenitized more readily than pearlite.

Data such as those in Fig. 6.12 suggest that at high enough temperatures,

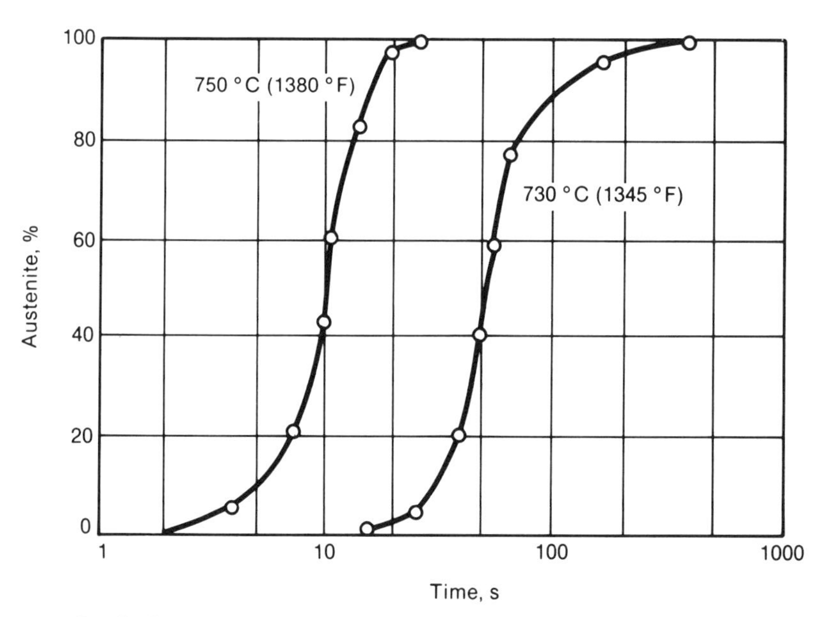

Fig. 6.12. Effect of austenitizing temperature on rate of austenite formation from pearlite in a eutectoid steel (from G. A. Roberts and R. F. Mehl, *Trans. ASM*, Vol 31, 1943, p. 613)

austenite forms in a fraction of a second. This fact is relied upon in surface hardening in which the workpiece surface or cross section is raised to a higher temperature than is normally attained in much slower, furnace-based processes. A large amount of effort has gone into the determination of the Ac_3 and Ac_{cm} temperature that signifies complete austenitization during *continuous* heating cycles such as those used in induction heating. Because the only "soaking" time available for phase transformation in these cases is that time after which the equilibrium transformation temperature (Ae_3 or Ae_{cm}) is exceeded, the continuous-heating transformation temperature (Ac_3 or Ac_{cm}) is always *above* the equilibrium one. This difference increases with heating rate, as might be expected—an effect shown in Fig. 6.10 for the Ac_3 temperature for 1042 carbon steel. Here, the Ac_3 temperature is the one at which it has been estimated that the austenite reaction is complete. As pointed out above, these data also show that the increase in transformation temperature depends on the initial microstructure. The fine quenched-and-tempered, or martensitic, microstructure revealed the least change in Ac_3 temperature as compared with the equilibrium Ae_3 temperature, whereas the same steel with an annealed microstructure exhibited the largest difference in Ac_3 as compared with the Ae_3 obtained with very low heating rates. Such a trend is readily explained by the fact that the diffusion distance to redistribute carbon is shorter in the former instance and longer in the latter microstructure, in which carbides are much larger.

Based on data such as those in Fig. 6.10, guidelines have been derived for required austenitizing temperatures for induction hardening of a wide range of steels (Table 6.6). Generally, these temperatures are approximately 100 °C (180 °F) above the equilibrium (low heating rate) austenitizing temperature primarily to reduce or eliminate totally the austenitizing soaking time during

Table 6.6. Recommended induction austenitizing temperatures for carbon and alloy steels(a)

Carbon content, %	Temperature for furnace heating, °F (°C)	Temperature for induction heating, °F (°C)
0.30	1550 to 1600 (845 to 870)	1650 to 1700 (900 to 925)
0.35	1525 to 1575 (830 to 855)	1650 (900)
0.40	1525 to 1575 (830 to 855)	1600 to 1650 (870 to 900)
0.45	1475 to 1550 (800 to 845)	1600 to 1650 (870 to 900)
0.50	1475 to 1500 (800 to 845)	1600 (870)
0.60	1475 to 1550 (800 to 845)	1550 to 1600 (845 to 870)
>0.60	1450 to 1510 (790 to 820)	1500 to 1550 (815 to 845)

(a) Free-machining and alloy grades are readily induction hardened. Alloy steels containing carbide-forming elements (e.g., niobium, titanium, vanadium, chromium, molybdenum,, and tungsten) should be austenitized at temperatures at least 55 to 100 °C (100 to 180 °F) higher than those indicated.

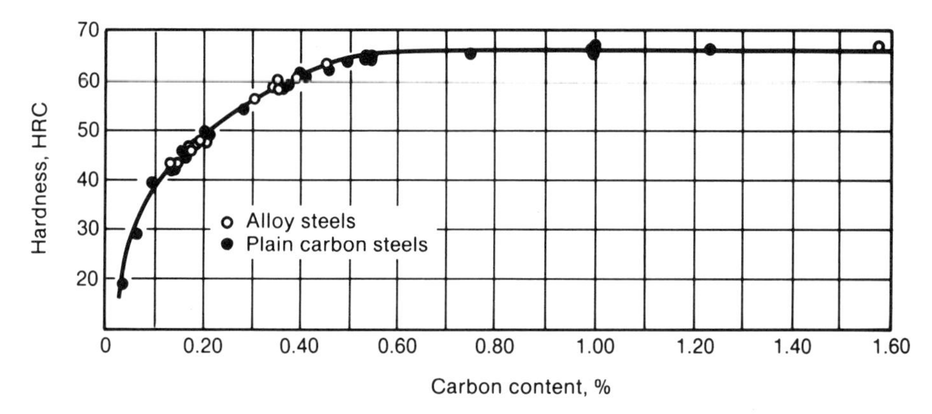

Fig. 6.13. Hardness of as-quenched martensite as a function of its carbon content (from J. L. Burns, T. L. Moore, and R. S. Archer, *Trans. ASM*, Vol 26, 1938, p 1; and W. P. Sykes and Z. Jeffries, *Trans. ASST*, Vol 12, 1927, p 871)

continuous induction heating cycles. However, they are still below the temperature at which undesirable austenite grain growth occurs rapidly. The recommended austenitizing temperatures are at least another 100 °C (180 °F) higher in alloys with strong carbide-forming elements (e.g., titanium, chromium, molybdenum, vanadium, or tungsten) than they are in carbon steels. These increases are a result of large increments in the transformation temperatures of alloy steels.

Property Requirements. Induction hardening is readily controlled to produce a wide range of final properties in finished products. The main properties which are of concern are case hardness and case depth. In as-quenched steels, fully hardened to 100% martensite, the hardness depends only on carbon content (Fig. 6.13). Thus, in many applications requiring a hard case and a soft core, an induction surface-hardened carbon steel can replace a surface- or through-hardened alloy steel. Often medium-carbon steels, such as 1045, provide the optimal blend of hardness and toughness at a modest cost.

For some products, only the surface-hardness characteristic is of importance with regard to service properties. This is particularly true in wear applications. Valve seats in automobile engines are a prime example of an application in which induction hardening has greatly improved wear resistance.

In applications such as torsional or bending fatigue, case depth and case hardness are both important. Figure 6.14 illustrates case-depth selection for a situation involving torsional fatigue. The drawing shows two different induction case-hardness patterns, each with a surface hardness of 52 HRC and a core hardness of 12 HRC. Assuming no stress concentrators, the applied stress is shown as a straight line from zero at the center to a maximum stress corresponding to 52 HRC at the surface. Case A provides a region in which the applied stresses exceed the yield strength; on the other hand, case B has a strength which equals or exceeds the applied stresses everywhere through the cross section. Thus, case B can be expected to provide better service properties.

The data in Fig. 6.14 suggest a means by which case-hardness patterns can be designed to meet service demands as specified by the applied stress patterns. For this purpose, the data in Table 6.7 give useful conversions between hardness and tensile and torsional yield strengths. The former conversions are important in bending fatigue, and the latter in torsional fatigue.

If hardness control were the only attractive feature of surface induction heat treatments, they would not be as popular as they are. What makes induction particularly useful is the fact that the process also introduces compressive residual stresses into the part surface. These residual stresses arise primarily from the density difference between the hard martensite layer and softer

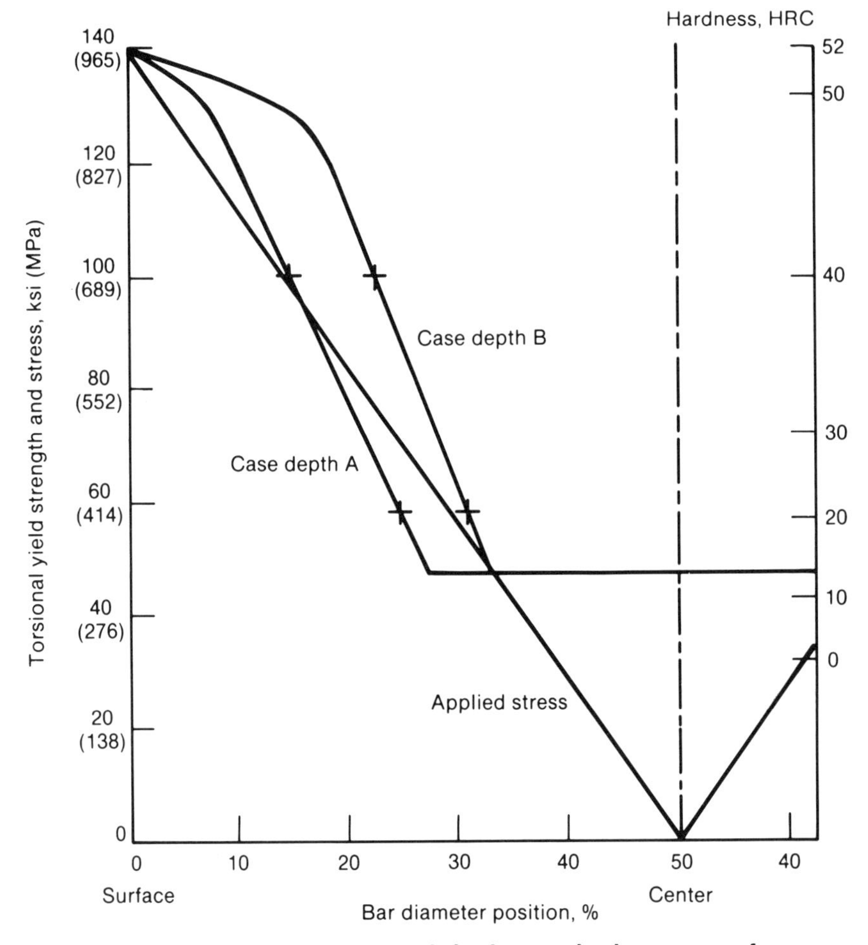

Fig. 6.14. Method of determining induction case-hardness pattern for shafts subjected to torsional loading (from G. A. Fett, *Metal Progress*, Vol 127, No. 2, February, 1985, p 49)

Table 6.7. Relationship between hardness and tensile and torsional strengths for hardened steels (from G. A. Fett, *Metal Progress*, Vol 127, No. 2, February, 1985, p 49)

Hardness		Strength, MPa (ksi)		
Rockwell C	Brinell	Tensile ultimate	Tensile yield	Torsional yield
52	514	1793 (260)	1613 (234)	965 (140)
50	481	1689 (245)	1524 (221)	917 (133)
40	371	1276 (185)	1151 (167)	689 (100)
30	286	965 (140)	841 (122)	503 (73)
20	226	758 (110)	614 (89)	365 (53)
10	187	655 (95)	503 (73)	303 (44)
0	150	517 (75)	365 (53)	221 (32)

interior layers. The magnitude of the residual stresses varies with the depth of hardening. It has been found in hardening of shafts, for example, that the level of the surface compressive residual stresses increases with the hardened depth. In addition, the depth to which the compressive residual stresses penetrate is usually about equal to the depth of the hardened layer.

The combination of a hard surface, compressive residual stresses, and a soft core results in excellent wear and fatigue resistance. Improvement in resistance to bending fatigue as a function of case depth in comparison with furnace treatment is shown in Fig. 6.15.

Equipment Selection—Frequency and Power Density. The equation for reference depth ($d \sim \sqrt{\rho/\mu f}$) can be used to estimate the optimum power-supply frequency for surface hardening by induction heating methods. Below the Curie temperature, d (cm) $\approx 5.8/\sqrt{f}$, or d (in.) $\approx 2.3/\sqrt{f}$, assuming $\mu = 100$. Above the Curie temperature (at which $\mu = 1$), d increases by a factor of 10. If heat conduction were not important, the sub-Curie reference-depth equation would give a reasonable approximation of the austenitized and hardened depth. Because of heat conduction, however, the case depth can be substantially greater, depending on the time allowed for conduction. To take this factor into account, therefore, a term proportional to the square root of the heating time τ is added to the reference depth to approximate the depth of austenitization and hardening, x_{case}:

$$x_{case}(\text{cm}) \approx (5.8/\sqrt{f}) + \sqrt{0.01\tau}$$

or

$$x_{case}(\text{in.}) \approx (2.3/\sqrt{f}) + \sqrt{0.0015\tau}$$

where the frequency f is expressed in hertz.

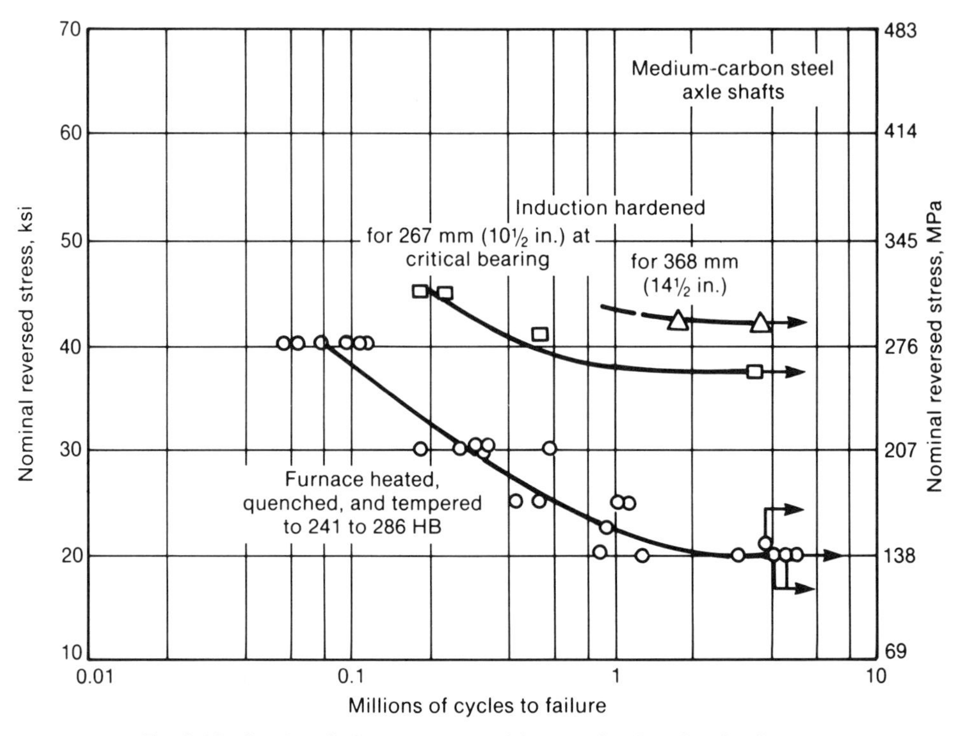

Fig. 6.15. Bending-fatigue response of furnace hardened and induction hardened medium-carbon steel tractor axles: shaft diameter, 7.0 cm (2.75 in.); fillet radius, 0.16 cm (0.063 in.) (from T. H. Spencer, *et al.*, *Induction Hardening and Tempering*, ASM, Metals Park, OH, 1964)

Power density also effects the depth of hardening.* For a given heating time and frequency, the case depth increases with power density. Because of this complicating factor, the usual practice in selection of frequency is to refer to tables (such as Table 6.8) that have been derived from years of experience with typical systems, rather than to use the above equations. Likewise, power densities for specific applications are also best determined by reference to nomographs derived from experience. An example is given in Table 6.9.

The information in Tables 6.8 and 6.9 is useful in selecting frequency and power density for surface hardening. However, there remains the problem of selecting heating time. In surface hardening, if the time is too long, the workpiece may reach austenitizing temperature at locations deeper than are necessary for the desired case thickness.

Two methods are available to establish the heating time for surface hardening operations. One involves a trial-and-error procedure in which the power and heating time are varied until the needed hardness and hardened depth are obtained. The hardened depth is determined metallographically by section-

*A theoretical treatment of the combined effects of heating time and power density is given in W. T. Shieh, *Metallurgical Transactions*, Vol 3, 1972, p 1433.

Table 6.8. Relationship between power-supply frequency and hardened depth obtained using induction heating

Frequency	Typical hardened depth
450 kHz	Less than 1.6 mm (0.06 in.)
10 kHz	1.6 to 3.2 mm (0.06 to 0.13 in.)
3 kHz	3.2 to 6.4 mm (0.13 to 0.25 in.)
1 kHz	6.4 to 9.5 mm (0.25 to 0.37 in.)
180 Hz	9.5 mm (0.37 in.) to full hardenability depth
60 Hz	Full hardenability depth

ing and acid etching. If the case is too shallow, the power input should be decreased and heating time increased. Conversely, if it is too deep, the power should be increased and the heating time decreased. Metallography will also show the presence of various microstructural constituents, such as martensite, which can be used as an indication of proper or improper austenitization.

A second method of determining optimum surface hardening parameters is through the use of nomographs such as those shown in Fig. 6.16 and 6.17. Figure 6.16 gives the relationship among generator frequency, applied power density, heating time, and hardened depth for steel shafts which are austenitized at temperatures between 850 and 900 °C (1560 and 1650 °F) using a single-shot induction heating method; this is a technique in which the entire area to be hardened is heated at one time by a coil which remains stationary relative to the part. From the figure, it is apparent that, for a given power density and heating time, shallow case depths require higher frequencies, or, at a fixed frequency, shallow case depths need higher power densities for shorter times. Figure 6.17 is a similar nomograph designed for surface hardening by a scanning method. In induction scanning processes, the coil moves relative to the part, thereby heating it in a progressive mode. This allows power supplies with low power ratings to be used for parts larger than would normally be possible. The curves in Fig. 6.17 give the approximate relationship between power density and heating time in the coil (coil length divided by scan rate) for various case depths. Note that different sets of curves apply depending on the power-supply frequency. Irrespective of which set is used, however, the greatest operating efficiencies are obtained by using conditions close to the steeper parts of the curves.

Localized Annealing of Steel Pipe Welds

The localized annealing of welded steel pipe represents another important application of the selective treatment capabilities of induction heating. In the actual fabrication process, steel strip is formed into a cylindrical shape, the seam is welded, and the tube is water quenched. This leaves a narrow zone of brittle martensite which must be stress relieved to enhance the toughness and subsequent formability. The amount of heat generated by welding affects the amount of heat required for subsequent induction annealing. Low-frequency welding operations provide diffuse heating and a wide heat-affected zone (HAZ), whereas high-frequency welding equipment produces a narrow HAZ.

Frequency requirements for induction annealing are typically found by

Table 6.9. Power densities required for surface hardening(a) (from T. H. Spencer, *et al.*, *Induction Hardening and Tempering*, ASM, Metals Park, OH, 1964)

Frequency, kHz	Depth of hardening(b), cm (in.)	Power density(c), kW/in.^{2}(d)		
		Low(e)	Optimum(f)	High(g)
500	0.038 to 0.114 (0.015 to 0.045)	7	10	12
	0.114 to 0.229 (0.045 to 0.090)	3	5	8
10	0.152 to 0.229 (0.060 to 0.090)	8	10	16
	0.229 to 0.305 (0.090 to 0.120)	5	10	15
	0.305 to 0.406 (0.120 to 0.160)	5	10	14
3	0.229 to 0.305 (0.090 to 0.120)	10	15	17
	0.305 to 0.406 (0.120 to 0.160)	5	14	16
	0.406 to 0.508 (0.160 to 0.200)	5	10	14
1	0.508 to 0.711 (0.200 to 0.280)	5	10	12
	0.711 to 0.914 (0.280 to 0.350)	5	10	12

(a) This table is based on use of proper frequency and normal over-all operating efficiency of equipment. Values given may be used for static and progressive methods of heating; however, for some applications, higher inputs can be used when hardening progressively. (b) For greater depth of hardening, a lower kilowatt input is used. (c) Kilowattage is read as maximum during heating cycle. (d) 1 kW/in.^{2} = 0.155 kW/cm^2. (e) Low kilowatt input may be used when generator capacity is limited. These kilowatt values may be used to calculate largest part hardened (single-shot method) with a given generator. (f) For best metallurgical results. (g) For higher production when generator capacity is available.

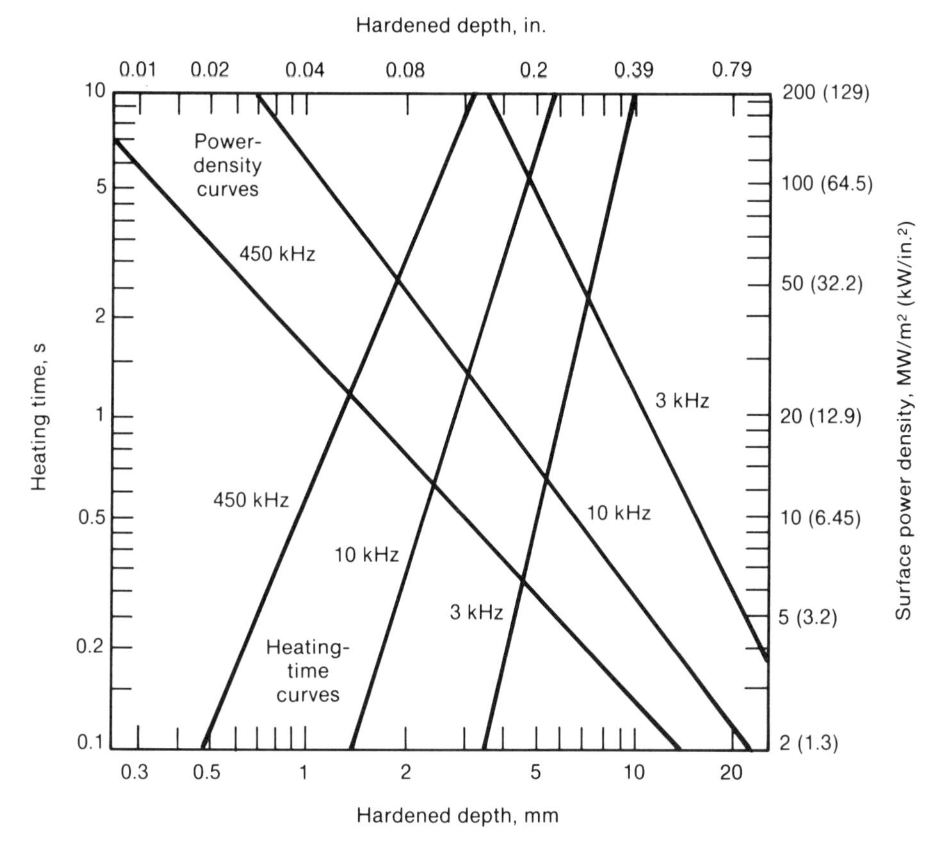

Fig. 6.16. Interrelationship among heating time, surface power density, and hardened depth for various induction generator frequencies (from M. G. Lozinskii, *Industrial Applications of Induction Heating*, Pergamon Press, London, 1969)

equating the wall thickness to the reference depth of steel above the Curie temperature. With this criterion, the usual frequency selections are 10,000 Hz for wall thicknesses up to 0.50 cm (0.20 in.), 3000 Hz for thicknesses between 0.32 and 0.95 cm (0.12 and 0.38 in.), and 1000 Hz for thicknesses over 0.95 cm (0.38 in.).

Localized heating through the thickness of the entire HAZ is provided by means of a split-return coil and flux concentrator in the form of a laminated core. The required power is then a function of the material flow rate (determined by the pipe speed, HAZ width, and wall thickness), the annealing temperature (which determines the heat content of the material), and the coupling efficiency. The calculation is simplified by reference to charts such as that shown in Fig. 6.18, which gives a factor by which the product of the line speed, HAZ width, and wall thickness should be multiplied to obtain the power in kilowatts. The derivation of Fig. 6.18 assumes a typical annealing temperature and air gap between the inductor and pipe.

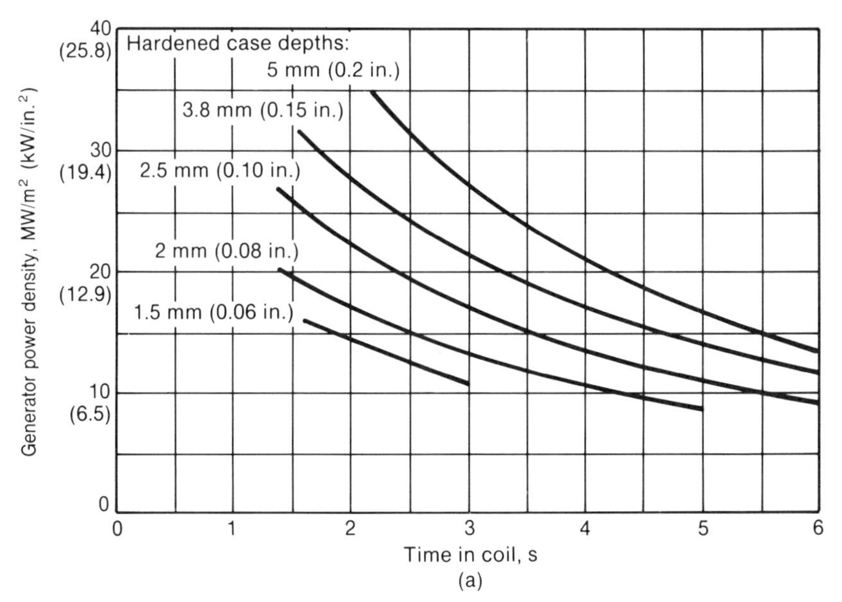

(a) Frequency, 3 kHz; minimum scanning rate, 6.5 mm/s (0.26 in./s); minimum shaft diameter, 25 mm (1 in.). (b) Frequency, 10 kHz; minimum scanning rate, 50 mm/s (2 in./s); minimum shaft diameter, 16 mm (0.63 in.). (c) Frequency, 450 kHz. ○—○: generator, power, 50 kW; case depth, 1.5 mm (0.059 in.). ○---○: generator power, 50 kW: case depth, 1.25 mm (0.049 in.). ●—●: generator power, 50 kW; case depth, 1.0 mm (0.039 in.). ●---●: generator power, 50 kW; case depth, 0.75 mm (0.030 in.). □—□: generator power, 25 kW; case depth, 1.5 mm (0.059 in.). □---□: generator power, 25 kW; case depth, 1.25 mm (0.049 in.). ■—■: generator power, 25 kW; case depth, 1.0 mm (0.039 in.). ■---■: generator power, 25 kW; case depth, 0.75 mm (0.030 in.). △—△: generator power, 10 kW; case depth, 1.5 mm (0.059 in.). △---△: generator power, 10 kW; case depth, 1.25 mm (0.049 in.). ▲—▲: generator power, 10 kW; case depth, 1.0 mm (0.039 in.). ▲---▲: generator power, 10 kW; case depth, 0.75 mm (0.030 in.).

Metal is SAE 1045 normalized steel. Coil length, 10 mm (0.39 in.). Two turns. Difference between coil ID and shaft OD, 3.5 mm (0.14 in.). Case depth determined by location at which microstructure is 50% martensite.

Fig. 6.17. Relationship among induction heating power density, time in coil, and case depth for progressive hardening of medium-carbon steel shafts under various conditions (from J. Davies and P. Simpson, *Induction Heating Handbook*, McGraw-Hill, Ltd., London, 1979)

DESIGN PROCEDURES FOR INDUCTION MELTING

As in heating prior to hot working and in heat treating, induction melting relies on electric currents that are induced in an electrical conductor (the charge to be melted) by suitably coupling it with a coil carrying an alternating current. Various coil and crucible designs are used for induction melting. The two most common are referred to as the coreless furnace (Fig. 6.19) and the channel furnace (Fig. 6.20).

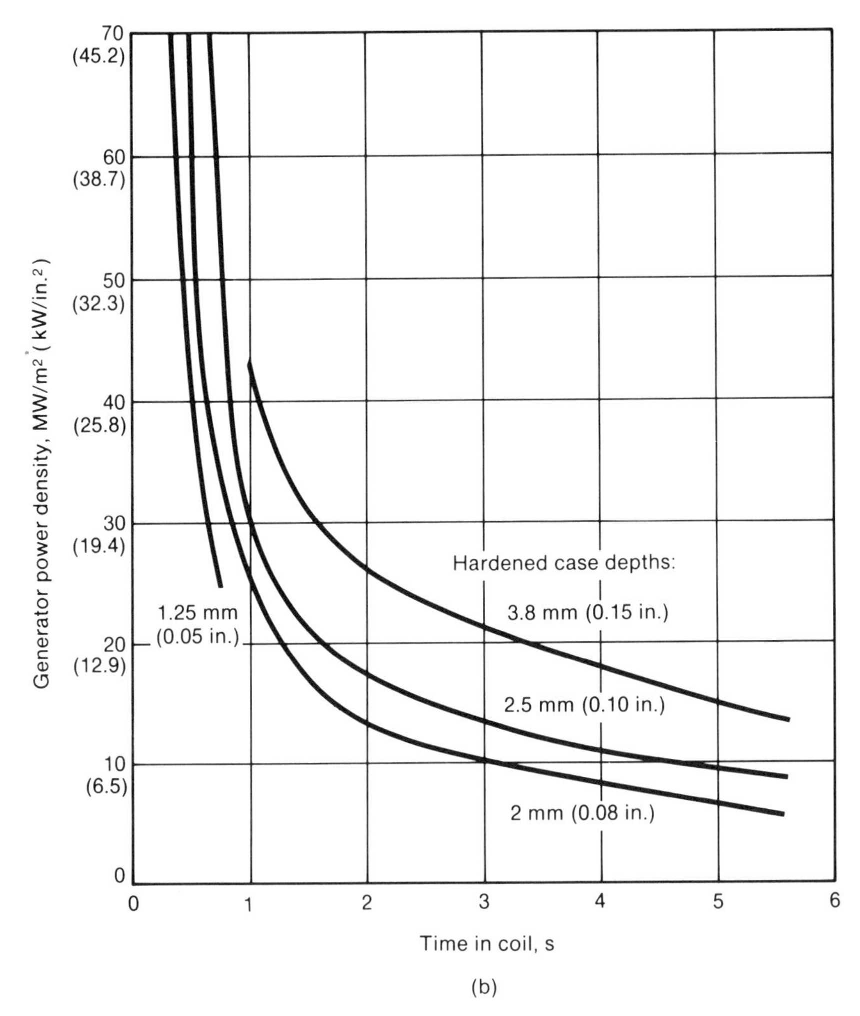

(b)

Fig. 6.17. continued

A coreless induction melting furnace consists of a refractory envelope to contain the metal surrounded by a coil. When the metal is molten, agitation occurs naturally. This stirring action is directly proportional to the power and inversely proportional to the square root of the frequency. Thus, with careful selection of frequency and power, the mixing and melting rate can be defined to give the best technical/commercial compromise. Coreless furnaces are classified according to the frequency of the ac power supply. The important types are line-frequency (50/60 Hz) and medium-frequency (180 Hz to 10 kHz) units. The line-frequency furnace is used primarily for high-tonnage applications (3 to 40 tons), whereas medium-frequency equipment finds its greatest use in applications ranging up to 5 tons.

Channel induction melting furnaces are primarily of the line-frequency type. The inductor in this design consists of a coil fitting over a core of magnetic steel laminations. The essential feature of the construction of the furnace is

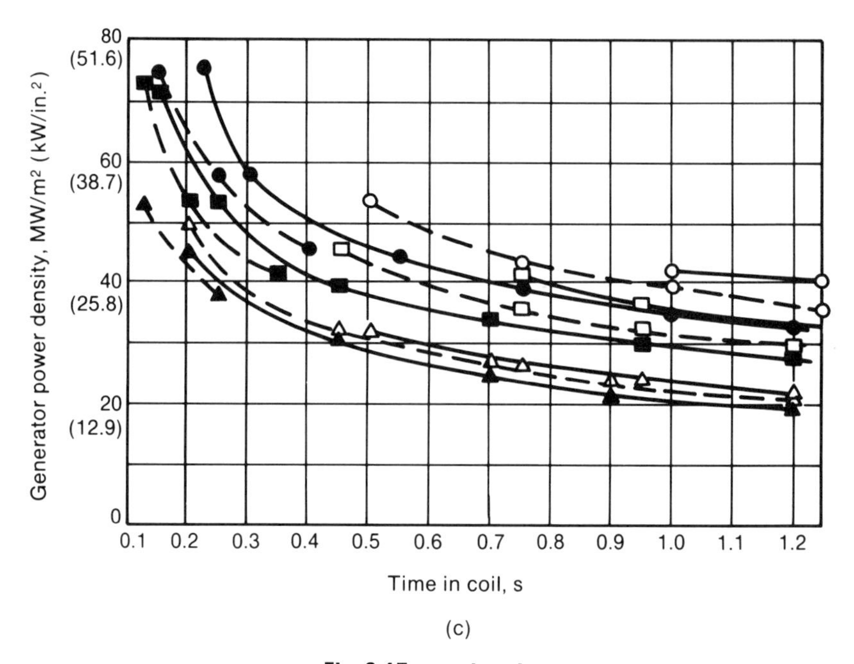

Fig. 6.17. continued

thus a small channel in the refractory vessel which surrounds the coil. This channel forms a continuous loop with the metal in the main part of the furnace body. By convection, the hot metal in the channel circulates into the main body of the charge in the furnace envelope to be replaced by colder metal. Unlike coreless induction melting, a source of primary molten metal is always required for start-up of the channel furnace. However, the major advantage of these units, which cover applications ranging from 1 to 50 tons, is the lower power input which can be tolerated for the furnace size. Surface turbulence within the main metal bath is considerably smaller, making this furnace more acceptable when gas pickup and volatile metal loss is a problem.

Induction melting units of both types are primarily used for refining and remelting of metals such as aluminum, copper, brass, bronze, iron and steel, and zinc. In addition to their use in the casting industry, channel-type induction melting furnaces are also currently being used for superheating and refining of blast furnace iron used subsequently in the basic oxygen furnace (BOF). Superheating involves heating of molten pig iron to a temperature considerably above its melting point. By doing this, more scrap steel can be melted in the BOF using the superheat contained in the charge from the induction furnace.

Design Considerations for Coreless Induction Melting Furnaces

Furnace Design. The main factors to consider in designing coreless induction melting furnaces are the crucible/refractory lining, coil geometry and flux guides, and the supporting frame. The refractory lining should be chosen with

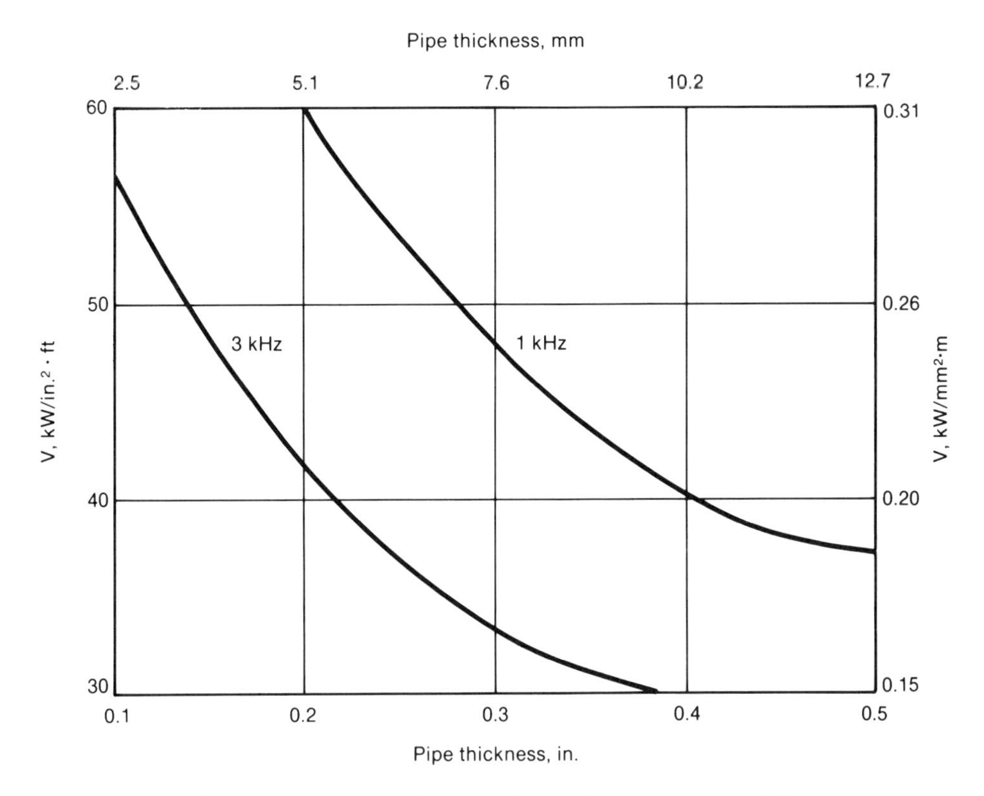

The required power in kilowatts is equal to the product of the line speed (in fpm), the pipe wall thickness (in in.), the width of the heated zone (in in.), and V as determined from the plot.

Fig. 6.18. Plot used to determine power requirements for induction welding of steel pipe (from G. F. Bobart, *Metal Progress*, Vol 94, No. 1, July, 1968, p 78)

due consideration of the metal to be heated. Lining materials are either acidic, basic, or neutral in nature. Acidic linings are suitable for high speed tool steels, plain carbon steels, low-alloy steels, cast irons, and nonferrous metals. Basic linings are used for stainless steels, high-manganese steels, and any steels in which special slag reactions are to be carried out. Basic refractory material is preferred for nickel-iron alloys. Lining materials commonly used are high-silica quartz (acidic), magnesite compounds (basic), and alumina or zirconia (neutral). Refractory installation and other furnace design features are discussed in more detail in Davies and Simpson.*

Frequency Selection. In induction melting, frequency selection is probably the most important design consideration. When choosing the power-supply frequency, the following factors should be weighed:

*J. Davies and P. Simpson, *Induction Heating Handbook*, McGraw-Hill, Ltd., London, 1979.

- Cost: Equipment cost increases with frequency. With the advent of high-efficiency solid-state power supplies, the expense of medium-frequency equipment relative to line-frequency systems is not excessive.
- Stirring action: The stirring action in a pool of liquid metal increases as the frequency is decreased. This agitation is desirable in recarburizing steel scrap or absorbing finely divided materials such as cast iron swarf, turnings, etc., with a minimum of metal loss. An excessively high frequency may produce insufficient stirring for proper alloying. On the other hand, excessive stirring often leads to shortened refractory-lining life and refractory inclusions in the metal.
- Starting: It is often difficult to start melting in an induction furnace when using a low frequency. For example, the melting rate is low in a line-frequency system (because of the poor ratio of particle size to skin depth) until 50 to 60% of the charge has been melted. To improve this situation, a small molten pool of 15 to 20% of the previous charge is left in the furnace. With medium frequencies, starting is not a problem.
- Efficiency: The poor starting at low frequencies reduces the efficiency of melting. This is partially offset by the dependence of converter efficiency on frequency. Modern solid-state systems, with automatic optimization of frequency and high efficiency, make medium frequencies increasingly attractive.

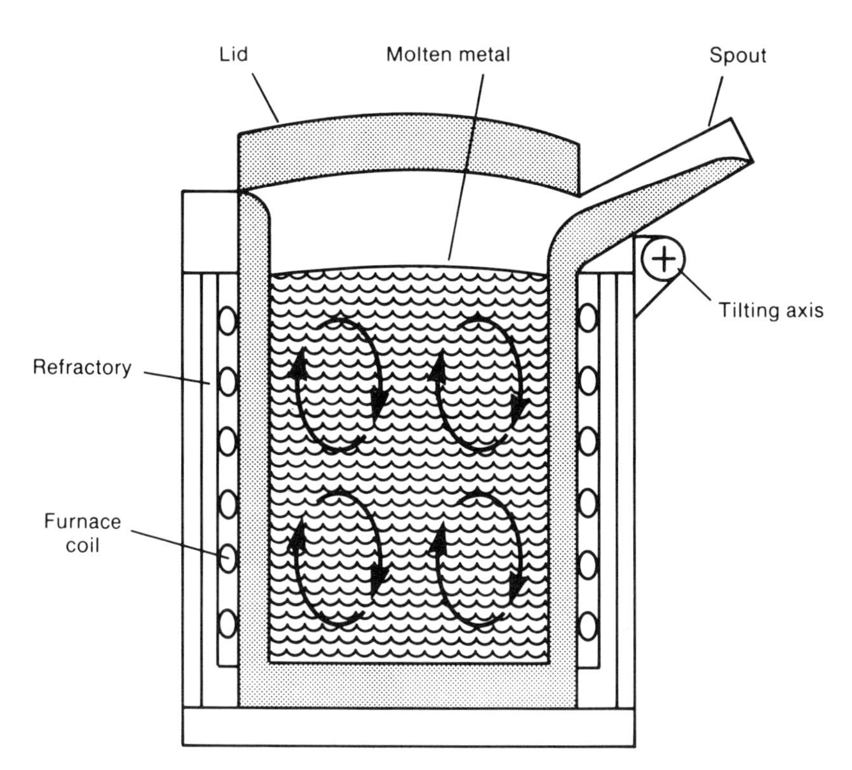

Fig. 6.19. Schematic illustration of a coreless induction melting furnace

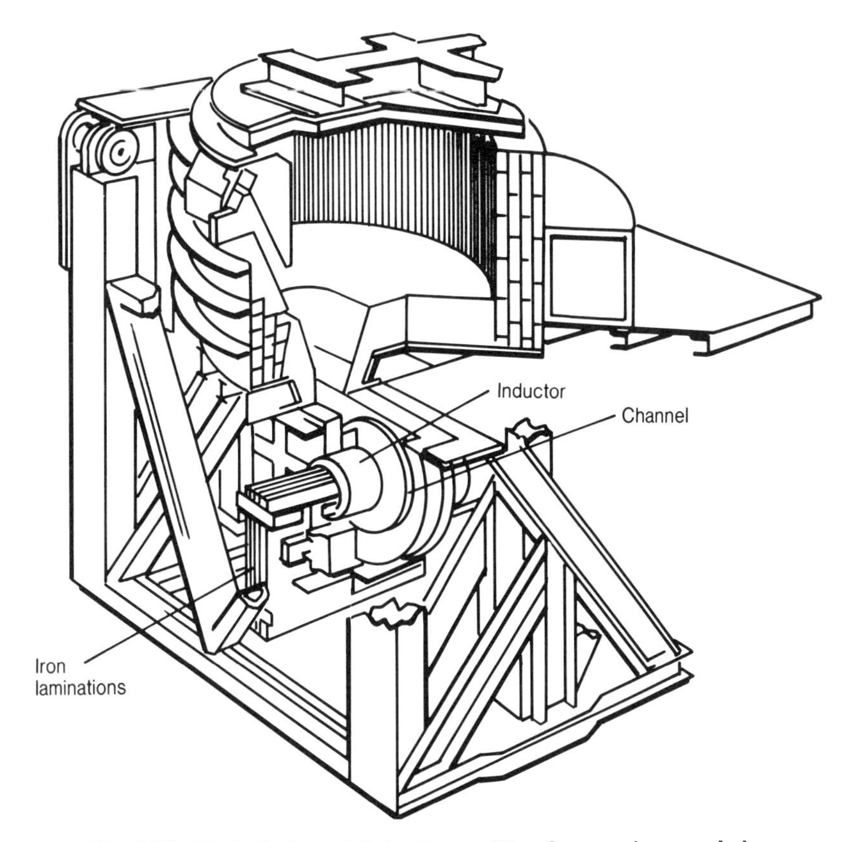

Fig. 6.20. Typical channel induction melting furnace (source: Inductotherm)

- Size of charge: Low frequencies are not suitable for small sizes of scrap because of the low ratio of particle size to reference depth until the melt has progressed sufficiently to absorb these materials into the molten metal. If there is a need for constant use of finely divided materials, starting with an empty furnace is preferable to using a medium-frequency furnace.
- Number of coil turns: The voltage induced in the workpiece or melt by induction heating is proportional to the ac frequency, the number of turns, and the strength of the field of magnetic induction (see Chapter 2). Hence, to maintain a given voltage, the number of coil turns must be increased if a lower frequency is employed. Thus, a low frequency requires a large number of turns with small conductors and small water passages. This leads to the increased risk of clogging and higher coil maintenance. A high frequency requires a large number of volts per turn and thus may lead to possible problems with arcing or electrical insulation.

Based on the above factors and on years of experience, it is possible to specify optimal frequencies for coreless induction melting as a function of furnace capacity. A useful summary of this information is given in Fig. 6.21.

Power Requirements. As with induction billet heating prior to hot working, power selection for coreless induction melting is largely based on the required output in tons of molten metal per hour. For example, the outputs of steel or cast iron that can be obtained with the various induction furnaces of one particular supplier are shown in Fig. 6.22 as a function of power rating. For the most part, doubling of the output in a given period of time requires doubling of the power input.

In practice, the power requirement for melting can be somewhat higher than that based on the heat capacity of the molten charge and the efficiency of coupling of the coil to the charge. For instance, about 500 kW·h per ton are needed to melt steel in a furnace with a capacity of 4 tons or more (Fig. 6.23), provided that the furnace is emptied once melting is complete. However, to allow for power-off time spent in charging, tapping, deslagging, etc., it may be necessary to allow a rating of 600 to 700 kW for every ton/hour of iron or steel required. Smaller furnaces will exhibit slightly higher energy consumption per ton. Energy consumption will also be higher if melting is done in a cold furnace lining.

Furnace Capacity. Furnace capacity should normally be at least equal to the desired batch size. Other factors may require the capacity to be increased above this figure. One of these factors is frequency as it affects stirring action. The large amount of stirring produced by line-frequency supplies means that

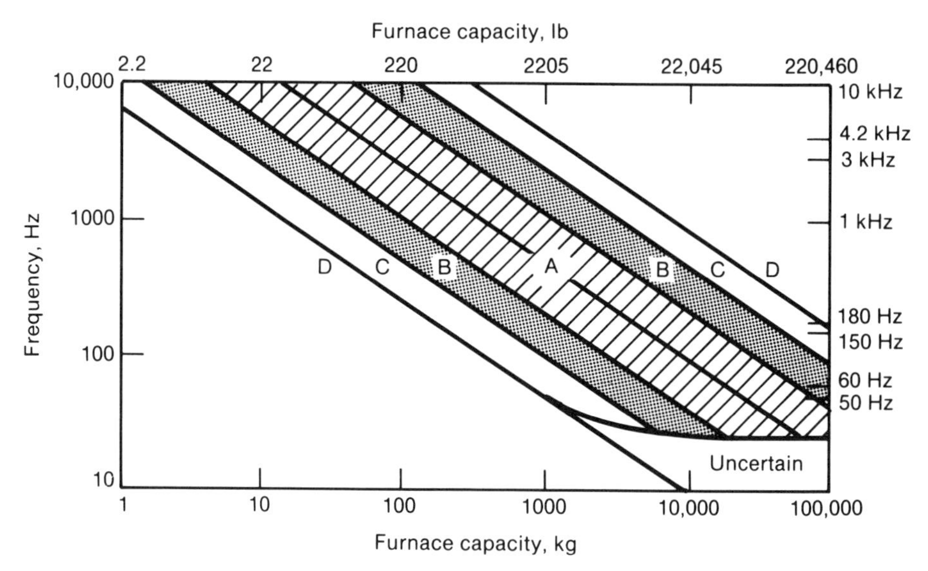

A = recommended frequency regime. B = acceptable frequency. C = furnace frequencies which have been used but which do not provide good results. D = unusable furnace frequencies.

Fig. 6.21. Selection of power-supply frequency for coreless induction melting furnaces as a function of furnace size (source: Inductotherm)

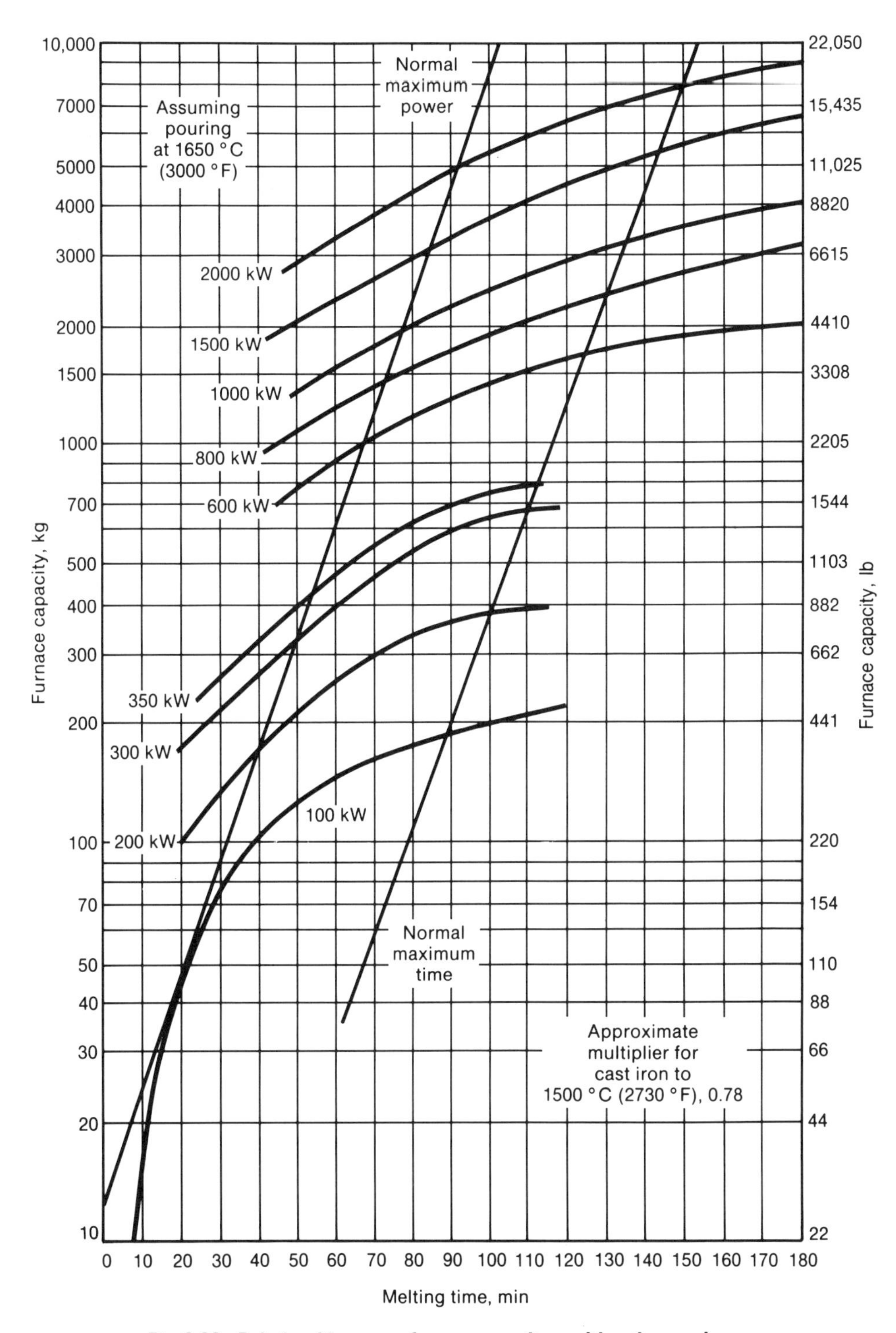

Fig. 6.22. Relationship among furnace capacity, melting time, and power requirements for coreless induction melting of irons and steels (Source: Radyne, Inc.)

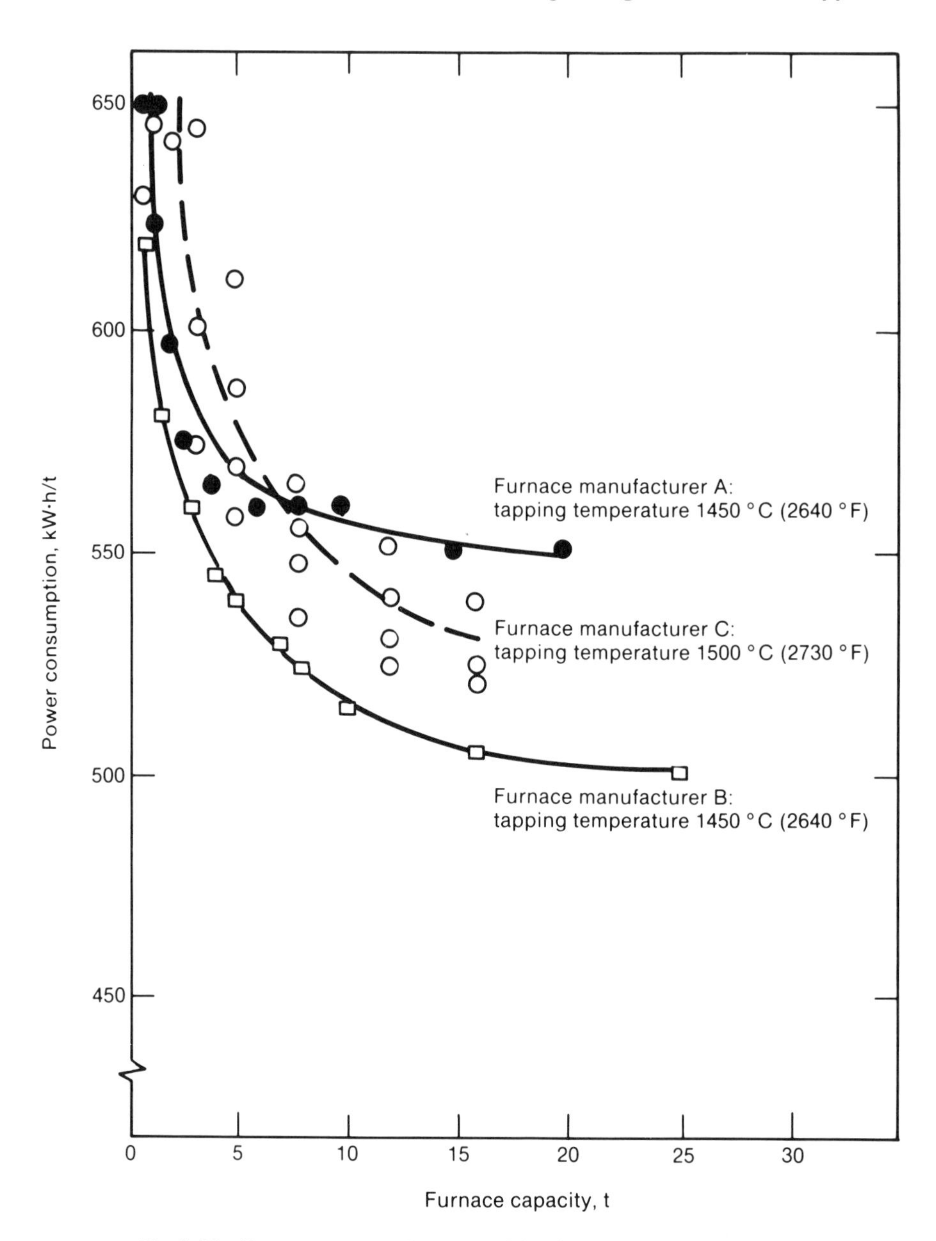

Fig. 6.23. Power consumption quoted by furnace manufacturers for melting of cast iron in line-frequency induction furnaces of various capacities (from W. A. Parsons and J. Powell, *Proc. Conf. on Electric Melting and Holding Furnaces in Iron Foundries*, University of Warwick, March, 1980, p 18-1).

the ratio of power to furnace capacity must be kept low when this kind of equipment is used. For example, if an output of 2.5 tons of molten steel per hour is obtained from a 6-ton-capacity furnace supplied with 1500 kW of line-frequency power, then the same output can be obtained from a 3-ton, 150-Hz

Table 6.10. Normal maximum furnace power ratings (from "Medium Frequency Coreless Furnaces for Iron Melting," British Electricity Council, Brochure EC 4402, November, 1983)

Furnace capacity, metric tons	Power rating, kW, at operating frequency of: 200 Hz	600 Hz	1000 Hz
1	...	500 to 700	600 to 800
2	...	1000 to 1400	1200 to 1600
3	1500 to 1800	1500 to 2100	1800 to 2400
4	2000 to 2400	2000 to 2800	...
5	2500 to 3000	2500 to 3500	...
6	3000 to 3600	3000 to 4200	...

furnace of equivalent power rating. Typical power ratings for coreless furnaces of several different frequencies are given in Table 6.10.

Operational Factors. Besides the above design considerations, several operational factors also impact the selection and economics of coreless induction furnaces. These include furnace utilization, furnace tapping time and frequency, molten heel practice,* slagging practice, lids and fumes extraction, charge handling, and single-shift versus multishift operation. Each of these is discussed in detail in a paper by Parsons and Powell.†

Design Considerations for Channel Induction Melting Furnaces

Because channel induction melting furnaces constrain metal to flow in a narrow channel, problems associated with bath turbulence are not encountered. Hence, such furnaces are almost always powered by line-frequency supplies. Furthermore, because the channel and the throat of the channel are the only parts subjected to intense heat input, this type of furnace is employed primarily as a holding or storage vessel for metal previously melted in an arc furnace, cupola, or coreless induction furnace.

The major design features for channel furnaces are furnace capacity, furnace construction, and holding-power requirements. Capacity is determined by use, which typically falls into one of four categories:

- Continuous charging from elsewhere (duplexing). In this case, the furnace is used to homogenize the composition and temperature of the melt. A capacity of about one hour's supply is usually sufficient.

*"Molten heel" refers to the molten pool left in the furnace after tapping.

†W. A. Parsons and J. Powell, "Design and Operational Factors Affecting Energy Consumptions," *Electric Melting and Holding Furnaces in Iron Foundries*, Proceedings of Conference Held at University of Warwick, March, 1980, p 18-1.

- Use as a storage vessel. When metal is melted during off-peak hours for use during the day, capacity usually must be large—on the order of the requirements for an entire day.
- Charging from another vessel. As mentioned above, charging from an arc furnace, cupola, or coreless induction furnace is quite common. In these cases, the capacity must be the same or greater than that of the other furnace.
- Use as a melting vessel. The size here depends on working needs, of course.

Furnace construction may take a variety of forms, including a vertical drum, a horizontal drum, or a drum with a rectangular bath which is either shallow or deep. In all cases, the channel in which metal is melted is located at the bottom of the holding vessel.

Holding power requirements depend to a certain extent on furnace construction. Horizontal drum furnaces have the lowest body heat losses. Semi-drum furnaces with rectangular baths have slightly higher body losses, and the vertical drum furnace has the highest losses. Despite these differences, required furnace power for metal holding is generally proportional to the tonnage of the furnace charge. Figure 6.24 shows a typical relationship for holding of irons and steel in channel induction furnaces.

DESIGN OF INDUCTION PIPE WELDING OPERATIONS

In induction welding of tube and pipe products, strip is formed between a set of specially designed rolls, the seam is brought together under a small amount of pressure, and a current is induced along the seam to bring about welding. Thus, there are important features of both mechanical and electrical design.

Mechanical Design Features

Mechanical design considerations are related to the incoming strip, forming rolls, seam guide, squeeze rolls, and work coil and impeder. Each of these is discussed separately.

Incoming Strip. The characteristics of the incoming material are very important in obtaining consistent results. The edges of the strip must be free of indentations, projections, inclusions, or laminations. If any of these are present there will probably be a discontinuity in the weld. Similar problems may result if the strip exhibits a width variation of as little as 0.25%, a tolerance which is very practical on modern slitting lines. Burrs from slitting should be situated such that they are on the inner surface of the tube after forming; this minimizes wear of the forming rolls. Problems related to burrs and nonuniform width can both be overcome by edge dressing on the tube mill. Dressing also is helpful in removing oxides prior to forming. On steel, rust on the edges is not deleterious. On the other hand, the edges of aluminum, copper, and brass strip should be free of oxides because very large voltages are required to break down the contact resistance in these situations. When this barrier is suddenly broken, inconsistent welding results are obtained.

Forming. The forming of strip prior to induction welding is somewhat different from the processing which precedes electric-resistance welding. In the latter

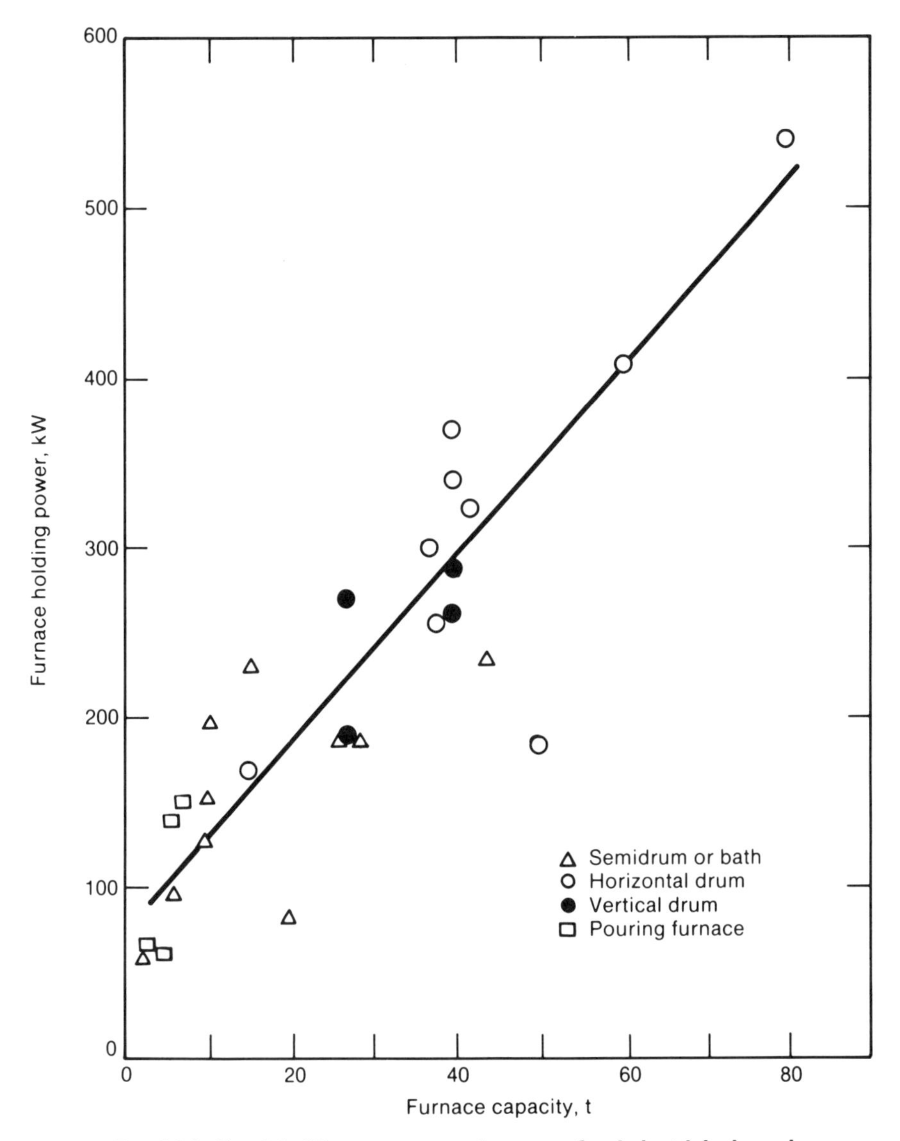

Fig. 6.24. Total holding power requirements for industrial channel induction furnaces of various capacities (from W. A. Parsons and J. Powell, *Proc. Conf. on Electric Melting and Holding Furnaces in Iron Foundries*, University of Warwick, March, 1980, p 18-1; and *British Foundryman*, Vol 76, No. 10, October, 1981, p 212)

situation, the strip is formed into a somewhat oval configuration, after which two welding electrodes on either side of the seam come into contact with the edges, thereby pressing them together during the actual welding process. The current path is through the seam from one electrode to the other.

Best results in induction welding are realized when the strip is formed such that the mating edges of the strip are parallel to one another at the first point of contact. Thus, a forming sequence which leads to a basically round profile with a slightly flattened portion at the seam is very desirable. This shape is usually achieved by overforming of the strip edges during the initial stages of the forming operation.

Seam Guide. A seam guide is a thin blade positioned immediately after the forming stands and before the welding station. Its purpose is to keep the weld positioned at the top so that excess weld bead on the outer surface of the tube is easily removed downstream by trimming tools. Tool steel seam guides are common, but they cause sparking and deterioration of strip edges prior to welding. Ceramic blades avoid these problems and have the added advantage of being electrically nonconducting, thus eliminating the flow of induced eddy currents between the strip and the blade.

Squeeze Rolls. Following forming, squeeze rolls are used to press the edges of the strip together with a well-controlled force and to shape the tube into its final welded form. Various arrangements of squeeze rolls are available to perform these functions, each with its advantages and disadvantages. These arrangements, some of which are depicted in Fig. 6.25, include:

- Two side pressure rolls alone
- Two side pressure rolls with one or two supplementary top rolls
- Two side pressure rolls with a bottom support roll and one or two top rolls
- Three pressure rolls.

Work Coil and Impeder. The inductor for tube welding is usually a simple solenoid coil (Fig. 6.26). With such a coil, the induced eddy currents form a complete circuit by flowing around the back of the tube and then along the open vee-shape edges to the point where the tube weld bead ends. The currents are more highly concentrated at this point than anywhere else, with the result that more heat is developed here. This makes it possible to weld the edges together without wasting a large amount of energy elsewhere.

The number of coil turns and the coil geometry are determined by the generator used (and thus considerations related to the tuning of the appropriate tank circuit) and by the production requirements. Guidelines for clearance between the coil and the tube are given in Fig. 6.27. Both the over-all coil

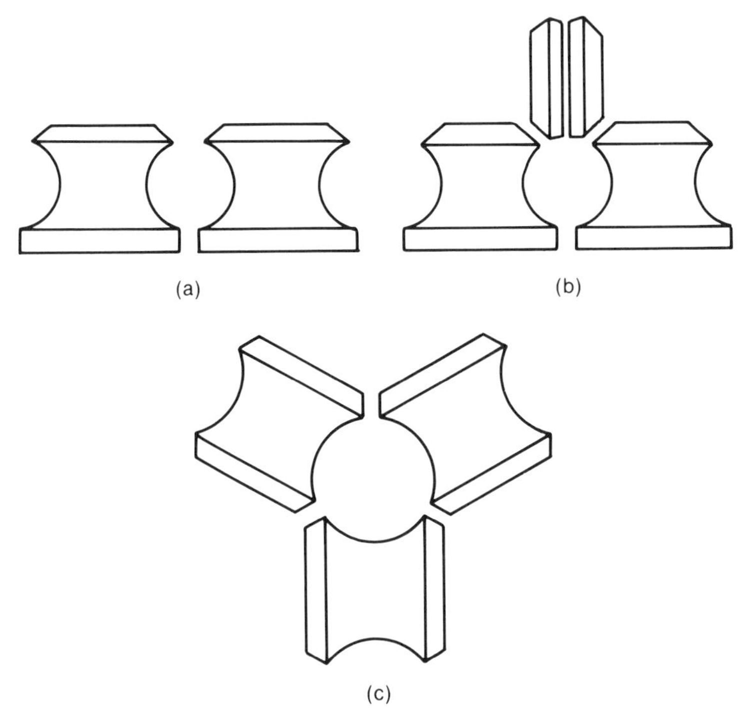

Fig. 6.25. Squeeze-roll arrangements for induction tube welding: (a) two-roll; (b) two-roll with supplementary roll; (c) three-roll (from J. Davies and P. Simpson, *Induction Heating Handbook*, McGraw-Hill, Ltd., London, 1979)

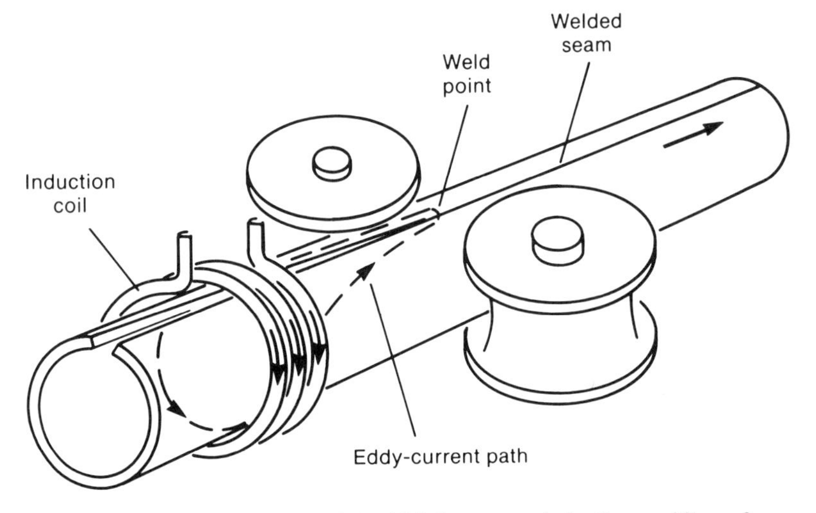

Fig. 6.26. Eddy-current path in high-frequency induction welding of tubular products (from C. A. Tudbury, *Basics of Induction Heating*, Vol 1, John F. Rider, Inc., New York, 1960)

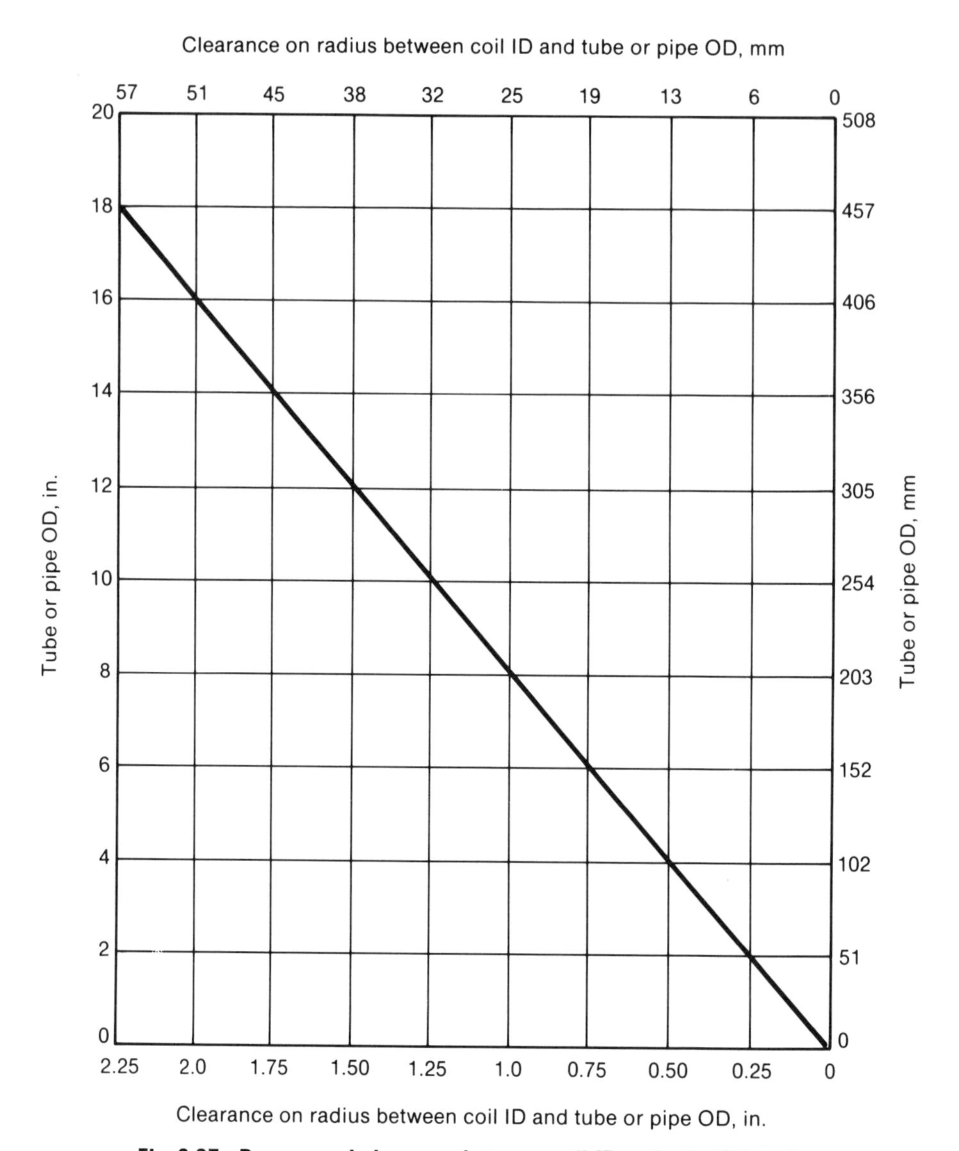

Fig. 6.27. Recommend clearance between coil ID and tube OD during induction welding of tubular products (from J. A. Redmond, *Induction/Resistance Tube and Pipe Welding*, Westinghouse Electric Corp., Baltimore)

length and the distance between the weld point and the nearest part of the coil should be approximately equal to the internal diameter of the coil. If the inductor is a relatively long distance from the weld point, most of the current returns through the wall of the tube and not through the weld point, resulting in decreased efficiency and a wide heat-affected zone. Similarly, if the inductor is relatively close to the weld point, the current through the weld point is high, but because of the low resistance of the short path, the total power input into the tube is small, again leading to low working efficiencies.

In order to obtain good heating efficiencies, it is necessary to use an impeder. An impeder is a flux concentrator placed inside the tube which improves the heating pattern. When the impeder is placed under the open vee, the current that would otherwise tend to spread out over the surface of the tube is caused to concentrate at the edges. The increased current density at the edges increases their temperature, and thus efficiency is greatly improved.

Impeders are made of ferrite or powdered iron to reduce eddy-current losses in the impeder. They should be placed as close to the vee opening as practical and should be held in a fixed position to prevent current variations. Often they are enclosed in a nonconducting tube so that they can be water cooled. The downstream end of an impeder is placed near the weld point, because this is where the induced eddy currents reverse direction. Impeder lengths for welding steel tubes of various sizes are listed in Table 6.11.

When nonferrous materials are welded, impeders are usually not needed. It has been found that the increased welding speeds obtained with impeders in these cases do not warrant their use because of the typically low power consumption for such materials.

Electrical Design Features

As with other induction heating applications, selection of frequency and power level are the two prime factors in power-supply selection.

Frequency Selection. As in pipe seam annealing, frequency requirements are found by equating the wall thickness to the reference depth d. In all cases, $\mu = 1$ in the equation for d, because steel is welded well above the Curie temperature. For typical wall thicknesses (i.e., up to 0.38 cm, or 0.15 in.), the

Table 6.11. Relationship between outer diameter and impeder length for induction welding of steel tubes (from J. A. Redmond, *Induction/Resistance Tube and Pipe Welding*, Westinghouse Electric Corp., Baltimore)

Nominal tube size, cm (in.)	Impeder length, cm (in.)
1.3 (0.5)	25.4 (10)
2.5 (1)	25.4 (10)
3.2 (1.25)	25.4 (10)
3.8 (1.5)	30.5 (12)
5.1 (2)	38.1 (15)
6.4 (2.5)	38.1 (15)
11.4 (4.5)	38.1 (15)

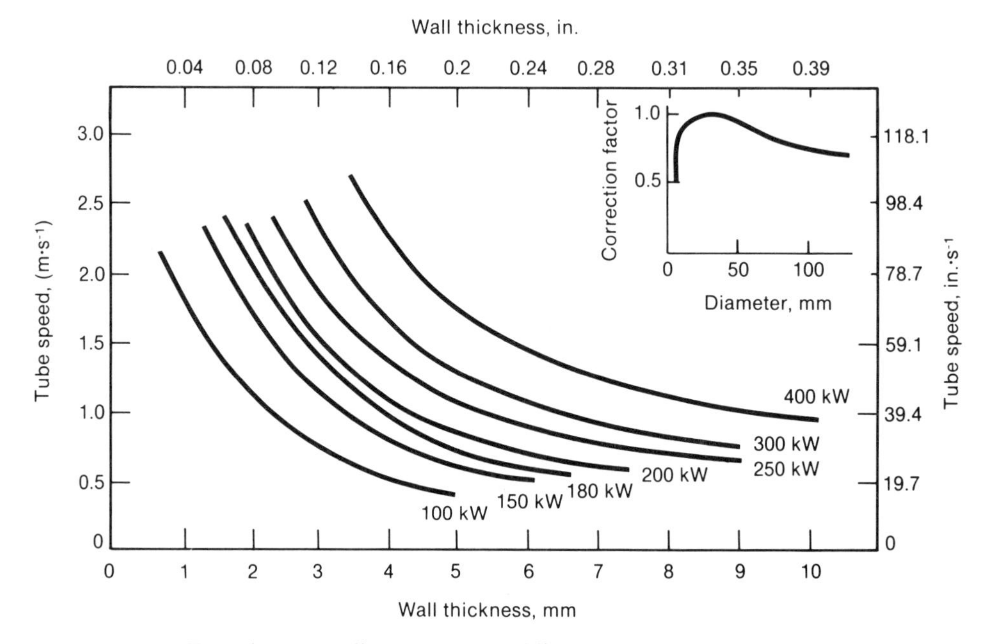

For other outer diameters, the welding speed should be multiplied by the correction factor given in the inset. For other metals, the speed should be multiplied by 1.4 (aluminum), 1.1 (brass), 0.8 (copper), or 0.9 (stainless steel).

Fig. 6.28. Welding speed as a function of wall thickness and input power for fabrication of 25-mm- (1-in.-) diam steel tubing (from Inductron, Ltd.; and J. Davies and P. Simpson, *Induction Heating Handbook*, McGraw-Hill, Ltd., London, 1979)

required generator frequency is in the radio-frequency range ($f \geq 100$ kHz). The most common frequency is 450 kHz.

If too low a frequency is selected, too much metal will be heated, and the edges will be upset an excessive amount by the squeeze rolls. If the frequency is too high, the outside portion of the skelp edges will overheat and the ID portion of the weld will be underheated. To compensate for the latter situation, the welding speed is held down to allow heat conduction to smooth out the temperature gradient.

Power Selection. For a given wall thickness and tube OD, selection of the power rating of the induction generator determines the welding speeds that can be achieved. As in other situations, the product of the mass flow rate (of the material that is heated) and the heat content imposes a lower limit on power requirements. The actual power needed is difficult to determine theoretically, however, because of coupling losses and losses associated with heating of the tube away from the fusion zone. Thus, it is easiest to refer to charts such as the one in Fig. 6.28 to determine the interrelationship of tube geom-

Table 6.12. Chemical compositions and melting ranges of alloys commonly used in induction soldering (from J. Libsch and P. Capolongo, *Lepel Review*, Vol 1 , No. 5, p 1)

Alloy	Composition, wt %			Melting range, °C (°F)	
No.	Tin	Lead	Silver	Liquidus	Solidus
1	99.8	...	...	230 (450)	230 (450)
2	62	38	...	185 (361)	185 (361)
3	60	40	...	190 (370)	185 (361)
4	50	50	...	215 (420)	185 (361)
5	40	60	...	240 (460)	185 (361)
6	30	70	...	260 (500)	185 (361)
7	...	97.5	2.5	310 (590)	310 (590)
8	...	96.5	3.5	315 (603)	310 (590)

etry, tube speed, and power-supply rating. Note that this figure applies only to generator frequencies between approximately 400 and 500 kHz.

DESIGN OF INDUCTION BRAZING AND SOLDERING OPERATIONS

Brazing and soldering involve melting of an alloy between the surfaces of metal parts to be joined. If the metal surfaces are clean, intimate contact is established and the melted material alloys with each surface, forming a joint upon solidification during cooling. The two methods of joining differ primarily in the type and melting temperature of the alloy used to form the joint.

In soldering, low-temperature alloys, generally containing lead and tin, permit joints of limited strength to be made at temperatures below 425 °C (800 °F). Soldering with these alloys (Table 6.12) is often termed soft soldering; typical metals that are joined by this technique are copper and copper alloys and aluminum and its alloys. Thorough cleaning prior to and during heating is basic for good soldered joints. Failures are often traceable directly to poor cleaning and inadequate fluxing. Surfaces to be joined should be chemically cleaned (i.e., to make them free of heat treatment scale, corrosion products, grease, etc.) prior to heating, and the joint areas should be fluxed as soon as possible to avoid contamination from handling or exposure.

Brazing alloys melt at considerably higher temperatures, and provide high-strength joints which resist reasonably elevated temperatures without failure. Products such as structural frames, musical instruments, jewelry, and ice skates are frequently fabricated by brazing. The metals joined include carbon and alloy steels, stainless steels, cast iron, copper and copper alloys, and nickel and nickel alloys. Brazing alloys (Table 6.13) are selected on the bases of

Table 6.13. Chemical compositions, melting ranges, and colors of alloys commonly used in induction brazing (from J. Libsch and P. Capolongo, *Lepel Review*, Vol 1, No. 5, p 1)

Alloy	Composition, wt %					Solidus,	Liquidus,	
No.	Silver	Copper	Zinc	Cadmium	Other	°C (°F)	°C (°F)	Color
1	72	28	...	...	...	780 (1435)	780 (1435)	White
2	15	80	...	...	5 P	640 (1185)	785 (1445)	Gray
3	50	15.5	16.5	18	...	625 (1160)	635 (1175)	Yellow
4	45	15	16	24	...	605 (1125)	620 (1145)	Yellow
5	35	26	21	18	...	605 (1125)	700 (1295)	Yellow
6	50	15.5	15.5	16	3 Ni	630 (1170)	690 (1270)	Yellow
7	54	40	5	...	1 Ni	725 (1340)	855 (1575)	Off white
8	65	20	15	...	...	670 (1235)	720 (1325)	White
9	10	52	38	...	...	765 (1410)	850 (1565)	Yellow
10	...	99.9 min	...	...	...	1085 (1981)	1085 (1981)	Copper
11	71.8	28	...	...	0.2 Li	...	...	White
12	85	...	...	...	15 Mn	960 (1760)	970 (1780)	...
13	...	Rem	...	...	37.5 Au	965 (1770)	995 (1825)	...
14	...	...	...	...	(a)	1010 (1850)	1065 (1950)	...
15	...	...	...	...	(b)	575 (1070)	580 (1080)	...

(a) 65 to 75 Ni, 13 to 20 Cr, 3 to 5 B, 10 max Fe + Si + C. (b) 11 to 13 Si, rem Al.

melting-temperature range, ability to wet the metallic surfaces to be joined, and the tendency to oxidize and/or volatilize. Additional selection criteria frequently include the strength and ductility of the alloy and unfavorable metallurgical reaction with the metals to be joined. As in soldering, cleaning and fluxing of the surfaces to be joined are very important.

One of the most important considerations in soldering and brazing, whether it is done by induction or other means, is joint design. This includes joint thickness as well as joint geometry. For a given joining alloy, there exists an optimum thickness for which the joint possesses a maximum shear strength (Fig. 6.29). Some common joint designs are shown in Fig. 6.30, 6.31, and 6.32. The designs in Fig. 6.32 are particularly useful for applications in which stress concentrations can cause problems in fatigue loading.

Most induction soldering and brazing applications use radio-frequency (~450 kHz) power supplies because of the high rate of power input and the ability to localize heat in this frequency range. Some large ferrous pieces are brazed with medium-frequency (3 to 10 kHz) power supplies; in these cases, though, furnace heating prior to brazing may be competitive on a technical or economic basis.

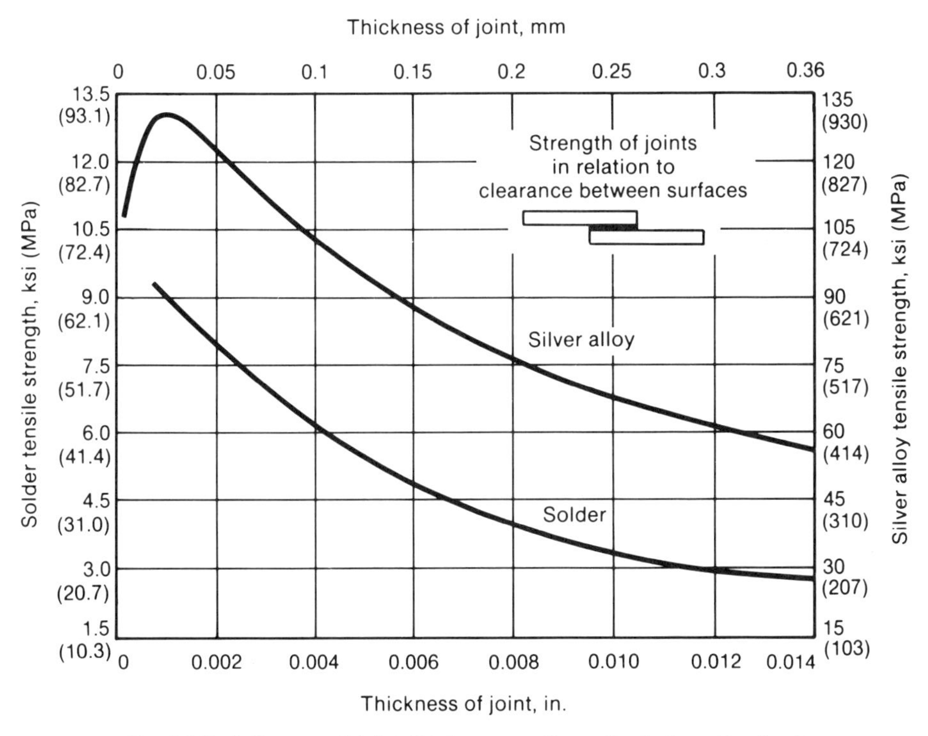

Fig. 6.29. Influence of joint thickness on theoretical strength of soldered and brazed joints (from F. W. Curtis, *High Frequency Induction Heating*, McGraw-Hill, New York, 1950)

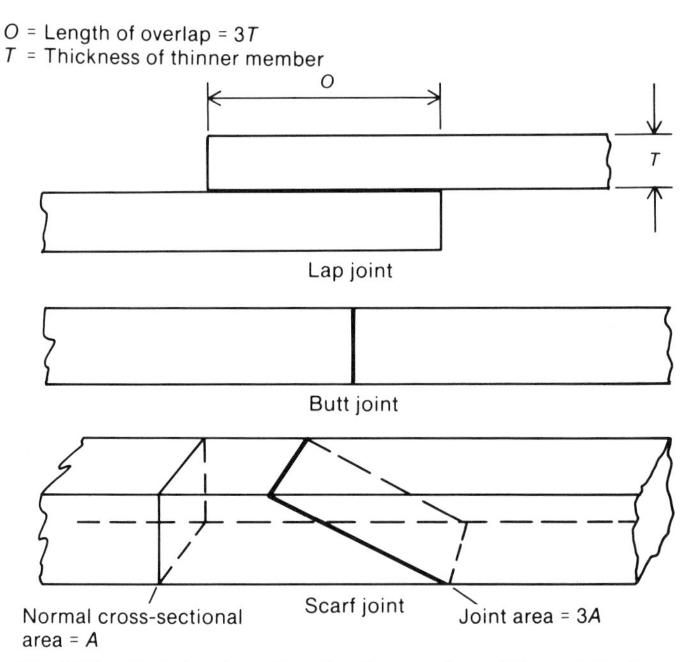

Fig. 6.30. Basic brazing joints: lap, butt, and scarf (from J. Davies and P. Simpson, *Induction Heating Handbook*, McGraw-Hill, Ltd., London, 1979)

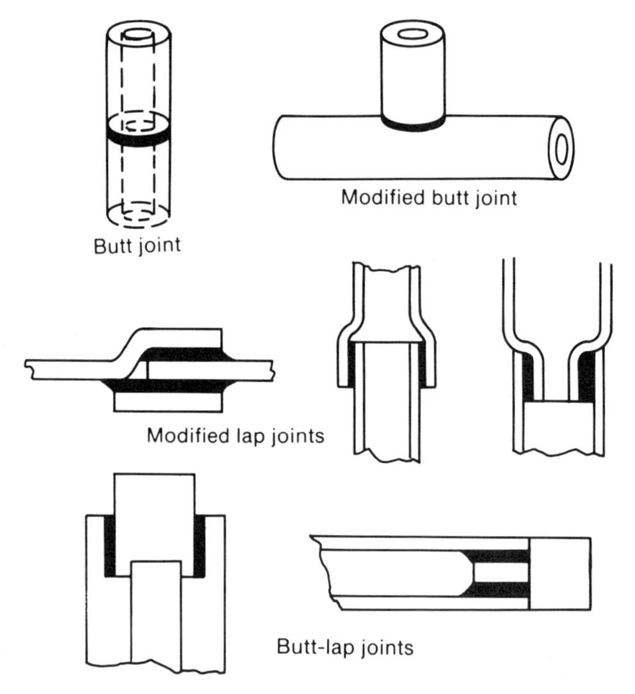

Fig. 6.31. Modified brazing joints (from J. Davies and P. Simpson, *Induction Heating Handbook*, McGraw-Hill, Ltd., London, 1979)

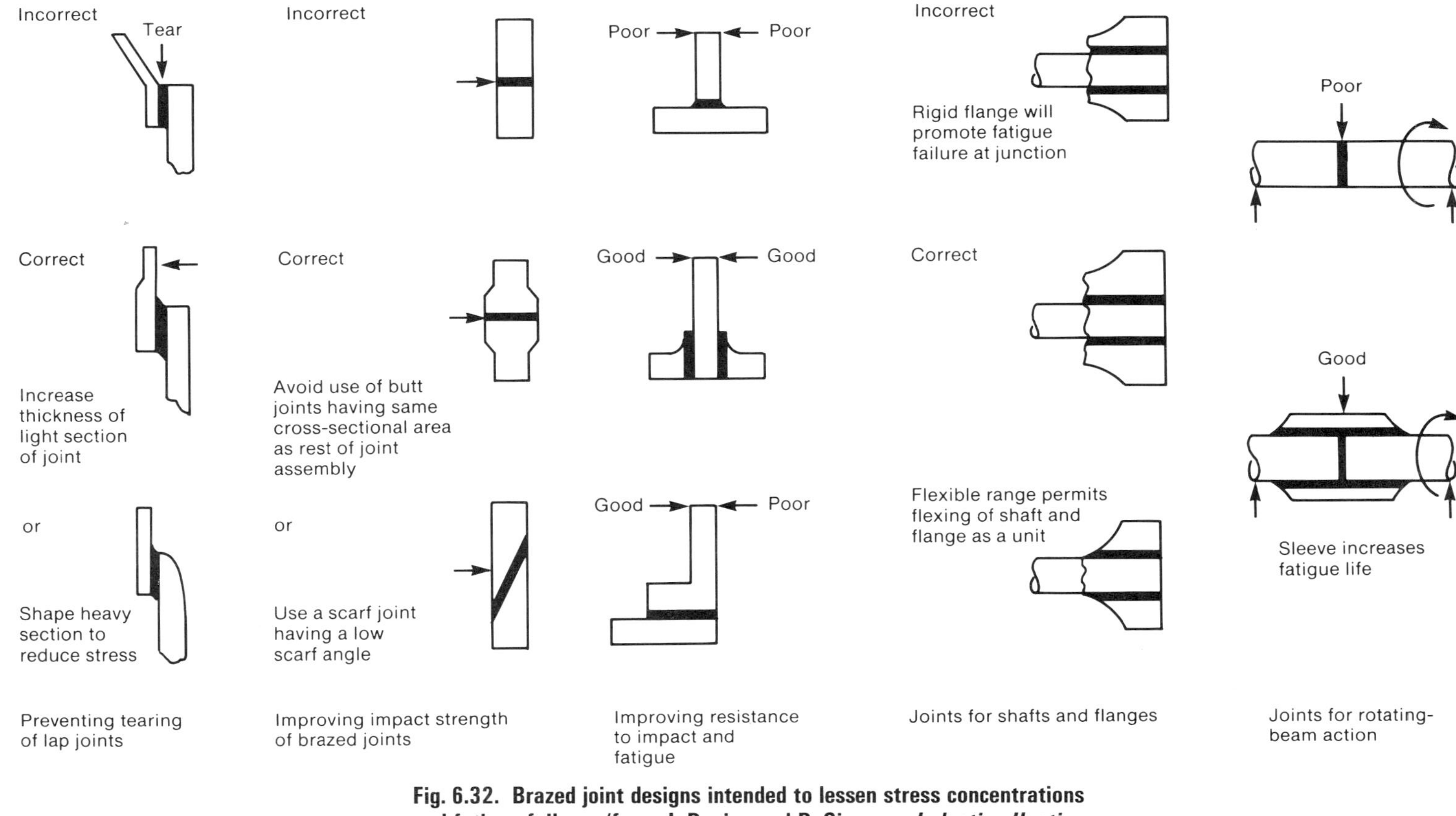

Fig. 6.32. Brazed joint designs intended to lessen stress concentrations and fatigue failures (from J. Davies and P. Simpson, *Induction Heating Handbook*, McGraw-Hill, Ltd., London, 1979)

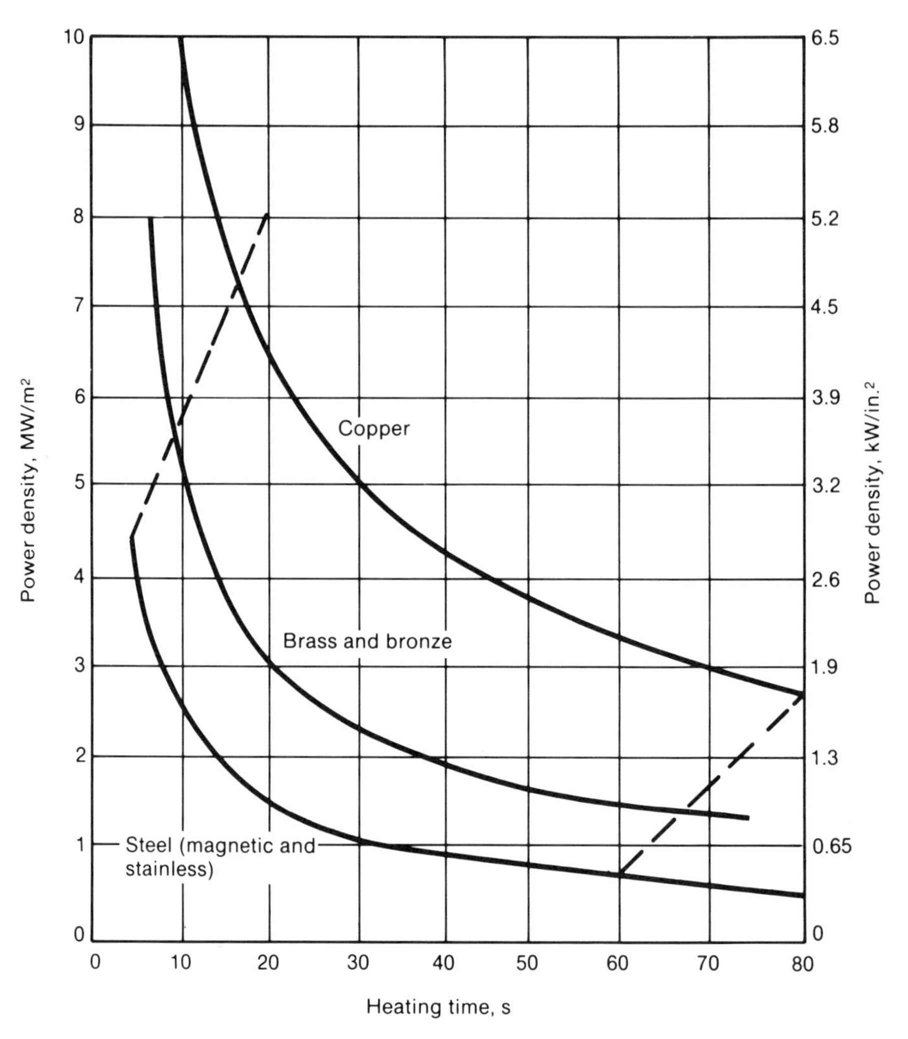

Fig. 6.33. Recommended power-density ranges (between broken lines) as a function of heating time for radio-frequency soft soldering at 188 °C (370 °F) (from J. Davies and P. Simpson, *Induction Heating Handbook*, McGraw-Hill, Ltd., London, 1979)

Power densities for induction soldering and brazing are similar to or slightly lower than those for surface heat treating of steel, mentioned earlier in this chapter. This is because of the somewhat lower temperatures involved in the former processes. Guidelines for power-density selection are given in Fig. 6.33 and 6.34. These should be considered only as approximate values, the best values being determined by actual trials. Improper selection of power level is a frequent source of difficulty. Often it is a result of excessive power density, which leads to nonuniform temperatures in the joint area. Improper power-level selection is especially troublesome when one of the materials being heated has poor thermal conductivity (e.g., stainless steel).

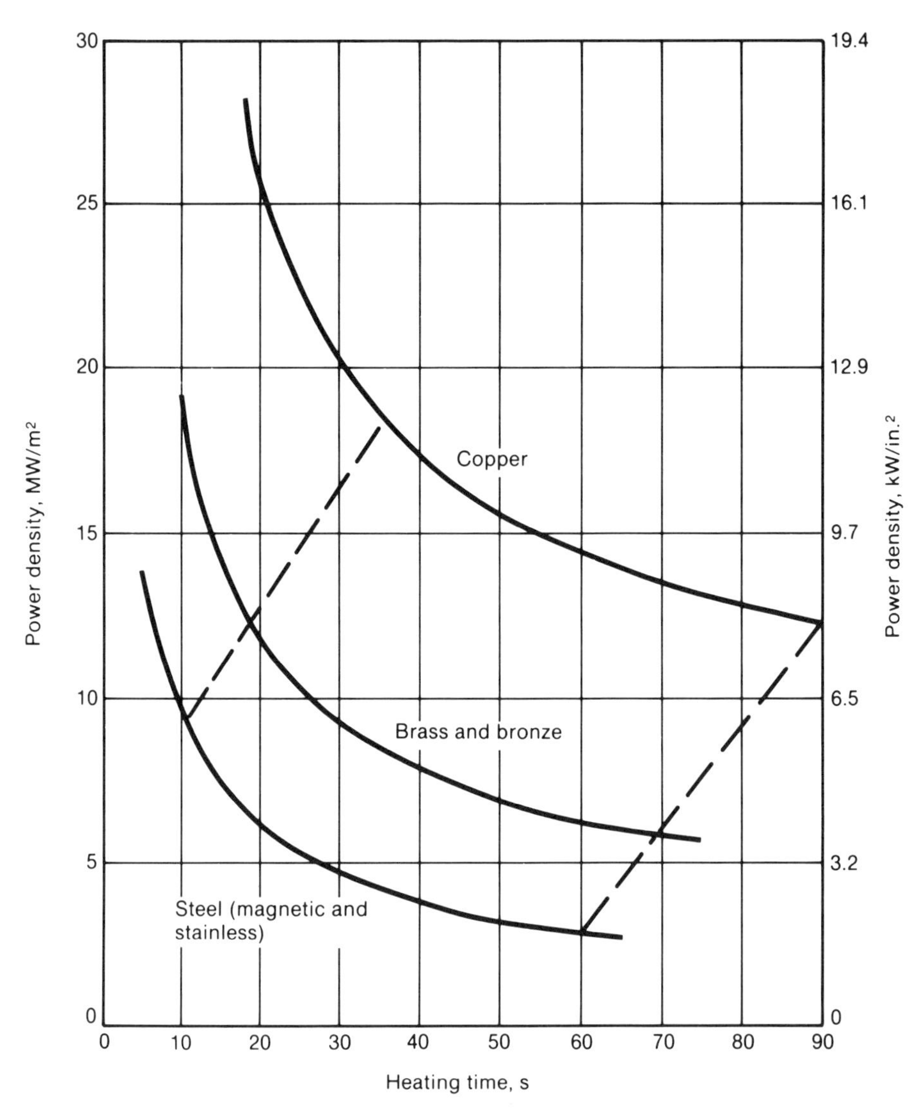

Fig. 6.34. Recommended power-density ranges (between broken lines) as a function of heating time for radio-frequency silver soldering at 700 °C (1290 °F). (From J. Davies and P. Simpson, *Induction Heating Handbook*, McGraw-Hill, Ltd., London, 1979)

Chapter 7

Fundamentals of Process Control

In Chapter 6, the basic design features of a variety of induction heating and melting applications were described. These features included selection of converter frequency and power capacity. Assuming that these features are chosen properly, the successful operation of the equipment to obtain reproducible results from one part to another or from one heat to another depends to a large extent on judicious control of input power, heating time, and part handling. Because induction systems are based on an electrotechnology, the coupling of the heating equipment to the electric-based control system is readily facilitated. The ease of control in these situations can be contrasted to nonelectric-based ones, such as fossil-fuel heating, in which problems such as those associated with thermal inertia of the furnace make quick response, process control, and high-speed automation more difficult.

The important elements of almost all process-control systems for induction heating include the following:

- A device or method for measuring workpiece temperature. This often consists of an instrument to be given a direct indication of surface or subsurface temperature (e.g., a thermocouple or radiation pyrometer). It can also consist of some indirect measurement that can be correlated to the temperature or heating pattern in the induction heated workpiece, such as measurement of coil impedance or of coil voltage and current.
- A controller for comparing the temperature of the workpiece to a preset desired temperature.
- A power-control device for regulating the power of the induction generator, or a device for controlling the heating time.
- A controller for related processing functions (e.g., part handling).

The above components of the control system must all operate in real time. That is to say, they must be capable of instantaneous measurement or response during the heating operation. Because they are electric based, this requirement is usually readily met. Selection of these process-control elements is discussed in this chapter. Attention is also focused on several other control techniques which are being implemented increasingly in industry. These include nondestructive inspection (NDI) methods and process simulation on computers. The first of these includes electromagnetic sorting of induction surface hardened parts, and lends itself readily to real-time control as well. Process simulation of induction heating is useful in the design stage as well as in establishing process parameters which lend themselves to automatic control. As such, any discussion of control technology for induction heating would be incomplete without some mention of this ever-expanding field.

TEMPERATURE MEASUREMENT

Methods of monitoring temperatures during induction heating have been reviewed extensively in the technical literature.* The most common techniques make use of thermocouples and radiation detectors. Despite the widespread use of each method, several major problems should be considered before a final selection is made. These include problems of poor workpiece surface condition, contact resistance, and response time for thermocouples. With radiation devices, emissivity variations are the major case of concern.

Thermocouples

Although thermocouples are not suitable for all applications, they provide good accuracy, measurement capability over a very broad temperature range, ruggedness, reliability, and low cost. Thermoelectric thermometry is a mature technology whose principles have been known since the early 1800's. It is based on the well-known relationship between a difference in junction temperatures and the resulting voltage (emf). In practice, the reference junction is held at a constant known value by various means—e.g., an ice bath, a controlled-temperature furnace, or an electrical method of simulating a known temperature. The temperature of the heated junction is determined by measuring the voltage and referring to calibration tables for the particular thermocouple materials. Thermocouples are of two basic types: contact and noncontact (or proximity).

Contact Thermocouples. There are various types of contact thermocouple arrangements the use of which permits accurate temperature measurements at fast response times. The simplest and probably the most reliable technique

---

*N. V. Ross, *Proc. Sixth Biennial IEEE Conference on Electric Heating*, IEEE, New York, 1963, p 29. J. B. Wareing, *High Temperature Technology*, 1983, Vol 1, No. 3, p 147. S. Zinn, *Heat Treating*, September, 1982, Vol 14, No. 9, p 28. H. Pattee, "State of the Art Assessment of Temperature Measurement Techniques", unpublished report, Battelle's Columbus Laboratories, Columbus, Ohio, 1984.

involves the direct attachment of a thermocouple to the part whose temperaturc must be determined. Typically this is done by percussion welding of the individual thermocouple elements to the part surface approximately 1 to 2 mm (0.04 to 0.08 in.) apart. This procedure is impractical for measuring workpiece temperatures in most induction heating processes because of the presence of an induction coil (single-shot and scanning applications) or the fact that the part is moving through the coil (scanning methods of induction heating). For this reason, contact thermocouples are normally used during initial trials to set up a particular process.

The two most common types of contact thermocouples are those in which the junction is welded to the specimen (referred to above) and those which are composed of two spring-loaded prods. In such an "open-prod" thermocouple (Fig. 7.1), the junction is made through the workpiece after contact is made. The prods can be made from any pair of thermocouple materials (e.g., chromel-alumel); the prods must be properly spaced to obtain the desired temperature-time response. Good electrical and thermal contact must also be established to produce accurate temperature measurements.

The open-prod thermocouple is most effective with metals which do not oxidize readily. However, by equipping the thermocouple assembly with a gas outlet port, it is possible to spray the ends of the prods and the contact area with a flux mixed with a combustible gas, thereby preventing scale formation

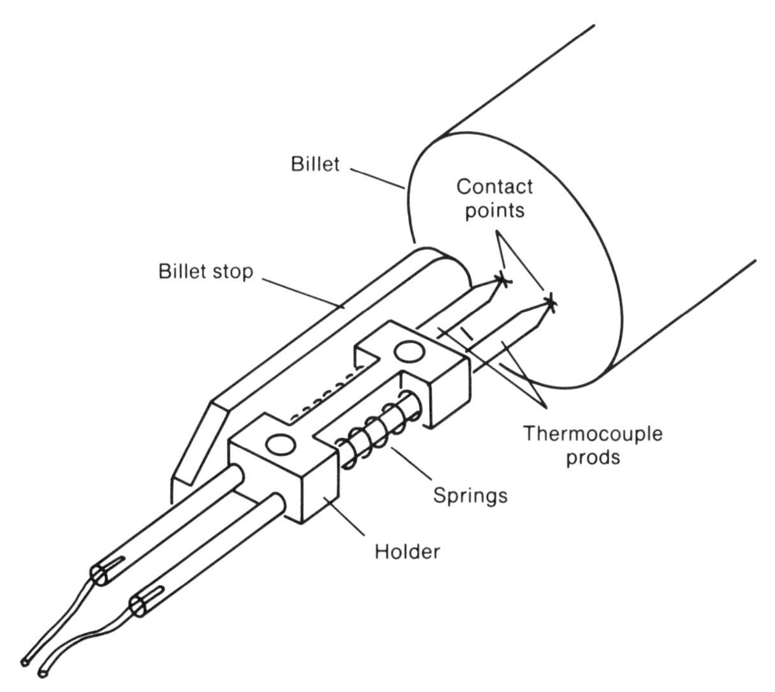

Fig. 7.1. "Open-prod" proximity thermocouple for making temperature measurements (from N. V. Ross, *Proc. Sixth Biennial IEEE Conference on Electric Heating*, IEEE, New York, 1963, p 29)

on both the part and the thermocouple itself. The open-prod thermocouple is not capable of measuring temperatures of metals harder than the thermocouple elements themselves. These harder metals include steels, titanium alloys, and nickel alloys. In these instances, noncontact thermocouples are appropriate.

Noncontact (Proximity) Thermocouples. To overcome the difficulties associated with scale formation, proximity thermocouples are often utilized. In this type of device, the thermocouple junction is attached to a disk of stainless steel or other high-temperature alloy. The disk (with the thermocouple attached to it) is welded on the inside and near the end of a stainless steel tube which acts as the temperature probe. By this means, the temperature-measuring device is isolated from the dirt and scale associated with the heating process. The distance between the disk and the workpiece surface is such that thermocouple temperatures do not unduly lag the actual surface temperatures. The disk is heated primarily by radiation from the hot surface; conduction through the entrapped gases and conduction through the tube also contribute to the total heat received by the disk.

Noncontact thermocouples are also widely used in melting applications such as those utilizing induction. In these instances the sensing element is enclosed in a protective sheath. Sheaths are usually made of a corrosion-resistant ceramic such as fused silica, magnesite, zircon-chromia, silicon nitride, or boron nitride. Of all the materials used to the present, boron nitride appears to offer a superior blend of corrosion/erosion resistance, shock resistance, machinability, and utility as a barrier material for sensor protection.

Thermocouple Combinations. Thermocouples are made of several different metal combinations. Selection of the particular design is usually based on the peak temperature to be measured. The most common couple materials are the following:

- Type "B": Pt-6Rh vs. Pt-30Rh. Used up to 1815 °C (3300 °F).
- Type "E": Ni-Cr (chromel) vs. Cu-Ni (constantan). Used up to 980 °C (1800 °F).
- Type "J": Fe vs. Cu-Ni (constantan). Used up to 760 °C (1400 °F).
- Type "K": Ni-Cr (chromel) vs. Ni-Al (alumel). Used up to 1150 °C (2100 °F).
- Type "R": Pt vs. Pt-13Rh. Used up to 1650 °C (3000 °F).
- Type "S": Pt vs. Pt-10Rh. Used up to 1650 °C (3000 °F).
- Type "T": Cu vs. Cu-Ni (constantan). Used up to 400 °C (750 °F).

There are also some very-high-temperature thermocouples based on tungsten. For example, tungsten-rhenium thermocouples find use at temperatures up to approximately 2200 °C (4000 °F).

Special Considerations for Thermocouples. Because temperature measurements made with thermocouples rely on the development of a dc voltage, due consideration should be given to construction and placement of such sensing elements when used in induction heating applications. Thermocouple lead wires should be run in a separate, shielded conduit away from all other wiring and should not be run parallel to or in proximity to ac wire runs. When the lead wires are enclosed in a grounded sheath, thermocouple manufacturers usually recommend leaving the thermocouple junction exposed (if it is outside the sheath) or slightly away from the sheath (if it is enclosed in it), particularly in critical electrical environments.

Extension wires are often used in conjunction with the thermocouple lead wires to minimize the over-all cost of the sensor, especially for those couples composed of noble metals. Usually they are chosen to have the same or nearly the same thermoelectric properties as the thermocouple wires themselves. Table 7.1 gives comparative data on extension wires in common use. The advantages of these wires, besides that associated with cost, include improvement in mechanical or physical properties of the thermoelectric circuit. For instance, the use of stranded construction or smaller-diameter solid wire may increase the flexibility of a portion of the circuit. Extension wires may also be selected to adjust the electrical resistance of the circuit. However, care must be taken to minimize sources of error that arise when using extension wires. These errors include the following:

- Errors due to the disparity in thermal emf between thermocouples and nominally identical extension-wire components of types EX, JX, KX, and TX.
- Errors due to temperature differences between the thermocouple/extension wire junctions. Errors of this sort are most troublesome for extension wire types WX and SX.
- Errors due to reversed polarity at thermocouple/extension wire junctions or at extension wire/measuring instrument junctions. Although a single reversal of polarity in the assembly would be noticeable, an inadvertent double reversal may likewise produce measurement errors that might be more difficult to detect.

Further discussion of problems associated with extension wires is given in ASTM Special Technical Publication 470.*

Guidelines for Using Thermocouples in RF Induction Heating Applications. Unsheathed thermocouples are used for measuring temperatures in low- and medium-frequency induction heating jobs. Fields are rarely induced in either thermocouple or extension wires because the wire diameters are generally very small. With high-frequency systems, however, RF pickup may pose a problem. Pickup can be minimized by attention to the following guidelines:

---

*_Manual on the Use of Thermocouples in Temperature Measurement_, ASTM STP 470, R. P. Benedict, ed., American Society for Testing and Materials, Philadelphia, 1970.

Table 7.1. Extension wires for common thermocouples (from *Manual on the Use of Thermocouples in Temperature Measurement*, ASTM STP 470, R. P. Benedict, ed., American Society for Testing and Materials, Philadelphia, 1970)

Thermocouple type	Extension type	Alloy type		Temperature range, °C (°F)	Limits of error, °C (°F)		Magnetic response(a)	
		Positive element	Negative element		Standard	Special	P	N
E	EX	NiCr (Chromel)	Constantan	0 to 205 (32 to 400)	±1.5 (±3)	...	0	0
J	JX	Iron	Constantan	0 to 205 (32 to 400)	±2 (±4)	±1 (±2)	M	0
K	KX	NiCr (Chromel)	NiAl (Alumel)	0 to 205 (32 to 400)	±2 (±4)	...	0	M
T	TX	Copper	Constantan	−60 to 95 (−75 to 200)	±1 (±1.5)	±0.5 (±0.75)	0	0
K	WX	Iron	Copper-nickel alloy	25 to 205 (75 to 400)	±3 (±6)	...	M	0
R	SX	Copper	Copper alloy	25 to 205 (75 to 400)	±6.5 (±12)	...	0	0
S	SX	Copper	Copper alloy	25 to 205 (75 to 400)	±6.5 (±12)	...	0	0

(a) M denotes ferromagnetic alloy; 0 denotes nonferromagnetic alloy.

- The part being heated should be used to shield the thermocouple and its lead wire. For example, the thermocouple can be passed down the bore of a tubular part.
- Thermocouple and extension wires should be kept as far as possible from power-supply leads, induction coils, and transmission lines. Wires should be run at right angles to conductors if practical.
- Shields, if used, should be kept as far away as possible from conductors so that they will not be induction heated. A shield should be grounded at the point where its leads are closest to the active conductors. Shields should also be grounded only in one location to prevent the formation of ground loops.
- Typically, a 0.1-μF (microfarad) capacitor should be run to ground from each of the metering system's input terminals to bypass any interference from the induction power supply.

Radiation Detectors

The other popular means of rapid temperature measurement in induction heating applications relies on the use of radiation detectors. These devices provide a noncontact method of measuring and controlling the temperatures of hot surfaces during heating. In comparison with thermocouples, radiation detectors provide the following advantages:

- Because contact is not required, temperatures can be measured without interfering with the heating operation. Measurements can be made conveniently.
- There is no upper temperature limit.
- The response during measurement is very fast. The readings are accurate if the instrument is properly calibrated and maintained.
- The service life of the equipment is indefinite.

Insofar as disadvantages are concerned, radiation temperature detectors are expensive. Also, measurements require direct observation of the heated surface. The latter drawback can be overcome to a great extent by the use of modern fiber optics in many instances.

The two most common types of radiation devices are optical and infrared pyrometers.

Optical Pyrometers. Optical pyrometers were among the first instruments used for noncontact temperature sensing. In practice, the operator looks at the incandescent body through a telescope-type device that contains a wire filament in the same optical plane as the observed body. The intensity of light from the filament is adjusted until the filament disappears against the background. The current to the filament is measured, and temperature is obtained from a calibration of filament current versus the temperature of a black body.

Optical pyrometers are portable and versatile, but most are not highly accurate. The readings from such instruments must be corrected to reflect the emissivity of the workpiece. The emissivity of a metal depends on many factors, the most important of which is the condition of the workpiece surface. Unfortunately, these factors vary with temperature and heating conditions. In addition, smoke and water vapor in the air affect optical pyrometer readings. Because of such variables, use of fixed correction factors are not usually feasible. Nevertheless, at very high temperatures, typical of many induction heating applications, they are usually considered accurate enough for process control.

Infrared Pyrometers. Infrared pyrometers are rapidly replacing optical pyrometers for many applications. They are more accurate (0.5 to 1% of full scale as opposed to ±2% of full scale for optical pyrometers), and they can be used as portable sensors or as part of a permanent, continuous temperature monitoring and control system. In addition, the readout can be analog or digital. Although there are many variations, the infrared pyrometer basically operates by comparing the radiation emitted by the hot target with that emitted by an internally controlled reference source. The output is proportional to the difference in radiation between the variable source and the fixed reference.

There are three types of infrared pyrometers:

- Single-color or single-band pyrometers, which measure infrared radiation of a fixed wavelength. These are the most versatile pyrometers because they are suited to many applications. Their accuracy may be affected by the presence of dust, atmospheric gases, and other characteristics of the measuring environment, and the emissivity of the target object must be taken into account. Single-color pyrometers operating at shorter wavelengths tend to be more accurate at high temperatures.
- Broad-band pyrometers, which measure the total infrared radiation emitted by the target. The accuracy of this type of pyrometer is also adversely affected by the presence of dust, gases, and vapor. Similarly, errors in measurement can occur unless the emissivity of the target is taken into account.
- Two-band or two-color pyrometers, which measure the radiation emitted at two fixed and closely spaced wavelengths. The *ratio* of the two measurements is used to determine the temperature of the target. The adverse effects of the environment (dirt, gases, and vapors) are largely eliminated with this pyrometer. Moreover, because emissivity affects both measurements equally, its effect on accuracy is eliminated.

Of the three types, the two-color pyrometer is surely the most accurate. As are the other types of pyrometers, it is readily adaptable for measuring of temperatures in induction heating systems using a variety of optical systems for focusing and, as mentioned above, using fiber optics.

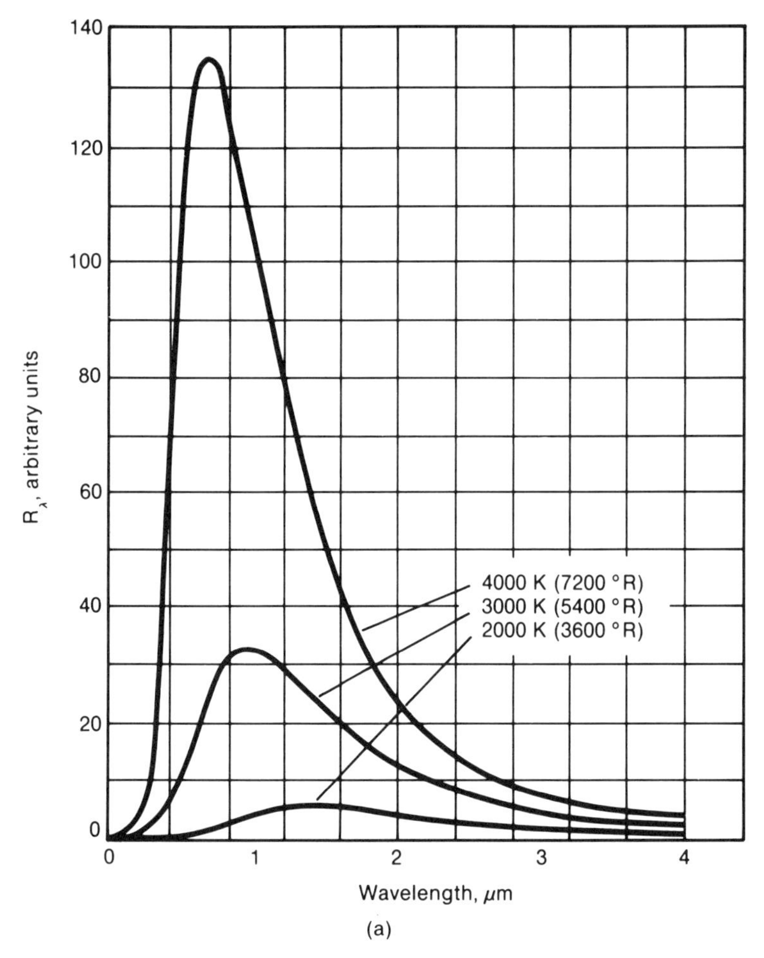

Fig. 7.2. Spectral-radiancy curves for (a) a black body at three different temperatures and (b) tungsten and a black body at 2000 K (3600 °R) (from D. Halliday and R. Resnick, *Physics*, Wiley, New York, 1967)

Theory of Operation of Infrared Pyrometers. As mentioned above, infrared pyrometers measure the radiation given off by the heated object. An internal calibration is used to convert the radiation level to temperature. The precise manner in which this is done can best be understood by reference to spectral-radiancy curves such as those given in Fig. 7.2. These curves represent the spectral radiancy, R_λ, or the amount of energy emitted by the body per unit time for wavelengths covering the interval from λ to $\lambda + d\lambda$. Single-color pyrometers measure the total radiation power, $R_\lambda d\lambda$, over the narrow wavelength band to which their pickup device is sensitive. *Infrared* pyrometers are designed for wavelengths *in excess* of those to which the naked eye is sensitive (0.4 to 0.7 μm, or 0.016 to 0.028 mil). Typical wavelengths employed usually lie in the range between 0.7 and 20 μm (0.028 and 0.79 mil).

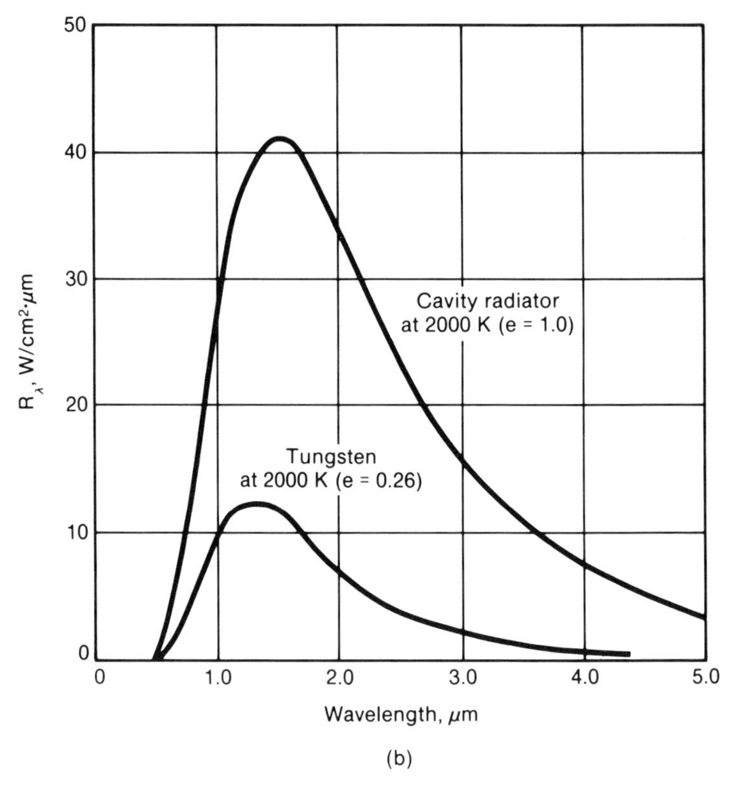

(b)

Fig. 7.2. continued

Upon examining the curves in Fig. 7.2(a) for a material with an emissivity e equal to 1.0 (a so-called black body), it is apparent that, for a given wavelength λ, the value of R_λ is very sensitive to temperature. Therefore, R_λ, or the voltage it induces in the detector in the pyrometer, is readily calibrated against temperature. For real materials, however, the emissivity is not unity (except for heavily scaled or oxidized surfaces); rather, it assumes a value between 0 and 1 (Table 7.2). Nonetheless, the relationship between the spectral radiancy of a real material and the idealized black body is quite simple: $R_\lambda(e \neq 1)/R_\lambda(e = 1) = e$. This relationship* is usually assumed to hold irrespective of λ for a given workpiece temperature (Fig. 7.2b) and is relied upon in the operation of single-color pyrometers. These instruments have an emissivity adjustment which permits the voltage generated to be multiplied by a factor of 1/e prior to its comparison with that which would be produced by a black body.

*In actuality, the emissivity e is defined as the ratio of the *total* radiancy of a heated object ($\int R_\lambda d\lambda$, integrated over the entire wavelength range) to the total radiancy of a black body heated to the same temperature. However, the ratio of the R_λ's at specific wavelengths is often taken as being equal to a constant (= e) irrespective of λ; this is a reasonable assumption for most engineering applications.

Table 7.2. Typical emissivities of metals (source: Ircon, Inc.)

Material	Emissivity: Smooth, polished	Emissivity: Smooth, oxidized	Material	Emissivity: Smooth, polished	Emissivity: Smooth, oxidized
Alumel	0.25	0.90	Nichrome	0.26	0.90
Aluminum	0.10	0.20	Nickel	0.15	0.90
Brass	0.10	0.70	Platinum	0.18	...
Carbon steel	0.25	0.75	Silicon	0.70	...
Chromel	0.25	0.90	Silver	0.03	0.80
Chromium	0.30	0.70	Stainless steel	0.25	0.85
Cobalt	0.25	0.75	Tantalum	0.10	0.70
Copper	0.04	0.70	Tin	0.22	0.60
Graphite	0.65	...	Tungsten	0.10	0.60
Iron	0.25	0.70	Vanadium	0.29	0.75
Lead	0.15	0.70	Zinc	0.07	0.50
Manganese	0.30	0.90	Zirconium	0.22	0.40
Molybdenum	0.28	...			

The temperature error due to an incorrect emissivity setting of a single-color pyrometer is readily calculated from the formula*:

$$\Delta T(\%) = \Delta e(\%)\left\{\frac{\lambda T}{C_2}\right\}$$

In this equation, ΔT and Δe denote the temperature and emissivity errors, respectively, and λ, T, and C_2 represent the pyrometer wavelength, workpiece temperature, and Planck's second radiation constant (0.0144 mK). As an example, consider the case in which the temperature of a steel bar at 1000 °C (1273 K) is being measured with a single-color pyrometer sensitive to radiation at 1 μm. Assume that e = 0.60, but that the emissivity setting on the pyrometer is 0.90. Thus, $\Delta e = 50\%$, and $\Delta T = (50)(1 \times 10^{-6})(1273)/(0.0144) = 4.42\% = 56\ K = 56\ °C$. The pyrometer would thus indicate a value which is *low* by 56 °C. It is obvious that low settings of emissivity produce temperature readouts which are higher than actual, and vice versa. Absolute temperature errors for a Δe of 10%, derived from the above equation, are plotted in Fig. 7.3. Those for other values of Δe are readily obtained by multiplying the numbers drawn from this plot by a factor equal to Δe (%)/10.

In a two-color pyrometer, the spectral radiancy from the workpiece surface is measured for two wavelengths simultaneously. At a given temperature, the *ratio* of these two readings is independent of the emissivity because all R_λ curves scale *linearly* with the master black-body curve by a factor of the emissivity. This ratio can then be calibrated against temperature.

*J. B. Wareing, *High Temperature Technology*, 1983, Vol 1, No. 3, p 147.

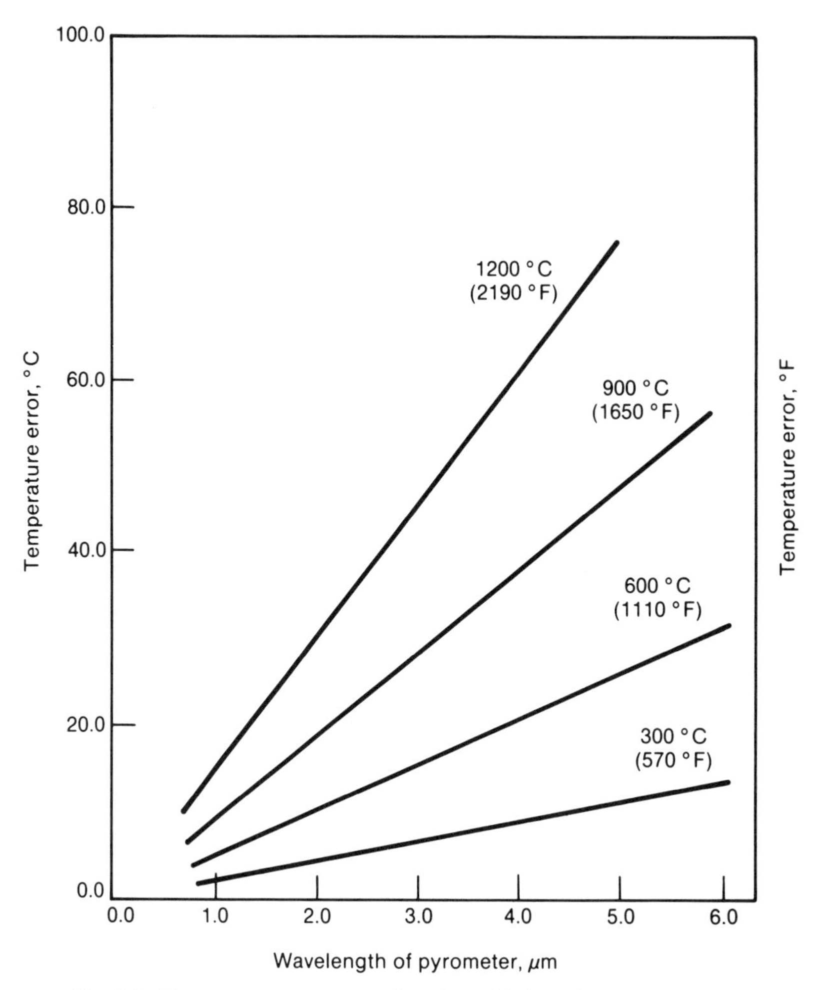

Fig. 7.3. Temperature error as a function of infrared pyrometer wavelength for a 10% error in emissivity adjustment (from J. B. Wareing, *High Temperature Technology*, Vol 1, No. 3, 1983, p 147)

Fiber Optics. Fiber optics are frequently used in conjunction with pyrometers to measure temperature.* The size and composition of such probes vary with application. Generally, they are less than 5 mm (0.2 in.) in diameter and can be fitted with lenses to sight through a gap in the induction coil at a very localized area. Long-focal-length lenses are also frequently employed to provide adequate distance between the workpiece and fiber optic and thus prevent overheating of the latter. Fiber optics capable of withstanding temperatures of approximately 500 °C (930 °F) are available. At higher temperatures, probes may be equipped with air jets to keep them cool and free of workpiece scale.

---

*J. Hansberry and R. Vanzetti, *Industrial Heating*, Vol 48, No. 5, May, 1981, p 6.

A typical optical fiber is usually constructed of silica. The device actually consists of a bundle of many hundreds of individual fibers contained within a flexible or rigid sheathing of either metallic or nonmetallic material. The end surfaces are highly polished to ensure a clearly defined angle of acceptance and to diminish reflectance losses due to surface irregularities. Using a large number of fibers in one bundle allows the gathering and transmission of the signal while retaining mechanical flexibility.

Recently, sapphire and quartz have also been used for optical-fiber applications in conjunction with infrared pyrometry. One application, patented by Accufiber, makes use of a sapphire fiber probe. This probe is coated with iridium to form a black-body cavity. The radiation from this cavity when heated is readily measured using a standard radiation detector to yield a very accurate temperature measurement. The manufacturer quotes a use-temperature range of 500 to 2000 °C (930 to 3630 °F), which is considerably wider than those of many thermocouple systems.

Advantages of fiber optics, in addition to those already cited, include the following:

- Very large numerical aperture. Unfocused fibers can be used to capture radiation from a large portion of the workpiece if necessary.
- Adjustment of viewing angle. With the addition of small lenses, the viewing angle can be cut down substantially, allowing sightings to be taken over regions as small as 1 mm (0.04 in.) in diameter.
- Inert and rugged construction. Fiber optics are unaffected by electromagnetic induction fields and contain no moving parts which can fail during service.
- Wide range of temperature measurement. With the proper selection of fibers and an infrared pyrometer, a single system may be used to cover a full range of temperatures from 100 to 2000 °C (210 to 3630 °F) with good resolution, accuracy, and repeatability. It is very important to note that infrared radiation of wavelengths greater than 2.5 μm is readily absorbed by the fibers after just a few centimeters of travel. Thus pyrometers sensitive to lower-wavelength radiation are often required when using fiber optics. However, sapphire fibers pose much less of a problem than those made of silica in this regard.

Sighting of Pyrometers. In use, a reticle in the pyrometer's optical system which contains an aiming circle is superimposed on the view of the part whose temperature is being measured. The sensor also "sees" the target via a special mirror. The size of the aiming circle is matched to that of the sensor, which then averages the energy passing through the area defined by the circle and produces a signal relative to the average target temperature. However, if the aiming circle is passing energy from both part and background, the sensor will average the two signals and give an incorrect reading.

In some heating applications, the induction coil may obstruct the pyrometer's view of the part and prevent its full resolution. The temperature of the obstructing, cold coil turns would then be averaged with that of the hot part. To overcome this problem, a piece of paper with a mark on it is held against the coil turns. The sensor is then focused on the mark to provide the smallest possible focal spot at a point between the coil turns. The spot size at the part's surface will be larger by an amount dependent on its distance from the coil's inside diameter. The sensor will measure the part temperature over this slightly larger spot. For the technique to work, instrument optics must be capable of providing a focal spot at the coil that is smaller than the spacing between turns.

Other Temperature-Measuring Techniques

Thermocouples and pyrometers are the most widely used devices for temperature measurement in industrial heating situations. Other techniques are currently under development and may find some use in the future. These methods include those based on ultrasonics, eddy-current detection, and real-time measurement of coil current or impedance during induction heating. Each of these is briefly discussed below.

Ultrasonic Methods.* Ultrasonic temperature-measuring techniques make use of transducers which generate and detect sound waves in the workpiece. The velocity of sound for a given material is a function of the material's elastic modulus and density. Because these properties are functions of temperature, the speed of sound can be used to infer temperature. Transducers under evaluation to determine their ability to generate the required acoustic waves include varieties such as electromagnetic-acoustic, electromechanical, and high-intensity lasers. These *noncontact* transducers are all designed to overcome one of the prime difficulties associated with ultrasonic temperature measurement—the difficulty of getting the sound waves into and out of a heated workpiece. Mechanical-contact methods using cooled buffer rods have been used, but do not lend themselves readily to production applications.

A second problem associated with ultrasonic temperature measurement relates to temperature distributions in bodies with temperature gradients. If ultrasonic velocity measurements are used to infer temperature, the sound-wave velocity is usually some average along the path of propagation. To find the velocity at a point, and thus the temperature at that point, it is necessary to measure the velocity along a number of intersecting paths and then "invert" the information to establish the spatial velocity distribution. To this end, the method of computer-assisted tomography (CAT), originally developed in the field of medicine, is being investigated. One of the basic tomographic schemes, depicted schematically in Fig. 7.4, attempts to determine the average temperature within concentric rings ("annular pixels") in cylinders by measuring

*R. G. Watson, *Advances in Instrumentation*, Vol 37, Part 1, 1982, p 267.

wave-propagation times for an appropriate set of ultrasonic rays. Initial results obtained using noncontact laser-generated ultrasonic waves have been encouraging, although only a relatively coarse pixel size has been used to date.

A related CAT technique for estimating temperature profiles makes use of the change in resonant frequency as a function of temperature. By making several resonance measurements at a single point on a slab or billet, a graphic profile of internal temperature can be plotted.

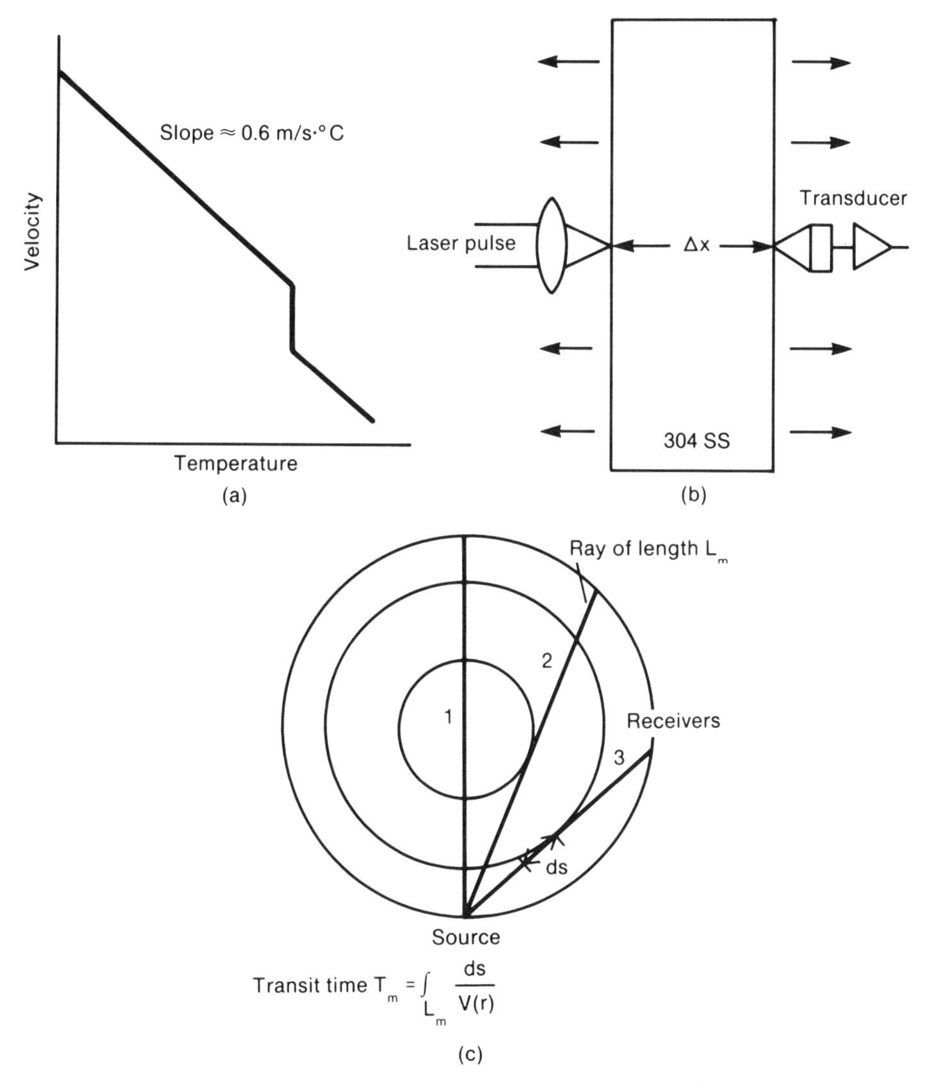

Fig. 7.4. Tomographic reconstruction of the ultrasonic velocity–temperature distribution in a solid bar: (a) schematic velocity-temperature curve for a uniformly heated bar, (b) schematic illustration of equipment setup, and (c) diagram illustrating method of data analysis (from R. Mehrabian and H. N. G. Wadley, *Journal of Metals*, Vol 37, No. 2, February, 1985, p 51)

Eddy-Current Techniques. Eddy-current detection of workpiece temperature during induction heating relies on the resistivity changes (usually increases) of metals with temperature. With the change in resistivity, the electrical load across an induction coil is modified. The most common means of detecting this is through measurements of the impedance of the coil and the workpiece resistance as reflected in the coil circuit, $Z = \sqrt{X_L^2 + (R_c + R_{eq})^2}$, where X_L is the inductive reactance of the coil, R_c is coil resistance, and R_{eq} is the resistance of the workpiece as reflected in the coil circuit.

Eddy-current techniques for temperature measurement typically utilize a second, or "inspection," induction coil brought into proximity with the work coil and workpiece. As for the work coil, the inspection coil is energized by an alternating current. However, the power put through the latter coil is negligible compared with that through the work coil. Thus, negligible heating of the workpiece can be brought about by the inspection technique itself.

The interactions between the inspection coil and the workpiece follow the same principles as those described in Chapter 2. The reflected workpiece resistance in the inspection circuit is a function of frequency through its influence on the reference depth. Reference depth is large at low frequencies and small at high frequencies. Hence, low inspection-coil frequencies are used to gage the average temperature (through workpiece electrical properties) to greater depths than high frequencies. In theory, multiple-excitation frequencies might be used to calculate the temperature *profile* within a heated billet. Care in frequency selection must be taken, however, to avoid interference from the field set up by the induction heating coil itself.

Several designs for the inspection coil (i.e., probe) that would be used in an eddy-current method of temperature measurement were investigated by Stutz.* Water-cooled probes wound with manganin wire were capable of withstanding exposure to steel workpieces heated to 1095 °C (2000 °F) with negligible drift; the probes were placed as close as 6 mm (0.25 in.) to the hot steel specimens. Various prototype geometries were also investigated to establish the temperature resolution that could be obtained with an eddy-current technique. It was found that, using the best probe examined, temperature differences of approximately 5 °C (10 °F) could be discerned in samples through heated to a nominal temperature of 540 °C (1000 °F).

Coil-Current Monitoring. A relatively new technique for obtaining an estimate of the temperature profile within an induction heated workpiece involves direct monitoring of the current through the work coil. It is similar to eddy-current techniques in that it relies on changes of impedance with temperature. One of the most interesting of such applications deals with case hardening of

*D. E. Stutz, unpublished research, Battelle's Columbus Laboratories, Columbus, OH, 1986.

steels by a single-shot method as described by Verhoeven, Downing, and Gibson.*

In case hardening, a surface layer is raised to austenitizing temperature, followed by quenching. Workpiece resistance, and therefore coil impedance, varies as a result of changes in resistivity with temperature as well as with phase changes. As discussed by Verhoeven and his colleagues, three distinct stages can be identified in the process of surface austenitizing by induction:

1. f stage: All induced current flows in ferromagnetic iron.
2. f/p stage: Current flows in both ferromagnetic and paramagnetic iron.
3. p stage: All current flows in paramagnetic iron.

Transitions from one stage to another are manifested by the coil current–time signature, a schematic of which is shown in Fig. 7.5. During the initial, or f, stage, the steel workpiece is heated without any phase transformations; the resistivity increases, the impedance increases, and the coil current drops. When the f/p stage begins, the relative magnetic permeability of the workpiece drops. Thus, the reference depth increases, the coil impedance decreases, and the coil current increases. The transition between the f and f/p stages is denoted by the time t_m. As more and more austenite is formed in the surface layers, the permeability decreases (at a rate faster than that at which resistivity increases), and the reference depth continues to increase, thereby further decreasing the coil impedance and increasing the coil current. The transition between the f/p and p stages is denoted by the time t_p. In the p stage, increased impedance (due to increased resistivity) would in theory cause a decrease in coil current. In practice, some materials exhibit a decrease and some an increase depending on the degree of end effects.

Verhoeven *et al.* also demonstrated that the I_c vs. t behavior can be correlated to the austenitized-and-hardened case depth, denoted by δ_{case}. This is done by examining the energy input into the workpiece:

$$\text{Energy input per unit area} = \int_{t_m}^{t} P_A \, dt$$

Here, t_m is used as a lower limit on the integration because this is the time at which the case first begins to form. P_A is the power density *into* the workpiece, or a quantity proportional to a factor of I_c^b where b varies from 1.54 to 2.0 during the transition from the f to the p stage. Using a value of 2.0 as an approximation, a bilinear correlation between energy input per unit area $\left(\approx \int_{t_m}^{t} I_c^2 \, dt\right)$ and case depth δ_{case} was found in the experiments carried out by these authors, as shown in Fig. 7.6.

*J. D. Verhoeven, H. L. Downing, and E. D. Gibson, *Journal of Heat Treating*, Vol 4, No. 3, June, 1986, p 253.

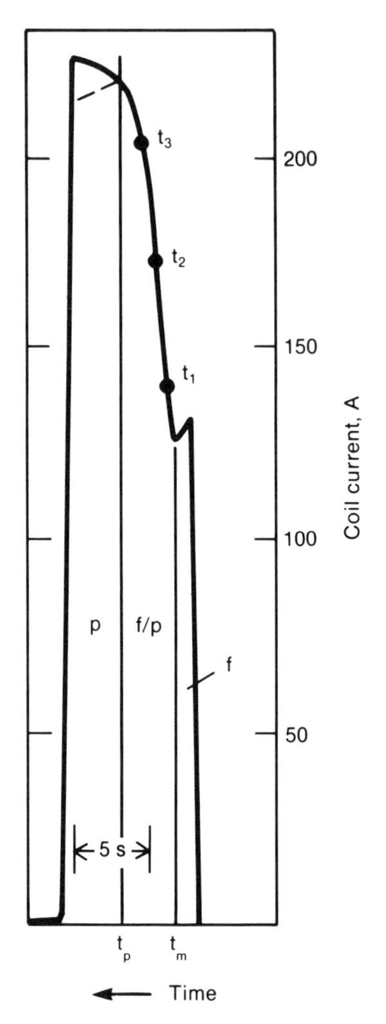

Fig. 7.5. Schematic coil-current-versus-time curve for a steel bar surface hardened by an induction heating method. The times t_m and t_p separate the three stages f, f/p, and p (from J. D. Verhoeven, H. L. Downing, and E. D. Gibson, *Journal of Heat Treating*, Vol 4, No. 3, June, 1986, p 253)

Load-Signature Analysis. A technique similar to coil-current monitoring, but useful for a variety of induction heating processes (including heat treating), is the one known as load-signature analysis. In this method, a number of characteristics of the induction heating circuit are monitored and compared with values known to give the proper heating cycle. These parameters usually include*:

---

*G. Mordwinkin, A. L. Vaughn, and P. Hassell, *Heat Treating*, Vol 18, No. 11, November, 1986, p 34; P. A. Hassell and G. Mordwinkin, *Industrial Heating*, Vol 53, No. 12, December, 1986, p 17.

- Current and voltage versus time
- Phase angle (between current and voltage) versus time
- Frequency of the applied voltage.

Figure 7.7 indicates how current (CT) and potential (PT) transformers are connected in either a constant-current (i.e., solid-state) or constant-voltage (i.e., motor-generator) heating circuit to get the data required to document the induction heating "signature."

Mordwinkin, Hassell, and their coworkers verified the usefulness of load-signature analysis through a number of experiments involving pipe heating and camshaft hardening. The former trials indicated that incorrect heating arising from gross geometric variations of the workpiece was readily discerned by signature analysis; however, none of the signature parameters listed above gave an unambiguous indication of heating differences associated with pipe warpage or the presence of through holes in the pipe wall. In the heat treating runs, load signatures were found to be distinctly different for sound versus cracked cams. Axial cam offsets in the coil of approximately 25% were also readily detected by this method. On the other hand, radial mispositioning of the workpiece showed little effect on the signature, particularly at low power-supply frequencies for which centering is often not critical anyway.

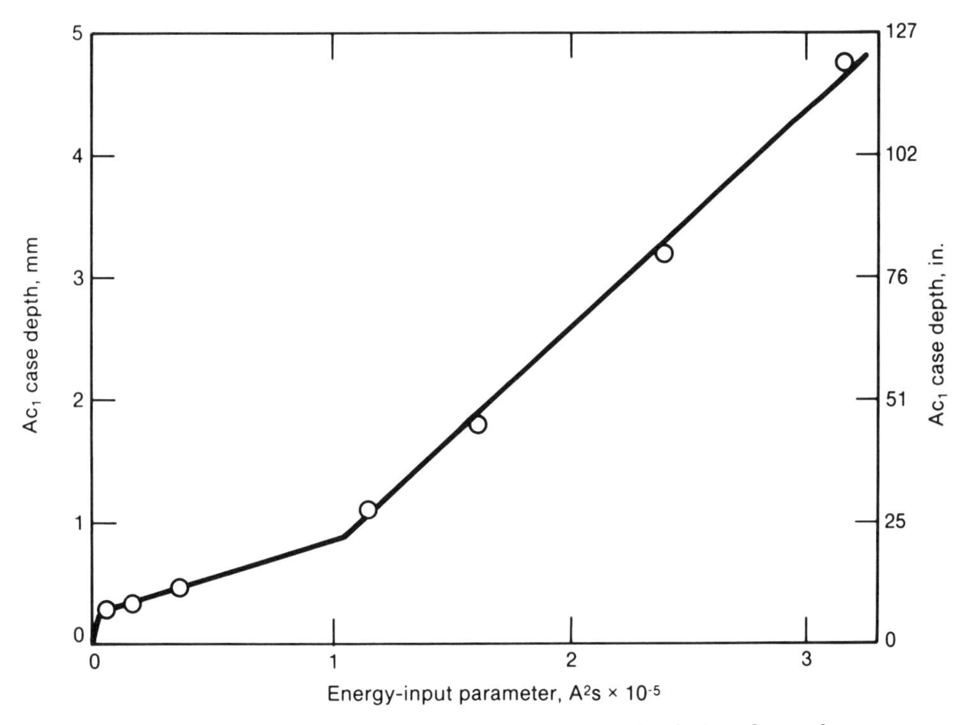

Fig. 7.6. Case depth obtained by induction surface hardening of a steel bar as a function of $\int I_c^2 dt$, where I_c and t denote induction coil current and time, respectively (from J. D. Verhoeven, H. L. Downing, and E. D. Gibson, *Journal of Heat Treating*, Vol 4, No. 3, June, 1986, p 253)

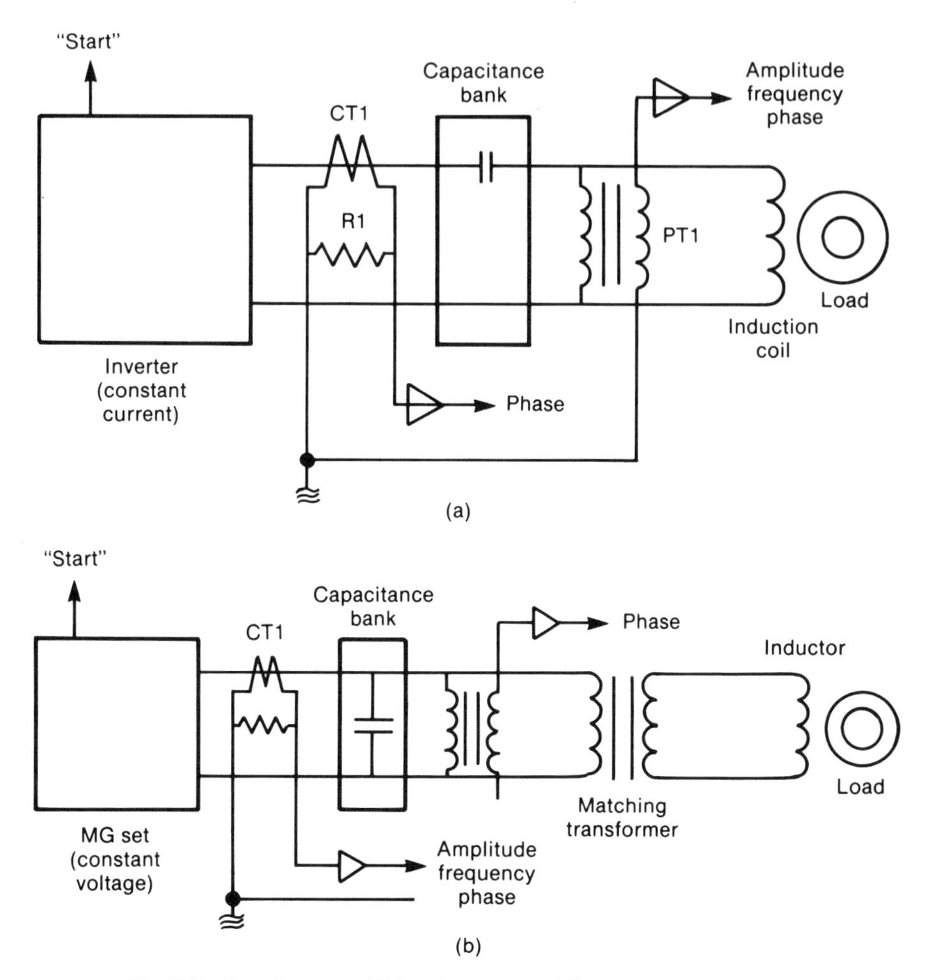

Fig. 7.7. Use of current (CT) and potential (PT) transformers to obtain induction heating "signatures" for systems utilizing (a) a constant-current (solid-state) power supply and (b) a constant-voltage (motor-generator) power supply (from G. Mordwinkin, A. L. Vaughn, and P. Hassell, *Heat Treating*, Vol 18, No. 11, November, 1986, p 34)

Other applications of load-signature analysis as discussed by Hassell *et al.* include determination of the influence of adjacent metal structures on heating patterns and troubleshooting of tank- or output-circuit components that might affect system performance. The latter include faulty tank-circuit capacitors, high-resistance or loose bus connections, and changes in high-current, low-voltage bus spacing. Malfunctions of induction generator controls for constant output of voltage, current, or power are also revealed by load-signature analysis, as are coil overheating (due to inadequate cooling-water flow or poor contact in clamp-type inductors) and deterioration of flux-modifier materials.

TEMPERATURE-CONTROL MODES

Control modes for induction heating systems may be either open loop or closed loop. Open-loop systems are the simpler of the two. In these systems, temperature control is obtained by regulating precisely the power input to the heating line while maintaining a fixed dwell time (single-shot induction heating) or feed rate (scanning induction heating techniques). By these means, every workpiece or portion of a workpiece receives a controlled amount of energy (kilowatt-seconds).

Closed-loop control systems make use of the electric signals developed by devices such as thermocouples and pyrometers in conjunction with a special controller. The controller compares the measured temperature with one preset on the controller. The temperature difference is then used to activate a power-control device attached to the induction power supply. Power regulation may then be direct (of the on/off type) or proportional through the use of contactors, silicon-controlled rectifiers (SCR's), or saturable-core reactors, among other devices.*

On/Off Control Mode. Many single-shot induction heating applications, such as preheating and heat treating, involve heating to a given temperature prior to further processing. The requirements of the temperature-sensing device and controller include accurate indication of workpiece temperature, minimum overshooting of the present temperature, and development of a signal that activates the operations following heating. For such applications, a temperature-monitoring device and a two-position (on/off) controller are sufficient. As its name implies, an on/off-type control mode calls for full heating power or none at all. During heating, the power remains fully on until the preset, or setpoint, temperature is reached and then switches off. Addition of timers and suitable relays, however, permits maintenance of temperature about the setpoint value for a period of time.

Proportional-Type Control Mode. Many progressive heating applications require a closely maintained temperature for a long period of time. Examples include hardening and tempering of steel bars and annealing of wire and strip, operations often carried out by induction scanning techniques. The growth of single crystals of germanium or silicon by very slow withdrawal of the seed crystal from a melt maintained within very close temperature limits is another important application in which tight temperature controls are required. In these and similar situations, another type of control mode, called proportional control, is usually employed.

Proportional control uses three types of actions (proportional, rate, and

---

*Anon., *Lepel Review*, Vol 1, No. 3, 1960, p 1; Anon., *Industrial Heating*, Vol 51, No. 3, March, 1984, p 29.

reset) which serve to minimize temperature fluctuations. Proportional action is the control mode by which an output, or control, signal is generated which is proportional to the magnitude of the difference between the workpiece and setpoint temperatures. Proportional action alone would function similarly to the on/off controller except for a modulation added to the control action. In other words, as the temperature of the workpiece reached the setpoint, the control would decrease the power input to the coil to prevent overheating or underheating.

The proportional action alone, however, has no way of sensing the amount of heat (or power setting) required to maintain a given setpoint temperature. Because of this, the temperature would tend to oscillate. In order to get the controller to equilibrate at the desired temperature, two other control actions, termed rate and reset, are employed. Rate action takes into account how fast the actual workpiece temperature is changing in relation to the setpoint. A large rate of change in workpiece temperature would result in a large change in the control signal, larger than if just a proportional action were used. Reset action takes into account the time the actual workpiece temperature is away from the setpoint. The longer the time, the larger the correction.

PROPORTIONAL CONTROLLERS AND HEAT-REGULATING DEVICES

When a proportional control mode is in use, the control signals from the various actions (proportional, rate, and reset) are added to obtain a final output signal from the temperature controller, which in turn drives the power-control device on the induction power supply. The exact type of output signal varies according to the type of temperature controller. However, there are three broad types: position proportioning, time proportioning, and current proportioning.

Position-proportioning controllers provide a variable dc voltage which can be used in controlling an RF oscillator. Time-proportioning controllers provide a continuous pulsed output signal of constant amplitude; proportioning comes from the ratio of on time to off time or, more specifically, from the number and width of the output pulses. This type of controller can be used with induction heating power-supply contactors, silicon-controlled rectifiers, and RF power supplies. Current-proportioning controllers produce a continuous dc output signal whose magnitude is determined by the deviation of the measured temperature from the setpoint temperature. They can be used with saturable-core reactors and SCR's, for example.

A typical setup using a position-proportioning controller and an RF power supply is shown schematically in Fig. 7.8. In operation, a deviation of workpiece temperature (A) from the setpoint changes the position of the controller slidewire (B), resulting in an error signal (a low-current, dc signal) applied across the position-proportioning controller (C). Here the error is continu-

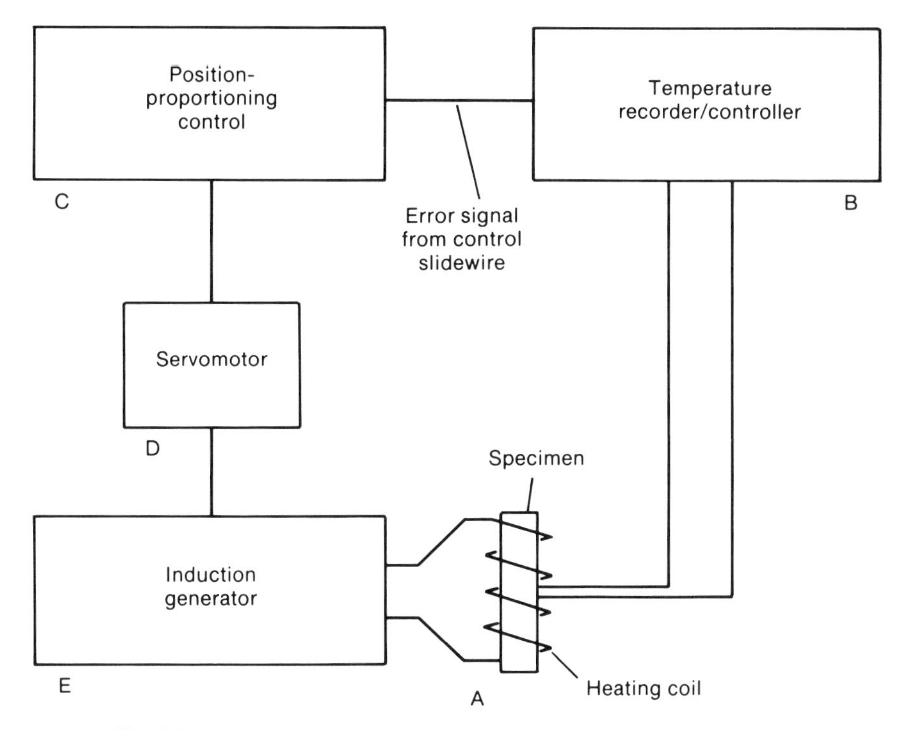

Fig. 7.8. Block diagram illustrating use of position-proportioning control and servomotor for temperature control (from Anon., *Lepel Review*, Vol 1, No. 3, 1960)

ously analyzed, and a corrective action required to restore the specimen temperature to the setpoint is applied through proportional, rate, and reset functions. The output from the proportioning controller is then fed into a converter and amplified to drive a servomotor (D) to regulate the power of the radio-frequency induction generator. Power regulation in the generator can be accomplished by a thyratron bias arrangement employing either phase shifting or variation of the grid voltage in accordance with the operation of the servomotor through a potentiometer.

Induction power-supply contactors can work in response to on/off as well as time-proportioning signals from a temperature controller; in essence, they consist of large mechanical relays with heavy-duty contacts that open or close the circuit between the power line and the induction heating generator. Silicon-controlled rectifier devices are relatively small for the amount of power they can control and are available in different forms. They can produce full-power pulses for lengths of time proportional to the signal from the control instrument.

Saturable-core reactors are used to control the voltage input into radio-frequency induction power supplies and hence the power available to the induction heating coil. Basically they consist of an iron-core transformer that regulates power flow by controlling the magnetic field in its iron core. This is done by impressing the temperature-control current across the control windings. If this current is low (i.e., if the temperature deviation from the setpoint is low), the core is not saturated, the inductance of the main winding is high, and a large voltage drop occurs across it. This results in a low *output* voltage drop from the induction power supply. Conversely, if the control current is high (because of a high temperature deviation), the core becomes saturated, the inductance of the main winding is low, and a small voltage drop occurs across the main winding and a large one across the output terminals.

INTEGRATION OF CONTROL FUNCTIONS

Although temperature control is surely the most important control aspect of an induction heating installation, other aspects of process regulation are also important. These include control of part sequencing, part handling and positioning, and electric demand. Often, a sophisticated control device is used in modern induction installations to monitor these latter functions as well as to carry out temperature control. The names of some of these systems do not indicate that they are computers, but in a broad sense they all perform tasks of information handling in accordance with programmed instructions. Some of the more common systems include the following*:

- Programmable controllers. These units were originally designed to replace hard-wired, relay logic control devices. They have analog control capability (as is needed for temperature control) and tabular and sequential logic programs, and can manage other systems.
- Loop controllers. These devices can be made as single-loop or multiple-loop units with a shared display. They perform the same functions as analog controllers, but can also be programmed for adaptive control, computation, and interactive modes with time-dependent or sequential setpoints.
- Distributed control systems. These devices are designed to locate control and data handling at various processing sites with an operator's interface at a remote central location. Graphical displays of process equipment, trending of process variables, group displays of the real-time parameters of interrelated process functions, and simplified alarm tracking and acknowledgment are some of the unique characteristics of distributed control systems.
- Minicomputers. These machines have large memories and are used for sophisticated programs for optimizing, process modeling, scheduling, and management planning. Personal computers are finding a place at the lower end of this category.

At first, digital machines such as those listed above were used to do the same control jobs performed by the analog devices that they replaced; it was

*J. E. O'Neill and J. A. Moore, *Industrial Heating*, Vol 52, No. 10, October, 1985, p 11.

quickly found that they did these tasks better. The digital devices do not drift, they have excellent resolution, and their displays are more readable than those of analog units. An operator has more confidence in an integer or decimal value read on a display than in a number that must be defined by interpolation of a pointer position between divisions of a scale.

Benefits of digital control devices include the following:

- Control from multiple recipes. Values of setpoint temperature, heating times and rates, and soak times are readily adjusted.
- Better control of sequential operations, allowing manpower reductions and preventing human error.
- Automatic collection and analysis of operating data, permitting rapid detection of out-of-specification operations.
- Centralized control.

Several examples will serve to illustrate the flexibility of integrated control systems.

Heating of Steel Slabs

Probably the largest megawattage use of induction is for the reheating of steel slabs prior to hot rolling. The facility at McLouth Steel, designed by Ajax Magnethermic Corporation, illustrates the controls required to operate such a process. The plant is designed to reheat up to 600 tons per hour of steel slabs whose incoming temperature varies between ambient and 815 °C (1500 °F).* Slabs are delivered by crane to one of six heating stations from a continuous casting machine or the slab storage yard. Prior to being loaded into the induction heaters, the slabs are checked for surface defects. Then they are moved sequentially on transfer cars through three induction heating stations fed by 20-MW, 10-MW, and 5-MW line-frequency power supplies. The total time from slab delivery to arrival at the rolling mill is 54 min.

Because of the complexity of the slab heating system, an automated control system for the operation is mandatory. The over-all control scheme (Fig. 7.9) makes use of equipment for slab-handling control (digital), heater control (digital), static power switches (SCR's), slab-temperature controls (analog), and a process computer (digital). Because of the multiple lines and heating stations, the static power-switching controls are among the most important features. The heaters draw different loads, and, because temperature control demands switching of heaters off and on, phase-balance control among the different lines is essential. The application of a process computer for this kind of control is excellent because of its abilities to gather, store, and analyze complex data; to make rapid decisions; and to perform routine functions consis-

---

*N. V. Ross, "Megaton Capacity for Hot Rolling Utilizing Induction Heating," *Proc. Seminario Sobre Laminacao-81 Colam*, Rio De Janeiro, Brazil, September, 1981; G. B. Bijwaard and H. Sorokin, *IEEE Transactions on Industry Applications*, Vol IA-8, November/December, 1972, p 735.

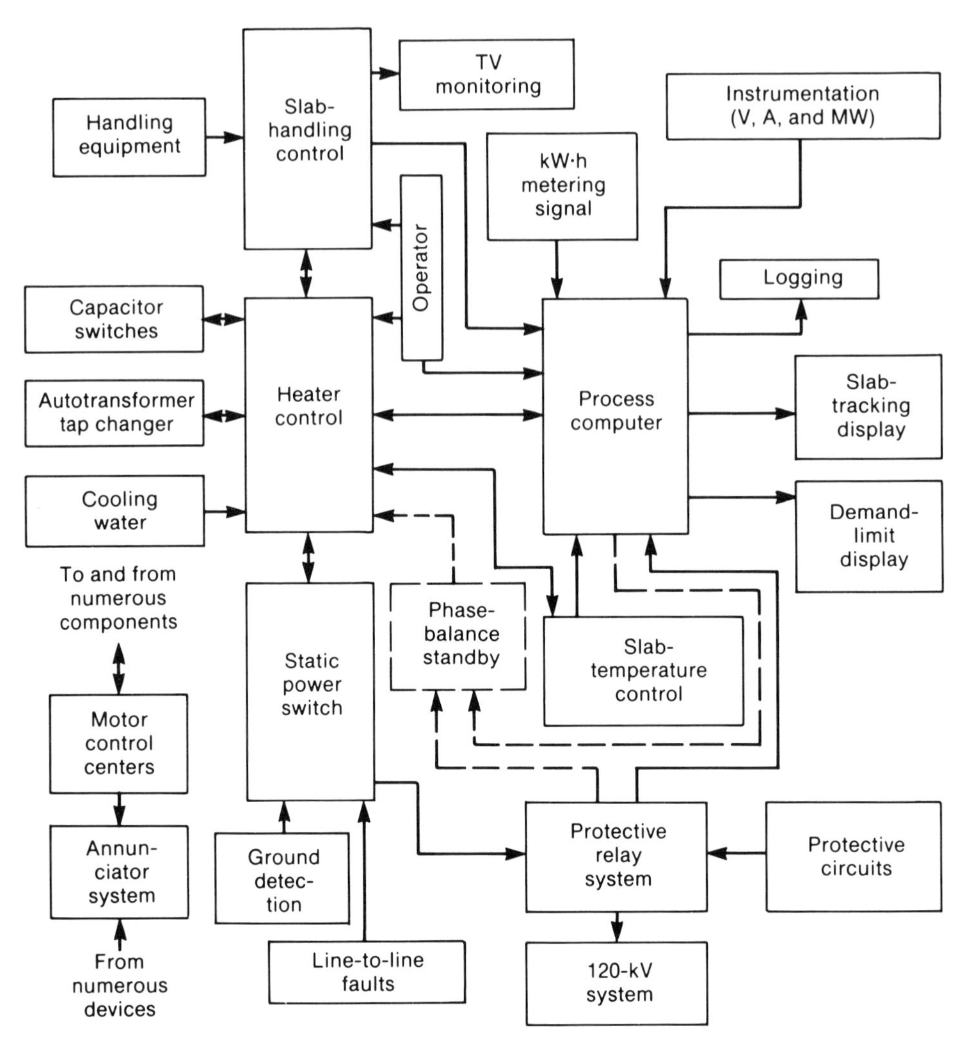

Fig. 7.9. Automatic handling and control scheme for an induction slab reheating facility (from G. B. Bijwaard and H. Sorokin, *IEEE Transactions on Industry Applications*, Vol IA-8, November/December, 1972, p 735)

tently. The ease with which the computer performs its slab-tracking, logging, phase-balance control, and demand-limit functions demonstrates the power of computer control in a time-sharing system.

Surface Hardening

Surface hardening of steel parts represents an operation in which many variables control final case depth and hardness and service characteristics. A numerically controlled (NC) system designed by Inductoheat (IPE Cheston) illustrates the state of the art for such applications.* This system can be used for parts as diverse as knuckle pivot pins, differential main shafts and drive

*R. N. Stauffer, *Manufacturing Engineering*, Vol 78, No. 5, May, 1977, p 44.

gears, transmission input and idler shafts, and rear-axle stub shafts. As many as 20 variables are controlled during a hardening-and-quenching cycle typically lasting 2 min. These include heating time, quenching time, scanning speed, power level, and part-rotation rate. The equipment can also be programmed to eliminate automatically the heating and quenching operations in those areas which are not to be hardened and to increase the scanning speed in these regions.

During setup of the induction hardening machine, control parameters are recorded on a special instrument panel. Once correct operation has been verified, this information is transferred to an NC tape which is read on a controller that incorporates solid-state timers and power-level controls. With this system, control of the hardness pattern from one run to the next is so good that quality-control checking can be eliminated. In previous systems, monitors that measured actual kilowatt-seconds of energy input were utilized; the machine was shut down at the end of a cycle if the energy level was not within a predetermined range. The accuracy and consistency obtained with the newer NC equipment eliminate the need for such monitoring.

The equipment also contains other automation features. For example, an externally mounted, multiple-position switch allows fast changing of autotransformer ratios when different parts or induction coils are installed on the machine. Air-operated upper tooling centers facilitate part loading and unloading. In addition, indexing and scanning of parts through the single-turn coils are done automatically with a drive on the part-holding fixture.

Vacuum Induction Melting

Vacuum induction melting and precision casting represent another area where computer or microprocessor control is very beneficial. Some of the functions for which such systems are being used include automation of melting and pouring sequences as well as programmed furnace temperature control and error diagnostics. Computers are also employed for alloying calculations and data retrieval and storage.

One of the more exacting operations for which computer control is attractive is precision casting following vacuum induction melting.* The quality of cast parts is mainly influenced by casting temperature, casting time, operating vacuum, and mold temperature. All of these parameters, except casting time (as determined by pouring speed), are relatively easy to measure and control. However, newer systems allow the computer to be "taught" the proper pouring speed. This is done by automatic recording of a series of manually operated pouring sequences. The ones that give the best castings are stored on an EPROM (Erasable Programmable Read Only Memory) for future use during automatic operation.

---

*F. Hugo, R. Schumann, W. Zenker, H. Bittenbrunn, and J. Mosch, *Metallurgy Plant and Technology*, Vol 8, No. 1, 1985, p 42.

Electric-Demand Control*

A final example of the use of process-control equipment relates to the control of electric demand in induction heating processes. Most commercial electric rate structures include two components—energy use and demand. The energy charge is based on the total kilowatt hours consumed, but the demand charge reflects the peak rate of energy consumption (i.e., the maximum *power* in kilowatts) drawn from the line. A reduction of energy usage will obviously lower the energy charge, but will not necessarily affect the demand charge. The converse is also true: it is possible to lower the demand charge without reducing the total energy consumed. Significant demand charges can be saved by coordinating the cyclical nature of certain electrical loads in order to smooth out the total plant demand profile. This is very readily done with a programmable controller or similar device. The controller program allows the electricity consumption in the plant to be tracked and higher demands to be forecast before they happen, and thus loads can be shed and restored without sacrificing over-all operations.

It should be realized that demand charges can be lowered only if the following conditions are met:

- The user's electric consumption rate has shown some degree of fluctuation or power peaks, such as those shown in Fig. 7.10.
- The precise time is known for shedding and restoring of loads.
- Significant loads are available for shedding during periods of high demand, without adversely affecting operations.

As an example, in a melting shop with several furnaces the melting operation closest to completion receives priority (to prevent overheating), while those at the beginning of their cycles can be shed temporarily. Such scheduling, and simultaneous monitoring and forecasting of electric demand, are easily carried out by programmable controllers.

DISTRIBUTED CONTROL

With the rising interest in flexible manufacturing and automation, induction heating equipment is increasingly becoming a part of larger manufacturing systems. Because induction heating is electric based, it is readily incorporated into processing schemes which make use of digital computers and controllers. This section briefly reviews the state of manufacturing control as it could apply to induction heating systems.

Concept of Distributed Control

The manufacture of a large number of products in automated factories can be quite complex. Because of this complexity, and in order to prevent major

*W. H. Sampson, *Industrial Heating*, Vol 49, No. 7, July, 1982, p 10.

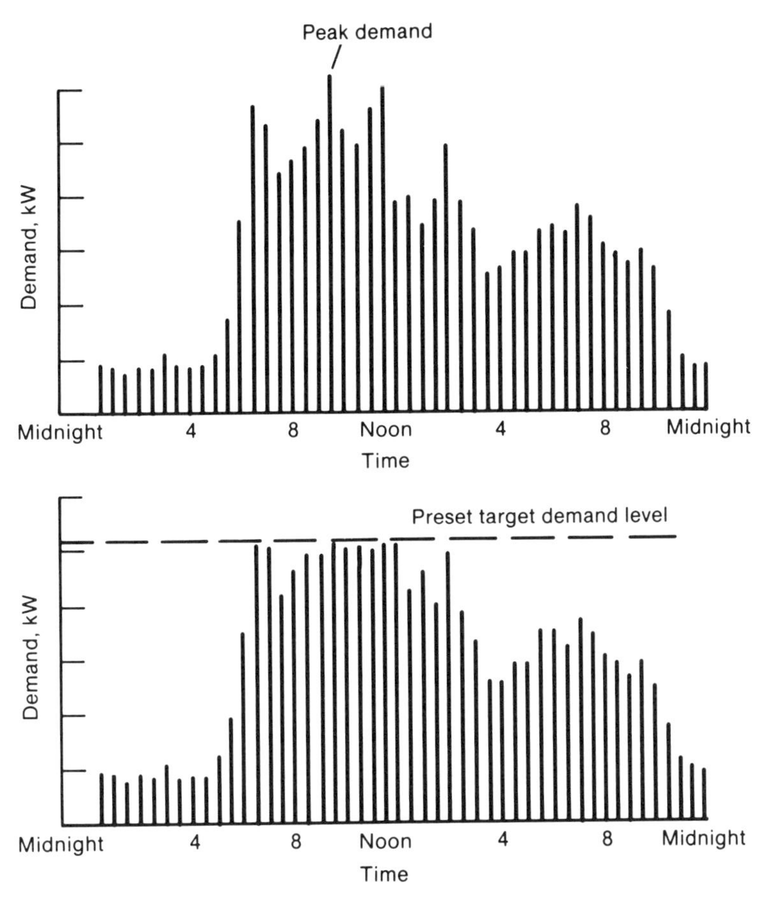

Fig. 7.10. Comparison of electrical load profiles without (left) and with (right) demand control (from W. H. Sampson, *Industrial Heating*, Vol 49, No. 7, July, 1982, p 10)

system failures and shutdowns, the control system that is most frequently implemented is a type known as "distributed." As discussed by McCurdy,* there are two major ideas behind distributed control. These are:

1. Each process in a manufacturing system is handled by an individual control device that is capable of operating independently. For example, the operation of an induction heating unit can be monitored *and* controlled by a separate programmable controller that will continue to do its job irrespective of what happens to other controllers, computers, etc., in the plant.
2. The individual controllers are connected through a common communications network. At the controller, or "baseboard," level, this network is known as a LAN (local area network). The LAN allows a host computer to gather information as required or to specify processes to be performed.

*D. W. McCurdy, *Industrial Heating* , Vol 53, No. 7, 1986, p 11.

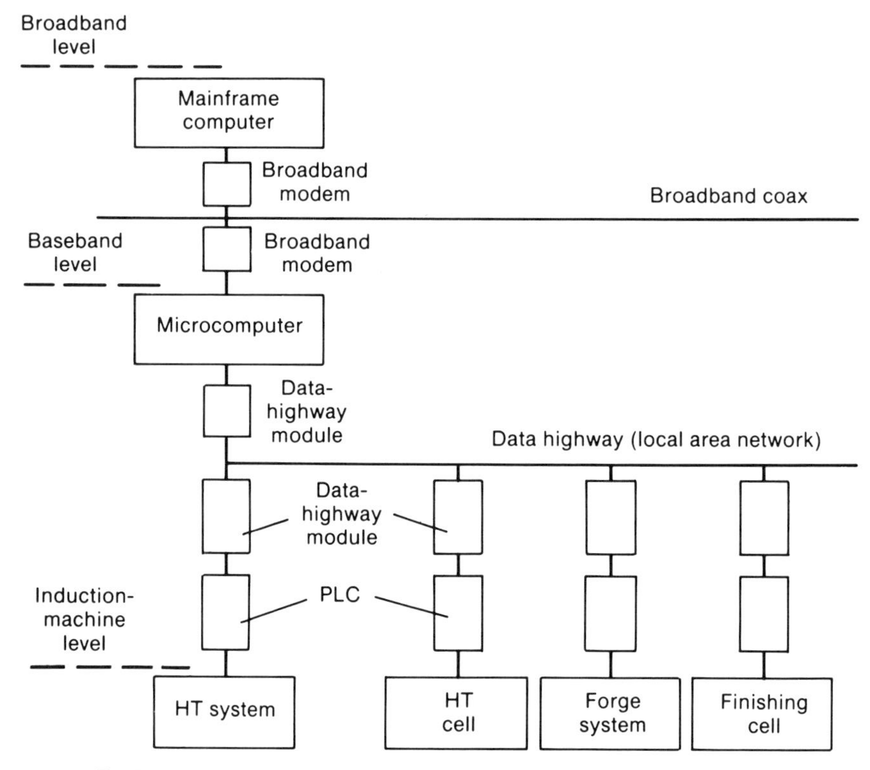

Fig. 7.11. Data-highway network for a system utilizing both baseband and broadband levels of communication (from G. F. Bobart, *Heat Treating*, Vol 18, No. 5, May, 1986, p 18)

In addition, the communications network often allows individual controllers to communicate among themselves without intervention or direction by a host computer.

The advantages of distributed control as compared with a centralized control system (in which a master device is required for operation) include the preservation of system integrity in the case of a (localized) hardware failure and a modular design which provides the ability to gradually expand processing capabilities without a major, one-time capital investment.

Induction heating cells can be tied into higher-level, or "broadband," plant-wide systems as well as be parts of local area networks. A schematic of how this can be accomplished is illustrated in Fig. 7.11.* The lower part of the over-all network utilizes baseband levels of communication. In the upper part of the diagram, the basic components of a plant-wide, broadband network are shown. The broadband network handles large amounts of concurrent communications and, in large operations, can provide a broad range of capabilities as well as many cost advantages. For example, functions such as order

*G. F. Bobart, *Heat Treating*, Vol 18, No. 5, May, 1986, p 18.

planning, production dispatch, cell control and reporting, and direct numerical control communications are typically carried out at the broadband level.

Interfacing/Connecting Control-System Components

The term "interfacing" refers to the means by which various digital (and analog) devices are connected to enable communications among them. The connection must accommodate operating voltage differences and, for control devices, the hardware and software characteristics of each device.

Hardware considerations in typical control systems generally center on the type of wiring or cable that is used. At this time there is no standard that is universally accepted. As described by McCurdy and Bobart, however, Electronic Industries Association Recommended Standard 422 (EIA RS 422) is evolving as the one most often relied on for connecting digital process controllers to each other and to computers. Usually, this consists of twisted pairs of wires that form the so-called data highway network. This method, or a modification of it, is attractive because it is inexpensive, it is suitable for use in adverse plant environments, and it is "multidroppable"—i.e., the same communications cable can be connected simultaneously to several controllers. There is also the added advantage that an RS 422 communications "port" is easily derived from the type found on most mini- and mainframe computers—an RS 232 port—using a relatively simple active converter. The RS 232 standard, however, is not suitable for typical plant environments, nor is it multidroppable.

At the broadband level of communications, coaxial cables are most often used to link higher-level system components. These cables permit simultaneous transmission of signals over multiple channels.

If wiring methods were the only concern in setting up a control system, process design and execution would be relatively simple. An equally if not more important design characteristic relates to the software, or instruction package, that controls the flow of information. Also known as "protocol," this software determines how information is sent and received over the network. This includes the amount of information sent at one time, error checking, addressing, and information coding. There are two basic types of communication*: synchronous and asynchronous. Synchronous communication requires that both the sender and the receiver (for example, two computers or one computer and a controller) have clocks which can be synchronized so that data may be sent at precise intervals. This requires expensive hardware, but allows fast and very reliable communications. In asynchronous communication, data are sent in an intermittent stream. Such a technique is simple, but speed is sacrificed. Asynchronous data transmission is the more common technique.

---

*R. G. Blocks, *Microprocessors in Heat Treating*, unpublished book used in ASM course, 1985.

In the communications network, several different pieces of equipment often must use the same channel; this applies especially to a data highway or baseband local area network which can handle only one signal at a time. Several methods have been developed and are described by Bobart. These include use of a token bus or token ring arrangement, whereby data flow is initiated only by the station holding the priority token. After transmission, the token is passed on to the next station, generally on a priority basis. Another method makes use of what is known as a collision-detection technique in which each station must wait a random period of time if more than one tries to transmit at the same time.

The problem of interfacing and communications can be particularly difficult in plant-wide or broadband systems comprised of equipment made by a number of different manufacturers. To help solve this problem, General Motors Corporation has developed a factory communications standard called Manufacturing Automation Protocol ("MAP") to enable programmable controllers, computers, CRT's, and LAN schemes made by various vendors to be efficiently interfaced. Most major U.S. manufacturing companies have accepted this specification and are cooperating in its further development.

MISCELLANEOUS CONTROL TECHNOLOGIES USED IN INDUCTION HEATING

With the advent of automation and the demand for controlled-quality products, increasing emphasis is being placed on nondestructive techniques for evaluating final product properties. Methods which allow for real-time *feedback* control of the actual induction heating process are of special interest. Several of these methods, pertinent to induction hardening processes, are discussed in this section.

Electromagnetic Sorting

Electromagnetic sorting processes make use of the dependence of magnetic properties of ferromagnetic steels on hardness, case depth, microstructure, residual stresses, etc. to verify proper heat treatment. These magnetic properties include permeability, coercive force, saturation, and remanence, each of which relates to phenomena during magnetization, as illustrated in Fig. 7.12. These drawings depict the relationship between magnetizing force (H) and the flux density of the field of magnetic induction (B). When a virgin specimen is used, the flux density B increases rather rapidly upon initial application of a magnetizing force H (Fig. 7.12a). Eventually it reaches a point at which any increase in H leads to no increase in B. The flux density at this latter stage denotes the saturation point. When the magnetizing force is reduced to zero (Fig. 7.12b), some residual magnetism is retained; this is referred to as remanence. When the magnetizing force is reversed and gradually increased in value, the flux continues to decrease (Fig. 7.12c). The value of H at which

B is equal to zero is referred to as the coercive force in the material. As the reversed field is further increased, the specimen again becomes saturated but in the opposite polarity. Upon increasing H again in the forward sense, the remaining portion of a closed loop, known as a hysteresis loop, is formed (Fig. 7.12d). If the magnetizing force and flux density are the result of an alternating current (either applied or induced), the hysteresis loop is traced out once for every current reversal.

The detailed hysteresis behavior itself, or one of the characteristics of the hysteresis loop, is relied on in electromagnetic sorting for evaluation of case depth (or average hardness). One of these characteristics is the value of the coercive force. In practice, a probe is placed on the test piece, and a coercive force measurement is made. Then the case depth may be read from a calibration curve previously prepared by destructive checks (i.e., metallographic sec-

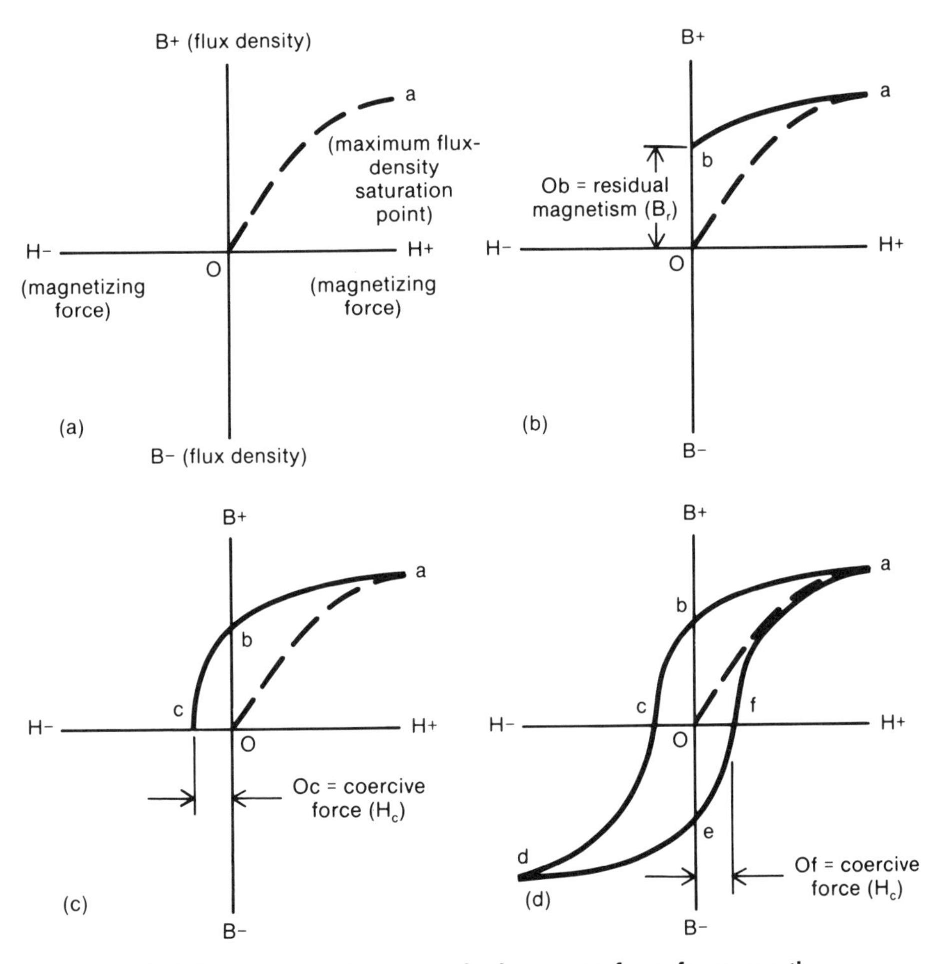

Fig. 7.12. Representative magnetization curves for a ferromagnetic material (from R. C. McMaster, *et al.*, *Metals Handbook*, 8th Ed., Vol 11, ASM, Metals Park, OH, 1976, p 93)

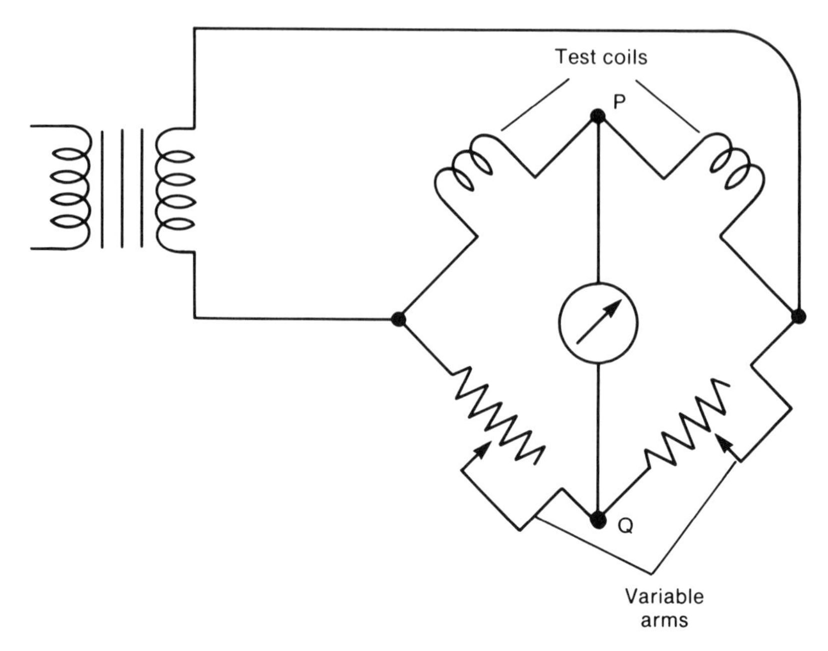

Fig. 7.13. A simple inductance bridge (from D. M. Lewis, *Magnetic and Electrical Methods of Nondestructive Testing*, George Allen and Unwin, Ltd., London, 1951)

tioning and examination) of samples with shallow and deep cases. This method is useful for steels with cases as deep as approximately 20 mm (0.8 in.).

Another technique, which is probably the most common nondestructive inspection test for hardness and case depth by electromagnetic means, uses a bridge-comparator system. The phenomenon which underlies this application is the fact that the magnetic permeability of a piece of steel is a function of material condition, such as the hardness level. The magnetic permeability is defined as the ratio of B to H. From the hysteresis loop, it can be seen that this ratio can assume values up to some maximum denoted by μ_{max}.

Bridge-comparator tests rely on various types of bridge circuits, nearly all of which are modifications of the simple inductance bridge (Fig. 7.13). The bridge includes reference and test coils which are identical in size and number of turns. For most applications, the bridge is energized by ac current at a frequency between 60 and 1000 Hz. Lower frequencies are typically used to check average or core hardness whereas higher frequencies are employed to discriminate between parts of different case depths.

When steel components are used as the cores of the coils, the over-all impedance (or inductance) measured across each of the coils depends on the magnetic characteristics (i.e., magnetic permeability) of their cores. For identical cores, the impedance of each arm will be the same, points P and Q will be at the same potential, and the indicator will yield a null reading. During the actual inspection operation, a reference component, or "standard," is

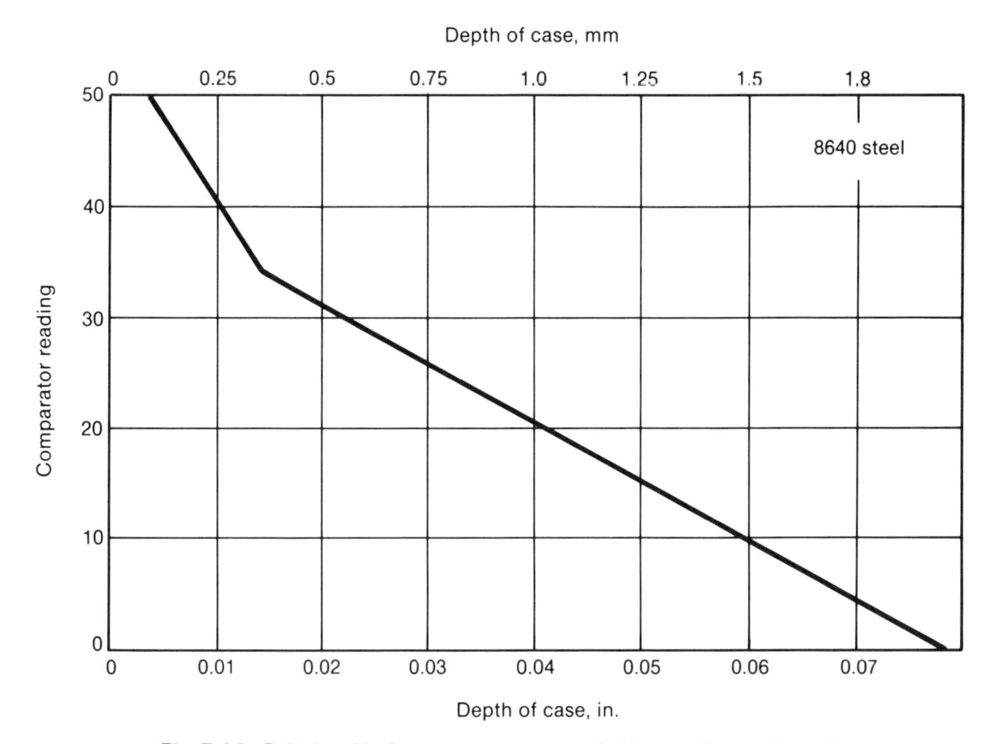

Fig. 7.14. Relationship between comparator bridge reading and depth of case obtained on induction hardened 8640 steel bars (from D. E. Bovey, *Instrument Practice*, Vol 2, No. 12, 1948, p 494; and *General Electric Review*, Vol 50, No. 11, 1947, p 45)

placed in one of the coils. The part, whose case depth or over-all average hardness is to be checked by comparison with the standard, is placed in the other coil. Dissimilarity in hardness patterns, and therefore impedance values, across the coils will produce a voltage difference across the indicator. This difference can be used to detect a difference in hardness or case depth through a suitable calibration based on destructive evaluation of several test samples. Figure 7.14 shows a typical calibration curve for induction case hardened 8640 specimens.

Many commercial sorting bridges also compare the hysteresis loops of the standard and test specimens on the assumption that, if they are completely alike, the values of B will be equal at all points in the ac cycle. Any instantaneous difference in the forms of the loops will result in phase displacements of the currents through the legs of the bridge which can be readily detected by an oscilloscope or similar equipment.

Resistivity Measurement of Case Depth

The use of resistivity measurements to estimate case hardened depth is similar to electromagnetic sorting techniques in that an electrical property which is structure-sensitive is correlated with case depth.* The major difference between the two methods lies in the type of electrical circuit utilized—a dc circuit in the former instance and an ac circuit for electromagnetic sorting. In essence, the average resistivity of the surface layers of case hardened steels is determined by attaching a pair of test probes to the surface. A pulsed dc current is passed through the probes and into the workpiece. Because the resistivity of martensite is greater than that of higher-temperature transformation products (e.g., pearlite, bainite), deeper case depths are indicated by higher average resistivity for a given probe geometry.

The resistivity measurement technique can be used to get an absolute indication of case depth by developing calibration curves. It can also be utilized as a sorting device in which the resistivity of a part is compared with that of a part known to have been properly heat treated.

Colorimetric Evaluation of Induction Hardening

Another, more recently developed, nondestructive inspection technique for evaluating the results of induction hardening operations is that referred to as the colorimetric method.† This test is used for flat parts which have been surface or locally heat treated. The teeth of gears and bandsaw blades are typical applications. Following induction hardening, various microstructural constituents are found in the heat treated parts. Layers at the outer surface are usually martensitic, whereas lower-temperature transformation products are found away from this zone. It has been noted that regions of varying microstructure (and hence peak temperature during the heating cycle) can be correlated with bands of various colors that are formed during processing. In turn, the extent and type of the colored bands can be correlated with the hardness pattern produced.

PROCESS SIMULATION

With the widespread use of high-speed digital computers, numerical simulation of induction heating processes is becoming increasingly common. With these programs, the effects of material variables (e.g., resistivity, permeability, etc.) and process conditions (e.g., power level) on heating patterns, residual stresses, and so forth, can be readily ascertained. Often this obviates the need to do costly and time-consuming experiments. In this way, coil designs and operating conditions are easily determined. However, at present, methods of treating the induction heating of only relatively simple geometries, such as round bars, tubes, sheets, plates, and slabs, have been developed. This is due to the complexity of the equations which describe the electrical and thermal

---

*W. R. Hain, *Heat Treating*, Vol 18, No. 8, August, 1986, p 35.
†A. Krilov, *NDT International*, Vol 37, No. 6, June, 1979, p 125.

aspects of the process. The important equations and some sample simulation results are briefly discussed below.

Problem Formulation

Simulation of induction heating processes includes input of required data and numerical solution of a set of differential equations subject to appropriate boundary conditions. Specifically, this comprises the following*:

- Definition of part shape and size. This operation can be done by hand or by use of engineering drawings that are stored in a computer memory bank. In addition, the part geometry must be broken into a number of discrete slices or elements to enable solution by techniques such as the finite-difference and finite-element methods. The discretization must be done with care to avoid elements which are too coarse (leading to solution errors) or too fine (causing high computation cost). Many times, elements of varying sizes are used, with the finer ones placed in regions in which higher current densities or temperature gradients are expected.
- Specification of coil shape.
- Input of material properties. This input falls under the categories of electrical and thermal properties. Electrical resistivity and relative magnetic permeability comprise the former, and thermal conductivity and emissivity are the more important parameters in the latter group. Resistivity and thermal-conductivity data as functions of temperature are readily available. The permeabilities of ferromagnetic irons and steels depend on local magnetic fields and must be taken into account. At high temperatures, radiation heat losses can be significant; therefore, the emissivity, which depends on surface condition (and thus on temperature through its effect on surface condition), must be included in the input data. Lastly, phase transformations and their kinetics must be described in the material data base. These data should include the effect of heating *rate* as well as the effect of temperature on phase changes, which affect other physical properties also.
- Method of determining eddy currents set up in the workpiece. This is probably the most difficult part of the simulation program. Methods of eddy-current calculation include analytical techniques, approximate analytical results, and numerical methods. Analytical techniques are available only for very simple geometries.† Calculations of eddy currents and magnetic fields, which are important in areas other than induction heating, are receiving ever-increasing attention in the technical literature.‡ Most treatments of

---

*V. E. Wood, unpublished research, Battelle's Columbus Laboratories, Columbus, OH, 1985.

†J. Davies and P. Simpson, *Induction Heating Handbook*, McGraw-Hill, Ltd., London, 1979.

‡A. Konrad, *IEEE Trans., Magnetics*, Vol 21, 1985, p 1805; C. J. Carpenter, *Proc. IEEE*, Vol 124, 1977, p 1026; T. Tortschanoff, *IEEE Trans., Magnetics*, Vol 20, 1984, p 1912; R. J. Lari and L. R. Turner, *IEEE Trans., Magnetics*, Vol 19, 1983, p 2474.

eddy currents in magnetic media assume that hysteresis losses can be neglected with respect to eddy-current losses. Sample calculations for a cylindrical geometry* suggest that the hysteresis losses can account for as much as 10% of the total losses—an amount which is not insignificant, particularly in steel heat treating applications.

- Thermal calculations. Determination of the temperature distribution within the workpiece, given the distribution of heat sources due to eddy currents or hysteresis, is a relatively standard problem in numerical analysis. The temperature dependence of the material properties must be included. Several simplifying approximations often used are that the eddy-current heat source is uniformly distributed in the sample at low frequencies and is localized at the surface at high frequencies.
- Boundary conditions. Boundary conditions have to be specified for both the electromagnetic and thermal portions of the problem. The most difficult of these involves the heat-transfer conditions at the workpiece surface. This is primarily a result of uncertainties associated with emissivity and radiation losses.

During computation, the eddy-current and temperature fields are calculated, subject to the boundary conditions, for a number of successive increments in time. For steels which undergo phase transformations (for example, during heating prior to hot working or during austenitization), the computation must also involve simultaneous estimation of the phase boundary within the workpiece.

Simulation of Surface Hardening

Computer results for surface hardening will serve to illustrate the power of process-simulation methods. The results are taken from the work of Melander,† who simulated static as well as scanning induction surface hardening of round bars of a steel similar to AISI 4142. Eddy currents were described using Maxwell's equation, and heat transfer was simulated using the one-dimensional (static hardening) or two-dimensional (scan hardening) heat-conduction equations with boundary conditions consisting of prescribed heat fluxes on the workpiece surface.

Phase transformations during the heating cycle were calculated with the use of isothermal transformation (reaustenitization) diagrams and a "staircase" model to account for the actual *continuous* heating involved in the induction experiments. The data were quantified using the expression due to Avrami‡ and Johnson and Mehl§:

$$V_k = 1 - \exp[-b_k(T) \cdot t^{n_k(T)}]$$

---

*J. D. Lavers, M. R. Ahmed, M. Cao, and S. Kalaichelvan, *IEEE Trans., Magnetics*, Vol 21, 1985, p 1850.

†M. Melander, *J. of Heat Treating*, Vol 4, No. 2, December, 1985, p 145.

‡M. Avrami, *J. Chem. Phys.*, Vol 7, 1939, p 1103.

§W. A. Johnson and R. F. Mehl, *Trans. AIME*, Vol 135, 1939, p 416.

where V_k is the volume fraction of phase k after holding for t seconds at a constant temperature T, and n_k is a constant. The diffusionless martensite transformation (on quenching) was modeled through the relation:

$$V_{martensite} = 1 - \exp[-\gamma(M_s - T)^{\beta}]$$

where $V_{martensite}$ is the martensite volume fraction, M_s is the "martensite start" temperature, and γ and β are experimentally determined constants.

The simulations conducted by Melander* also permitted estimation of residual stresses and hardness distributions. Residual stresses were derived from the stress-equilibrium equation, the flow rule, and an assumed elastic-plastic unloading. Hardnesses were estimated from regression equations relating Vickers hardness of a given phase to its chemical composition.

Several sample solutions obtained by Melander are shown on the next two pages in Fig. 7.15 and 7.16. Figure 7.15 compares measured and predicted temperature-versus-time histories and final hardness, martensite, and axial residual stress distributions in a 40-mm- (1.57-in.-) diam bar. The bar was heated in a single-shot mode with an 80-mm- (3.14-in.-) long solenoid inductor energized by a 300-kHz power supply. Following heating, the workpiece was allowed to air cool for 4.3 s and then water quenched. The simulation and experimental results exhibit excellent agreement.

Figure 7.16 also shows fairly good agreement between measurement and prediction for a bar that was progressively hardened. The 40-mm- (1.57-in.-) diam bar was hardened with a single-turn coil which was 18 mm (0.71 in.) wide and which traversed the workpiece at a rate of 3.47 mm/s (0.14 in./s). The bar was quenched with a polymer quenchant following induction heating.

*M. Melander, *ibid.*

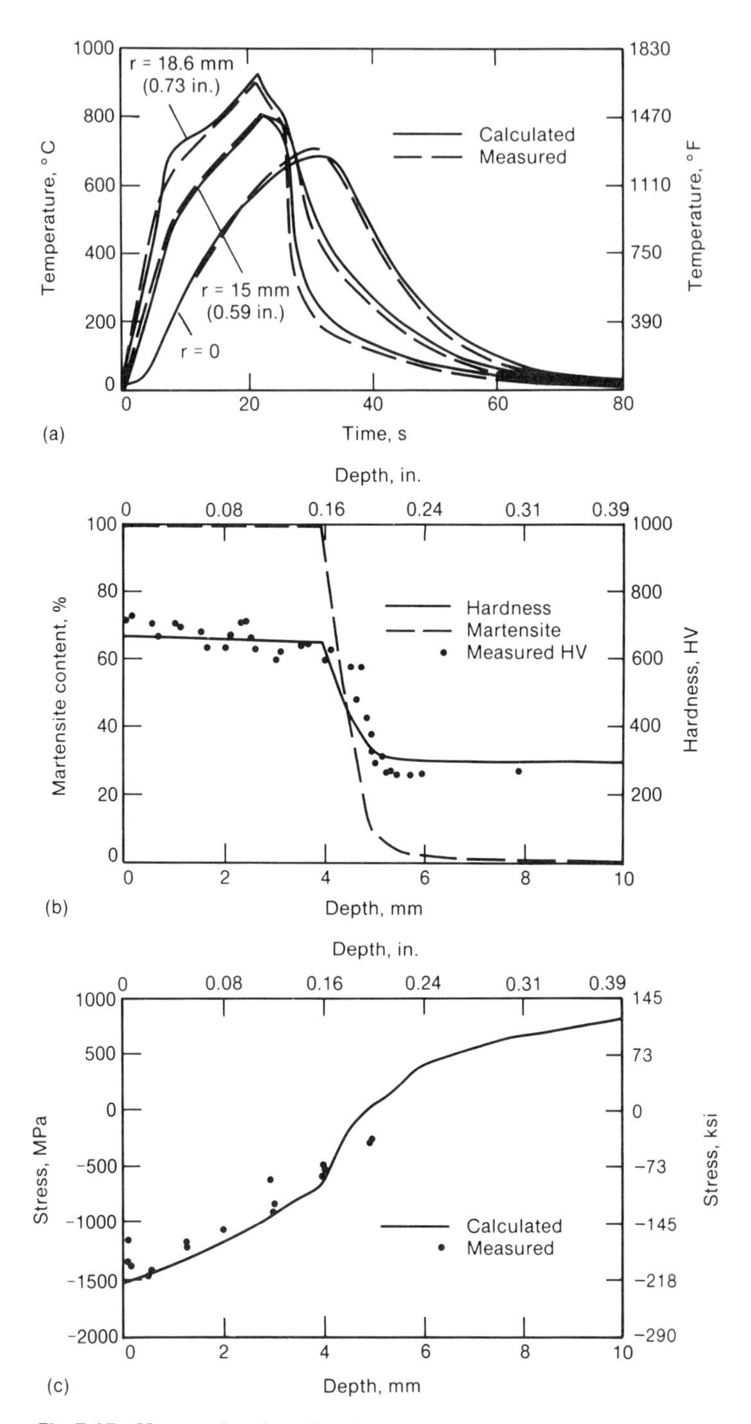

Fig. 7.15. Measured and predicted values of (a) temperature-versus-time history, (b) Vickers hardness and martensite-content profiles, and (c) axial residual stress pattern in AISI 4142 steel samples surface hardened by a single-shot induction heating technique (from M. Melander, *J. of Heat Treating*, Vol 4, No. 2, December, 1985, p 145)

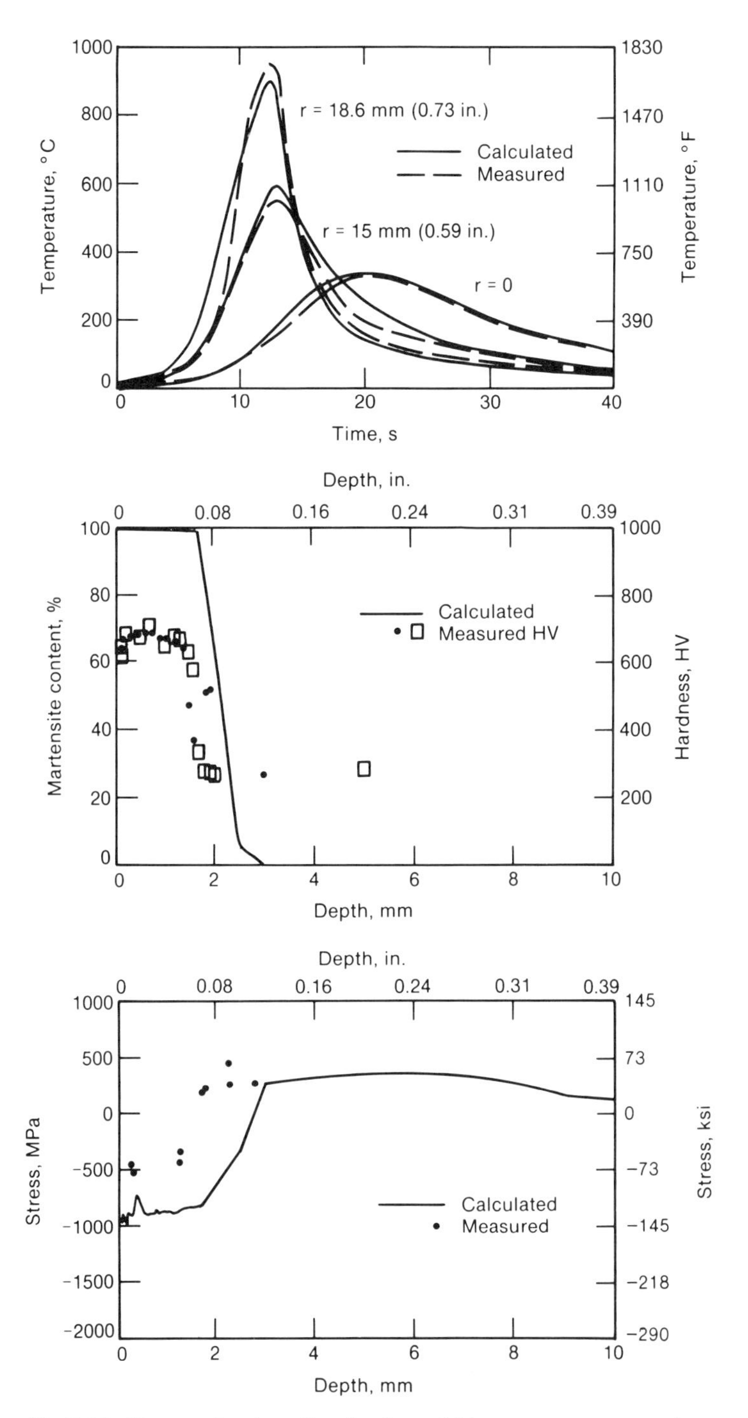

Fig. 7.16. Measured and predicted values of (a) temperature-versus-time history, (b) Vickers hardness and martensite-content profiles, and (c) axial residual stress pattern in AISI 4142 steel samples surface hardened by an induction scanning technique (from M. Melander, *J. of Heat Treating*, Vol 4, No. 2, December, 1985, p 145)

Chapter 8

Coil Design and Fabrication

In a sense, coil design for induction heating is built upon a large store of empirical data whose development has sprung from theoretical analyses of several rather simple inductor geometries such as the classical solenoidal coil. Because of this, coil design is generally based on experience. The objective of this chapter is to review the fundamental electrical considerations in design of inductors and to describe some of the most common coils in use. The actual construction of coils and the selection of power-supply leads are also treated.

BASIC DESIGN CONSIDERATIONS

In previous chapters, the analogy between a transformer and an inductor/workpiece combination has been described. The inductor is similar to the transformer primary, and the workpiece is equivalent to the secondary (Fig. 8.1). Therefore, several of the characteristics of transformers are useful in the development of guidelines for coil design.

One of the most important features of transformers lies in the fact that the efficiency of coupling between the windings is inversely proportional to the square of the distance between them. In addition, the current in the primary of the transformer multiplied by the number of primary turns is equal to the current in the secondary multiplied by the number of secondary turns. Because of these relationships, there are several conditions that should be kept in mind when designing any coil for induction heating:

1. The coil should be coupled to the part as closely as feasible for maximum energy transfer. It is desirable that the largest possible number of flux lines intersect the workpiece at the area to be heated. The denser the flux at this point, the higher will be the current generated in the part.
2. The greatest number of flux lines in a solenoid coil are toward the center of the coil. The flux lines are concentrated inside the coil, providing the maximum heating rate there.

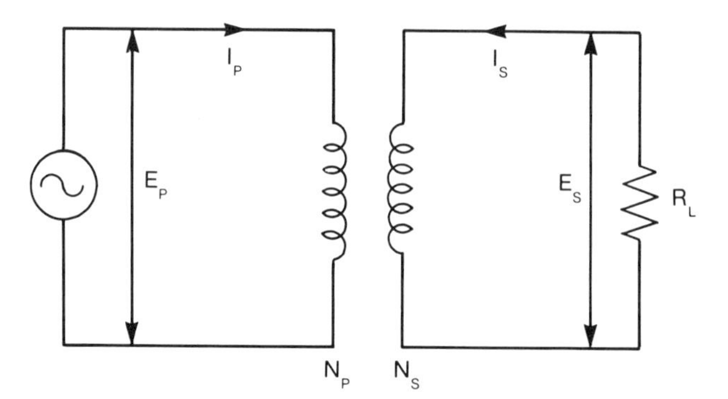

E_p = primary voltage (V); I_p = primary current (A); N_p = number of primary turns; I_s = secondary current (A); N_s = number of secondary turns; E_s = secondary voltage (V); R_L = load resistance (Ω).

Fig. 8.1. Electrical circuit illustrating the analogy between induction heating and the transformer principle

3. Because the flux is most concentrated close to the coil turns themselves and decreases farther from them, the geometric center of the coil is a weak flux path. Thus, if a part were to be placed off center in a coil, the area closer to the coil turns would intersect a greater number of flux lines and would therefore be heated at a higher rate, whereas the area of the part with less coupling would be heated at a lower rate; the resulting pattern is shown schematically in Fig. 8.2. This effect is more pronounced in high-frequency induction heating.
4. At the point where the leads and coil join, the magnetic field is weaker; therefore, the magnetic center of the inductor (along the axial direction) is not necessarily the geometric center. This effect is most apparent in

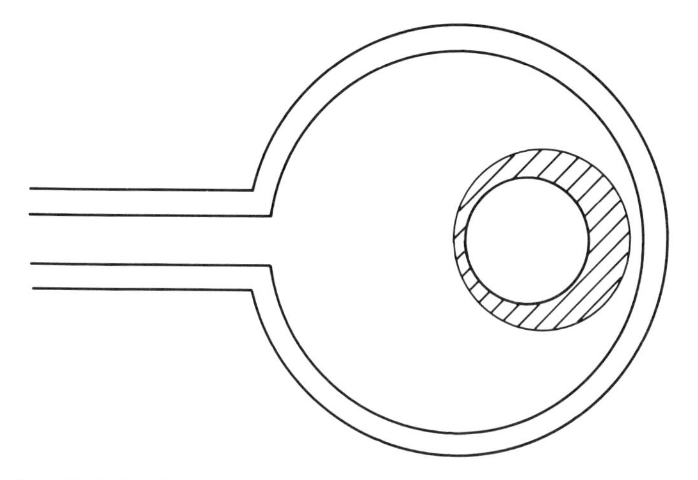

Fig. 8.2. Induction heating pattern produced in a round bar placed off center in a round induction coil

single-turn coils. As the number of coil turns increases and the flux from each turn is added to that from the previous turns, this condition becomes less important. Due to this phenomenon and the impracticability of always centering the part in the work coil, the part should be offset slightly toward this area. In addition, the part should be rotated, if practical, to provide uniform exposure.

5. The coil must be designed to prevent cancellation of the field of magnetic induction by opposite sides of the inductor. The coil on the left in Fig. 8.3 has no inductance, because the opposite sides of the inductor are too close to each other. Putting a loop in the inductor (coil at center) will provide some inductance. The coil will then heat a conducting material inserted in the opening. The design at the right provides added inductance and is more representative of good coil design.

Because of the above principles, some coils can transfer power more readily to a load because of their ability to concentrate magnetic flux in the area to be heated. For example, three coils which provide a range of heating behaviors are the following:

- A helical solenoid with the part or area to be heated located within the coil and thus in the area of greatest magnetic flux.
- A pancake coil where the flux from only one surface intersects the workpiece.
- An internal coil for bore heating, where only the flux on the outside of the coil is utilized.

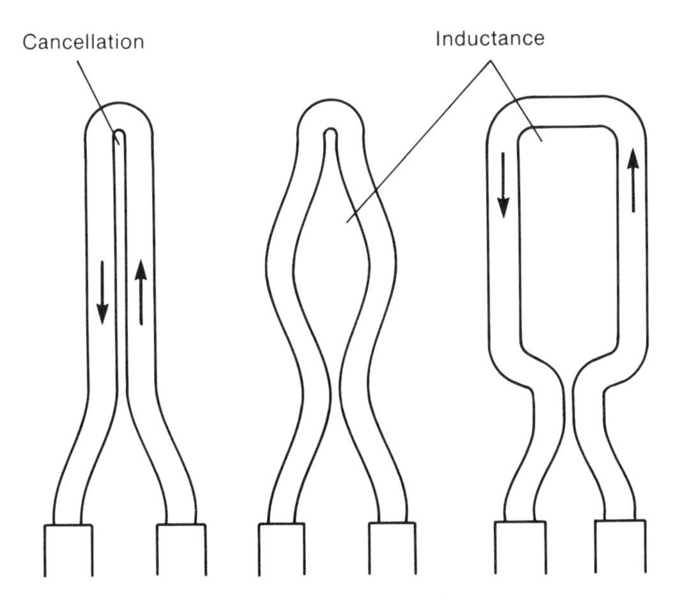

Fig. 8.3. Effect of coil design on inductance (from F. W. Curtis, *High Frequency Induction Heating*, McGraw-Hill, New York, 1950)

Table 8.1. Typical coupling efficiencies for induction coils

	Coupling efficiency at frequency of:			
	10 Hz		450 kHz	
Type of coil	Magnetic steel	Other metals	Magnetic steel	Other metals
Helical around workpiece	0.75	0.50	0.80	0.60
Pancake	0.35	0.25	0.50	0.30
Hairpin	0.45	0.30	0.60	0.40
One turn around workpiece	0.60	0.40	0.70	0.50
Channel	0.65	0.45	0.70	0.50
Internal	0.40	0.20	0.50	0.25

In general, helical coils used to heat round workpieces have the highest values of coil efficiency and internal coils have the lowest values (Table 8.1). Recall from Chapter 2 that the coil efficiency, η, is the fraction of the energy delivered to the coil which is transferred to the workpiece. This should not be confused with over-all system efficiency.

Besides coil efficiency, design considerations with regard to heating pattern, part motion relative to the coil, and production rate are also important. Because the heating pattern reflects the coil geometry, inductor shape is probably the most important of these factors. Quite often, the method by which the part is moved into or out of the coil can necessitate large modifications of the optimum design. The type of power supply and the production rate must also be kept in mind. If one part is needed every 30 s but a 50-s heating time is required, it is necessary to heat parts in multiples to meet the desired production rate. Keeping these concurrent needs in mind, it is important to look at a wide range of coil techniques to find the most appropriate one.

BASIC COIL DESIGNS

Low-Frequency Heating

Low-frequency induction heating is generally desired for through heating of metals, particularly those with large and relatively simple cross sections. Typical applications are round or round-cornered square (RCS) stock for forging or extrusion and slabs for hot rolling. In these cases, coil design is often quite simple, usually consisting of a solenoid coil or a variation of it that matches the basic workpiece cross-sectional shape (e.g., square, rectangular trapezoidal, etc.).

Low-frequency coils often have many turns. Accordingly, the coil usually forms the total tank inductance, and an autotransformer may be used to match the high coil impedance to that of the induction generator. In any case,

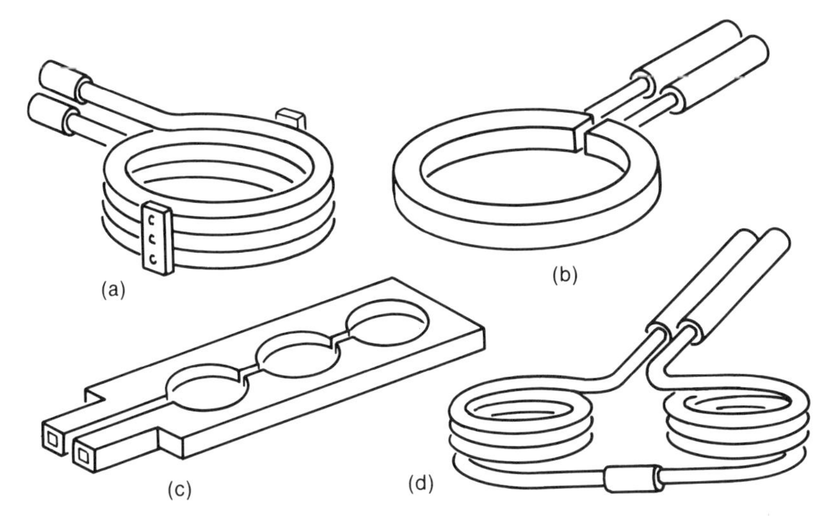

Fig. 8.4. Typical configurations for induction coils: (a) multiturn, single place; (b) single-turn, single-place; (c) single-turn, multiplace; (d) multiturn, multiplace (from F. W. Curtis, *High Frequency Induction Heating*, McGraw-Hill, New York, 1950)

the coil or transformer inductance must be high in order to reduce the number of tank capacitors needed to tune the resonant circuit at the operating frequency. Generally, the lower the frequency, the larger the coil or the greater the number of turns.* When low-inductance coils are occasionally required, isolation transformers can be used to match the coil impedance to that of the induction generator. However, these occasions are considerably fewer at the lower frequencies. In a practical sense, coils of this nature are generally purchased as part of an over-all system, with total responsibility for operation resting with the system supplier.

Medium-to-High-Frequency Coils

Simple solenoid coils, as well as variations of them, are also often relied on in medium-to-high-frequency applications such as heat treatment. These include single- and multiple-turn types. Figure 8.4 illustrates a few of the more common types based on the solenoid design. Figure 8.4(a) is a multiturn, single-place coil, so called because it is generally used for heating a single part at a time. A single-turn, single-place coil is also illustrated (Fig. 8.4b). Figure 8.4(c) shows a single-turn, multiplace coil. In this design, a single turn interacts with the workpiece at each part-heating location. Figure 8.4(d) shows a multiturn, multiplace coil.

*A computer program that can be used to calculate the required coil turns, tank-circuit capacitance, and transformer ratio for load matching and circuit tuning is given in Appendix D of S. L. Semiatin and D. E. Stutz, *Induction Heat Treatment of Steel*, ASM, Metals Park, OH, 1986.

More often than not, medium-to-high-frequency applications require specially configured or contoured coils with the coupling adjusted for heat uniformity. In the simplest cases, coils are bent or formed to the contours of the part (Fig. 8.5). They may be round (Fig. 8.5a), rectangular (Fig. 8.5b), or formed to meet a specific shape such as the cam coil (Fig. 8.5c). Pancake coils (Fig. 8.5d) are generally utilized when it is necessary to heat from one side only or where it is not possible to surround the part. Spiral coils (Fig. 8.5e) are generally used for heating bevel gears or tapered punches. Internal bores can be heated in some cases with multiturn inductors (Fig. 8.5f). It is important to note that, with the exception of the pancake and internal coils, the heated part is always in the center of the flux field.

Regardless of the final part contour, the most efficient coils are essentially modifications of the standard, round coil. A conveyor or channel coil, for example, can be looked at as a rectangular coil whose ends are bent to form "bridges" in order to permit parts to pass through on a continuous basis. The parts, however, always remain "inside" the channels where the flux is concentrated. Figure 8.6 illustrates similar situations in which the areas to be hardened are beside the center of the coil turns, and thus are kept in the area of heaviest flux.

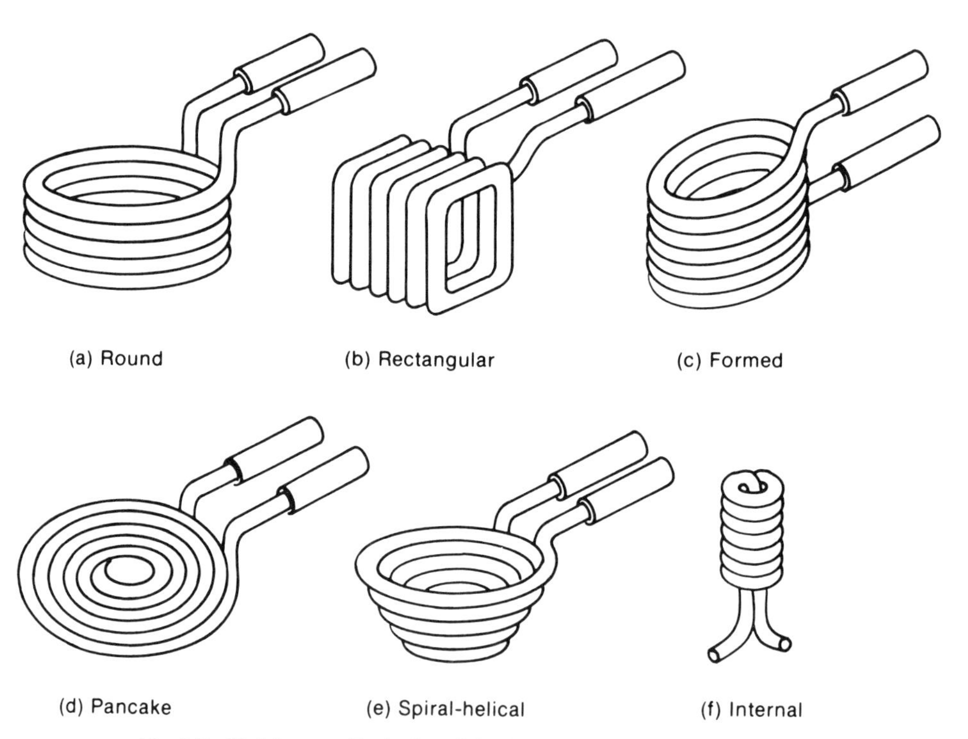

Fig. 8.5. Multiturn coils designed for heating parts of various shapes: (a) round; (b) rectangular; (c) formed; (d) pancake; (e) spiral-helical; (f) internal (from F. W. Curtis, *High Frequency Induction Heating*, McGraw-Hill, New York, 1950)

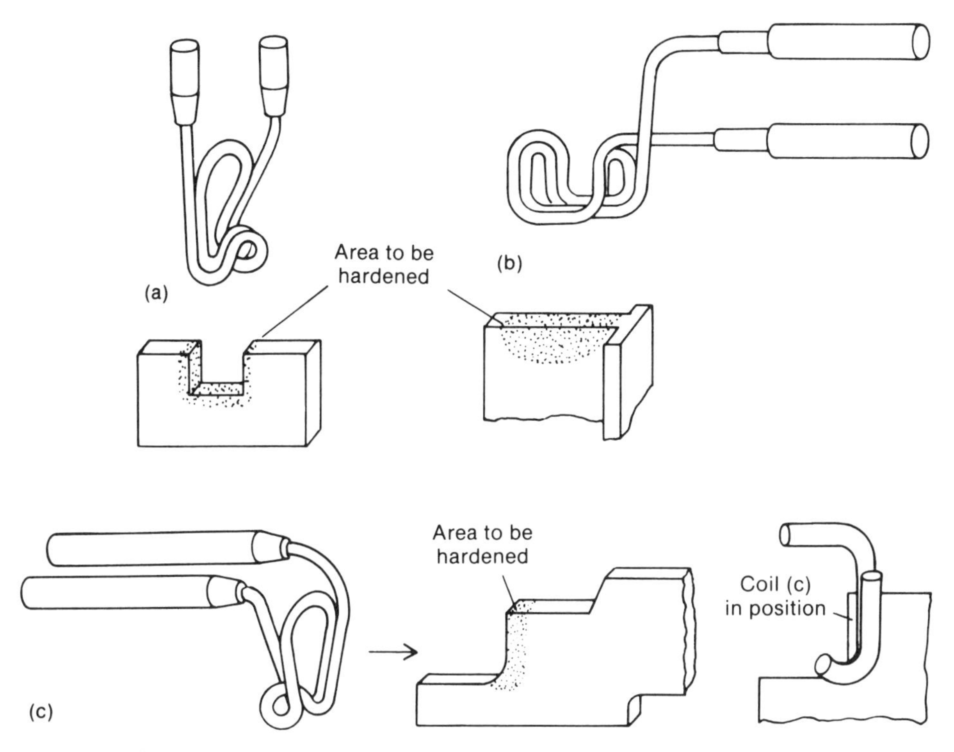

Fig. 8.6. Coil modifications for localized heating (from F. W. Curtis, *High Frequency Induction Heating*, McGraw-Hill, New York, 1950)

Internal Coils

Heating of internal bores, whether for hardening, tempering, or shrink fitting, is one of the major problems most commonly confronted. For all practical purposes, a bore with a 1.1-cm (0.44-in.) internal diameter is the smallest that can be heated with a 450-kHz power supply. At 10 kHz, the practical minimum ID is 2.5 cm (1.0 in.).

Tubing for internal coils should be made as thin as possible, and the bore should be located as close to the surface of the coil as is feasible. Because the current in the coil travels on the inside of the inductor, the true coupling of the maximum flux is from the ID of the coil to the bore of the part. Thus, the conductor cross section should be minimal, and the distance from the coil OD to the part (at 450 kHz) should approach 0.16 cm (0.062 in.). In Fig. 8.7(a), for example, the coupling distance is too great; coil modification improves the design, as shown in Fig. 8.7(b). Here, the coil tubing has been flattened to reduce the coupling distance, and the coil OD has been increased to reduce the spacing from coil to work.

More turns, or a finer pitch on an internal coil, will also increase the flux density. Accordingly, the space between the turns should be no more than one-half the diameter of the tubing, and the over-all height of the coil should not exceed twice its diameter. Figures 8.7(c) and 8.7(d) show special coil

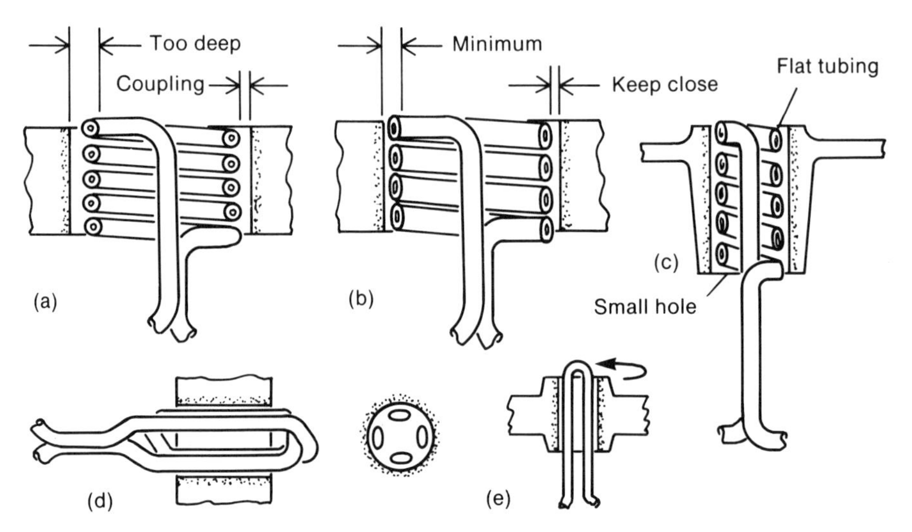

Fig. 8.7. Induction coils designed for internal (bore) heating (from F. W. Curtis, *High Frequency Induction Heating*, McGraw-Hill, New York, 1950)

designs for heating internal bores. The coil in Fig. 8.7(d) would normally produce a pattern of four vertical bands, and therefore the part should be rotated for uniformity of heating.

Internal coils, of necessity, utilize very small tubing or require restricted cooling paths. Further, due to their comparatively low efficiency, they may need very high generator power to produce shallow heating depths. In the case of trunnion cups used on universal joints, the pattern shown in Fig. 8.8 requires three turns of 0.32-cm (0.13-in.) square tubing on a small bore, uti-

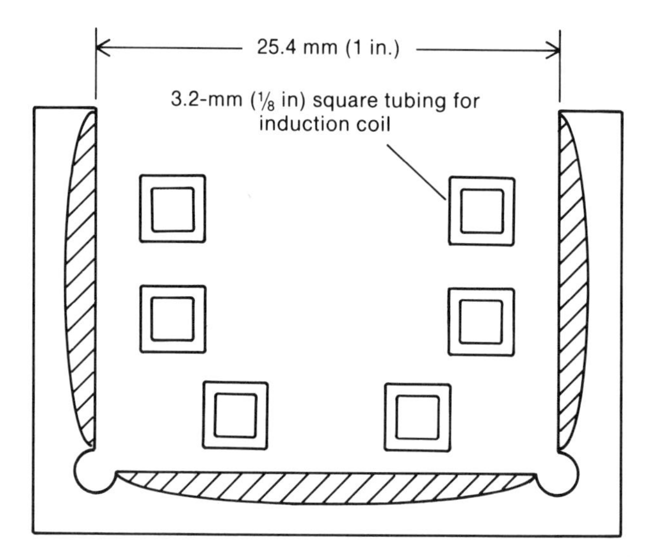

Fig. 8.8. Schematic illustration of induction-coil design for surface hardening of trunnion cups

lizing 60 kW of power. Because of the high currents utilized during the heating cycle as well as the heat radiated from the workpiece surface, more cooling than can be obtained using the normal coil water supply is required here. In a case of this nature, it is best to provide a separate, high-pressure water supply for the coil to achieve a satisfactory flow rate.

COMMON DESIGN MODIFICATIONS

Coil Characterization

Because magnetic flux tends to concentrate toward the center of the length of a solenoid work coil, the heating rate produced in this area is generally greater than that produced toward the ends. Further, if the part being heated is long, conduction and radiation remove heat from the ends at a greater rate. To achieve uniform heating along the part length, the coil must thus be modified to provide better uniformity. The technique of adjusting the coil turns, spacing, or coupling with the workpiece to achieve a uniform heating pattern is sometimes known as "characterizing" the coil.

There are several ways to modify the flux field. The coil can be decoupled in its center, increasing the distance from the part and reducing the flux in this area. Secondly, and more commonly, the number of turns in the center (turn density) can be reduced, producing the same effect. A similar approach—altering a solid single-turn inductor by increasing its bore diameter at the center—achieves the same result. Each of these techniques is described and illustrated in this section.

In Fig. 8.9(a), the coil turns have been modified to produce an even heating pattern on a tapered shaft. The closer turn spacing toward the end compensates for the decrease in coupling caused by the taper. This technique also permits "through the coil" loading or unloading to facilitate fixturing. A similar requirement in the heat treatment of a bevel gear is shown in Fig. 8.9(b). Here, because of the greater part taper, a spiral-helical coil is used. With a pancake coil, decoupling of the center turns provides a similar approach for uniformity.

Selection of Multiturn Vs. Single-turn Coils. Heating-pattern uniformity requirements and workpiece length are the two main considerations with regard to the selection of a multiturn vs. a single-turn induction coil. A fine-pitch, multiturn coil which is closely coupled to the workpiece develops a very uniform heating pattern. Similar uniformity can be achieved by opening up the coupling between the part and the coil so that the magnetic flux pattern intersecting the heated area is more uniform. However, this also decreases energy transfer. Where low heating rates are required, as in through heating for forging, this is acceptable. When high heating rates are needed, however, it is sometimes necessary to maintain close coupling. The pitch of the coil must be opened to prevent overloading of the generator.

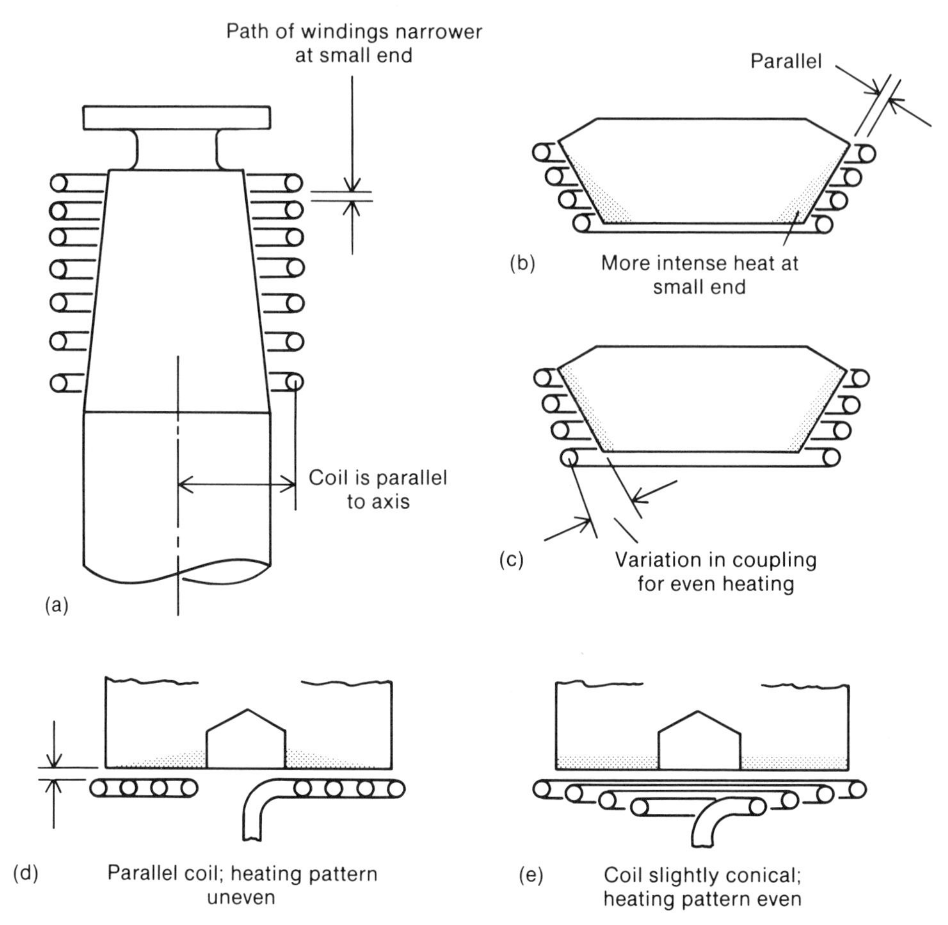

Fig. 8.9. Adjustment ("characterization") of induction heating patterns for several parts by varying the coupling distance or turn spacing (from F. W. Curtis, *High Frequency Induction Heating*, McGraw-Hill, New York, 1950)

Because the heating pattern is a mirror image of the coil, the high flux field adjacent to the coil turns will produce a spiral pattern on the part. This is called "barber poling" and can be eliminated by rotating the workpiece during heating. For most hardening operations, which are of short duration, rotational speeds producing *not less than* ten revolutions during the heating cycle should be used.

If part rotation is not feasible, heating uniformity can be increased by using flattened tubing, by putting an offset in the coil, or by attaching a liner to the coil. Flattened tubing should be placed so that its larger dimension is adjacent to the workpiece. The offsetting of coil turns (Fig. 8.10) provides an even horizontal heating pattern. Offsetting is most easily accomplished by annealing the coil after winding and pressing it between two boards in a vise. A coil liner is a sheet of copper which is soldered or brazed to the inside face of the

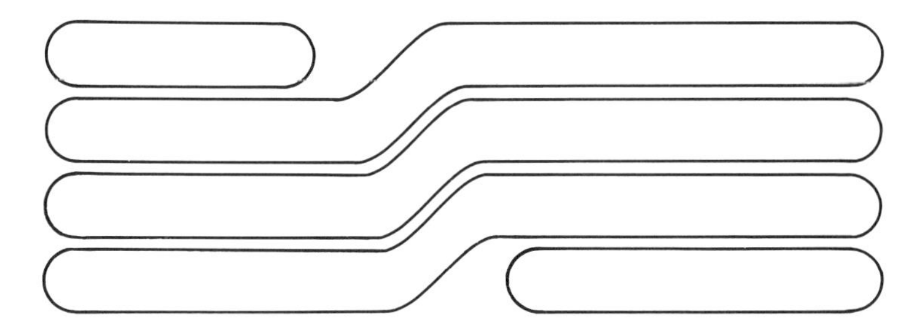

Fig. 8.10. Induction coil with an offset used to provide heating uniformity

coil. This liner expands the area over which the current travels. Thus, a wide field per turn can be created. The height (i.e., axial length) of this field can be modified to suit the application by controlling the dimensions of the liner. When a liner is used, the current path from the power supply passes through the connecting tubing (Fig. 8.11). Between the two connections, the tubing is used solely for conduction cooling of the liner.

In fabricating coils with liners, it is necessary only to tack braze the tubing to the liner at the first and last connection points, with further tacks being used solely for mechanical strength. The remainder of the common surfaces between tubing and liner can then be filled with a low-temperature solder for maximum heat conduction, because the coil-water temperature will never exceed the boiling point of water, which is well below the flow point of the solder. This may be necessary because the copper may be unable to conduct heat fast enough from the inside of the coil.

In multiturn coils, as the heated length increases, the number of turns generally should increase in proportion. In Fig. 8.12(a), the face width of the coil is in proportion to the coil diameter. In Fig. 8.12(b), the ratio of the coil diameter to face width is not suitable; the multiturn coil shown in Fig. 8.12(c) provides a more acceptable heat pattern. Multiturn coils of this type are generally utilized for large-diameter, single-shot heating, in which the quench medium can be sprayed between the coil turns (Fig. 8.12d).

When the length of the coil exceeds four to eight times its diameter, uniform heating at high power densities becomes difficult. In these instances, single-turn or multiturn coils which scan the length of the workpiece are often preferable. Multiturn coils generally improve the efficiency, and therefore the scanning rate, when a power source of a given rating is used. Single-turn coils are also effective for heating of bands which are narrow with respect to the part diameter, particularly for ferrous materials.

The relationship between diameter and optimum height (i.e., axial thickness) of a single-turn coil varies somewhat with size. A small coil can be made with a height equal to its diameter because the current is concentrated in a comparatively small area. With a larger coil, the height should not exceed one-half the diameter. As the coil opening increases, the ratio is reduced—i.e., a

5.1-cm- (2-in.-) ID coil should have a 1.91-cm (0.75-in.) maximum height, and a 10.2-cm- (4-in.-) ID coil should have a 2.5-cm (1.0-in.) height. Figure 8.13 shows some typical ratios.

Coupling Distance. Preferred coupling distance depends on the type of heating (single-shot or scanning) and the type of material (ferrous or nonferrous). In static surface heating, in which the part can be rotated but is not moved through the coil, a coupling distance of 0.15 cm (0.060 in.) from part to coil is recommended. For progressive heating or scanning, a coupling distance of 0.19 cm (0.075 in.) is usually necessary to allow for variations in workpiece straightness. For through heating of magnetic materials, multiturn inductors

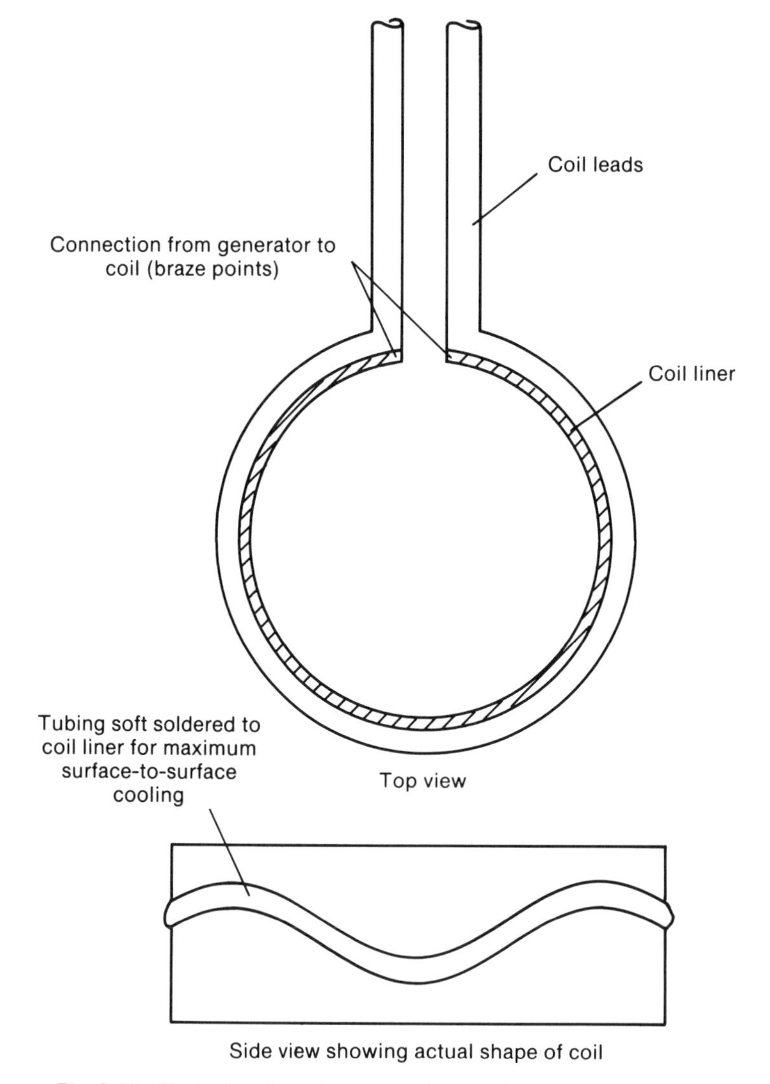

Fig. 8.11. Method of inserting a liner in a coil to widen the flux path

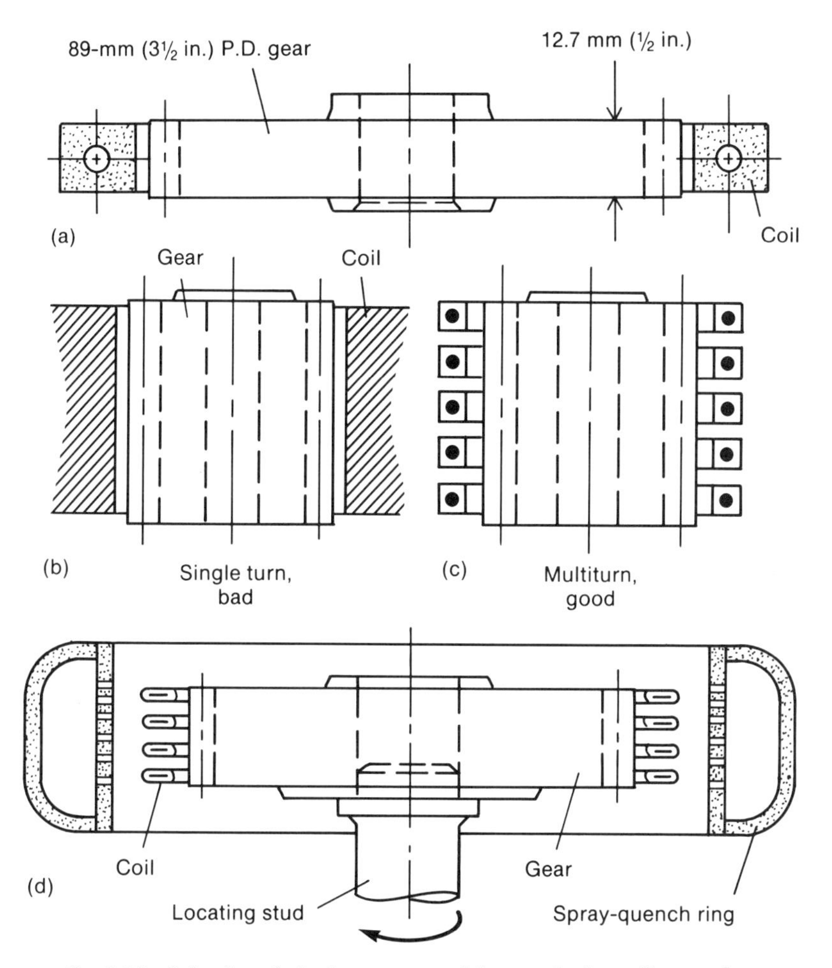

Fig. 8.12. Selection of single-turn vs. multiturn coils depending on the length-to-diameter ratio of the workpiece (from F. W. Curtis, *High Frequency Induction Heating*, McGraw-Hill, New York, 1950)

and slow power transfer are utilized. Coupling distances can be looser in these cases—on the order of 0.64 to 0.95 cm (0.25 to 0.38 in.). For nonferrous materials, coupling should be somewhat closer, usually between 0.16 and 0.32 cm (0.06 and 0.13 in.). It is important to remember, however, that process conditions and handling dictate coupling. If parts are not straight, coupling must decrease. At high frequencies, coil currents are lower and coupling must be increased. With low and medium frequencies, coil currents are considerably higher and decreased coupling can provide mechanical handling advantages. In general, where automated systems are used, coil coupling should be looser.

The coupling distances given above are primarily for heat treating applications in which close coupling is required. Coupling for through heating of billet stock (prior to hot working) must be increased significantly because

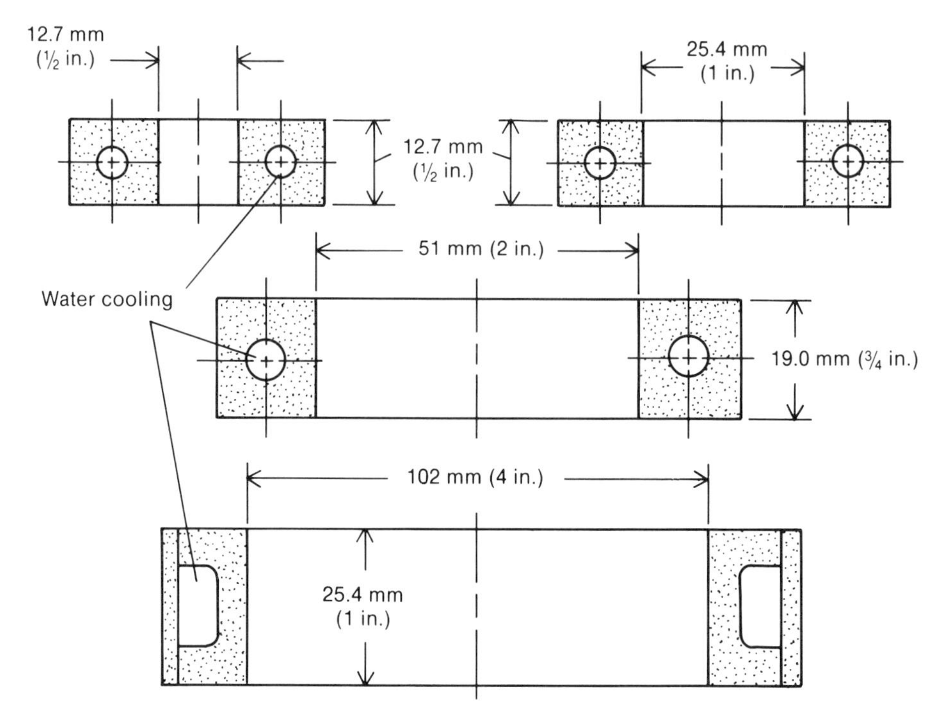

Fig. 8.13. Typical proportions of various single-turn coils (from F. W. Curtis, *High Frequency Induction Heating*, McGraw-Hill, New York, 1950)

thick refractory materials, as well as electrical insulation, must be incorporated into the inductor design. In most cases, the distance increases with the diameter of the part, typical values being 19, 32, and 44 mm (0.75, 1.25, and 1.75 in.) for billet-stock diameters of approximately 38, 102, and 152 mm (1.5, 4, and 6 in.), respectively.

Effects of Part Irregularities on Heating Patterns. With all coils, flux patterns are affected by changes in the cross section or mass of the part. As shown in Fig. 8.14, when the coil extends over the end of a shaftlike part, a deeper pattern is produced on the end. To reduce this effect, the coil must be brought to a point even with or slightly lower than the end of the shaft. The same condition exists in heating of a disk or a wheel. The depth of heating will be greater at the ends than in the middle if the coil overlaps the part. The coil can be shortened, or the diameter at the ends of the coil can be made greater than at the middle, thereby reducing the coupling at the former location.

Just as flux tends to couple heat to a greater depth at the end of a shaft, it will do the same at holes, long slots, or projections (Fig. 8.15). If the part contains a circular hole, an additional eddy-current path is produced which will cause heating at a rate considerably higher than that in the rest of the part. The addition of a copper slug to the hole can be used to effectively cor-

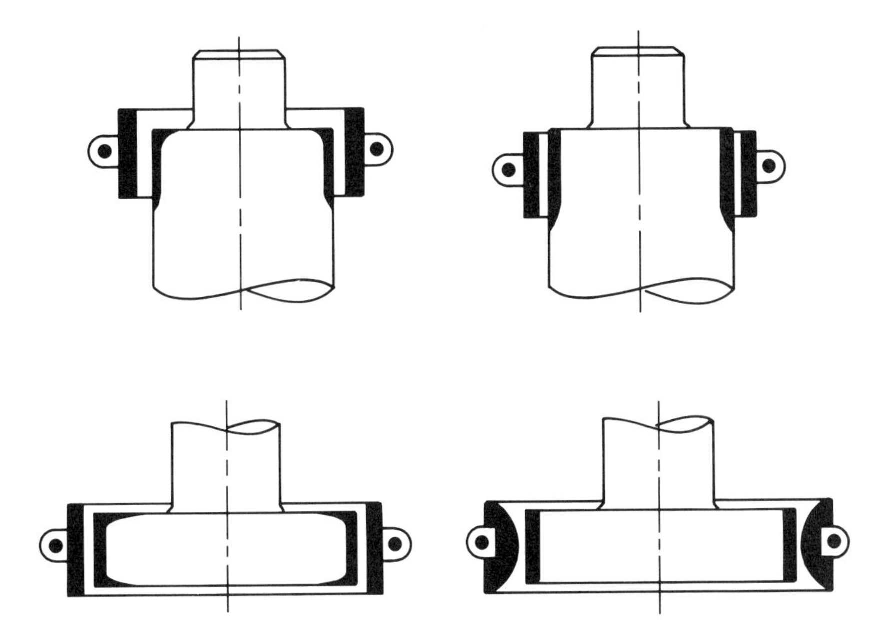

When the coil overlaps the edge (left), overheating will occur on the end surface. Coil placement should be slightly below the edge (right) for a more uniform heating pattern.

Fig. 8.14. Effect of coil placement on the heating pattern at the end of a workpiece (from F. W. Curtis, *High Frequency Induction Heating*, McGraw-Hill, New York, 1950)

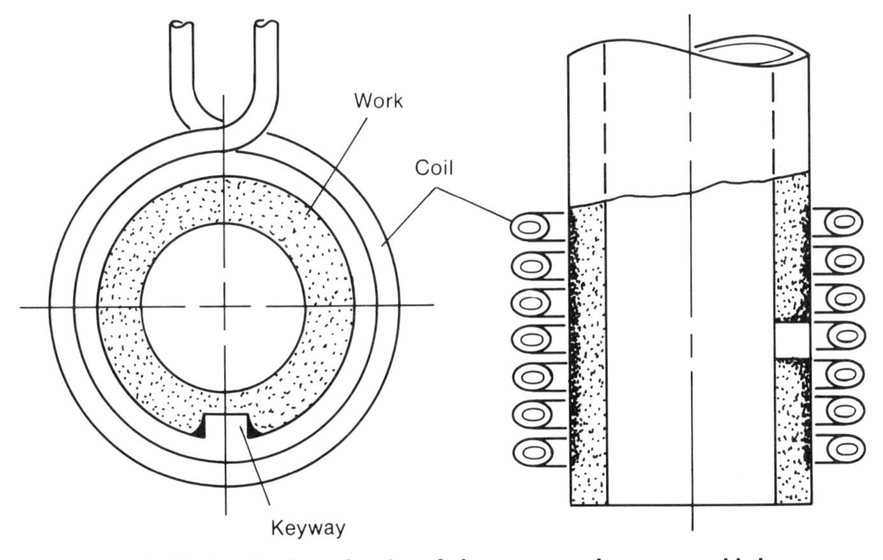

Fig. 8.15. Localized overheating of sharp corners, keyways, and holes most prevalent in high-frequency induction heating (from F. W. Curtis, *High Frequency Induction Heating*, McGraw-Hill, New York, 1950)

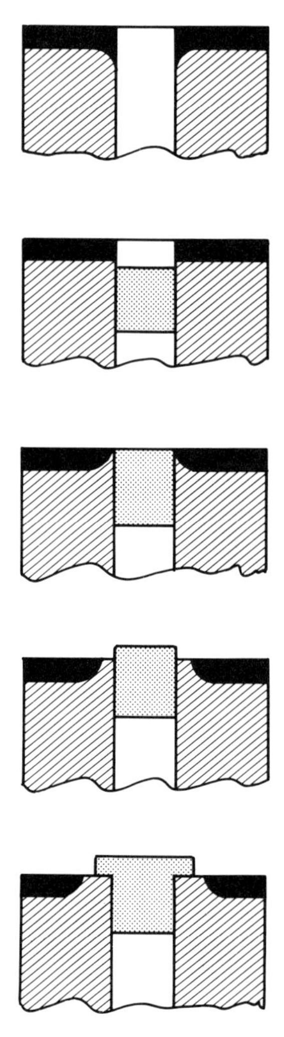

Fig. 8.16. Control of the heating pattern at a hole through use of copper slugs (from M. G. Lozinskii, *Industrial Applications of Induction Heating*, Pergamon Press, London, 1969)

rect or eliminate this problem. The position of the slug (Fig. 8.16) can control the resultant heating pattern. In addition, the slug will minimize hole distortion if the part must be quenched following heating. For slotted parts heated with solenoid coils (Fig. 8.17), the continuous current path is interrupted by the slot, and the current must then travel on the inside of the part to provide a closed circuit. This is the basis for concentrator coils, which are discussed below. It is of interest to note, however, that with the slot closed, the applied voltage of the work coil causes a higher current to flow. This is due to the fact that the resistive path, now around the periphery of the part, is considerably shorter. The increase in current then produces a considerably higher heating rate with the same coil.

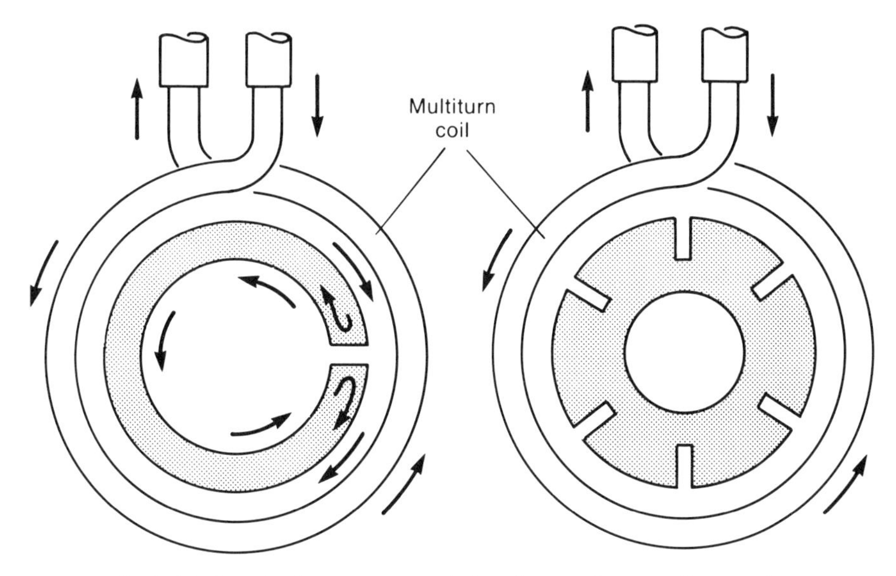

Fig. 8.17. Localized overheating of slots in certain parts that results from the tendency for induced currents to follow the part contour (from F. W. Curtis, *High Frequency Induction Heating*, McGraw-Hill, New York, 1950)

Flux Diverters

When two separate regions of a workpiece are to be heated, but are close together (Fig. 8.18), it is possible that the magnetic fields of adjacent coil turns will overlap, causing the entire bar to be heated. To avoid this problem, successive turns can be wound in opposite directions. By this means, the inter-

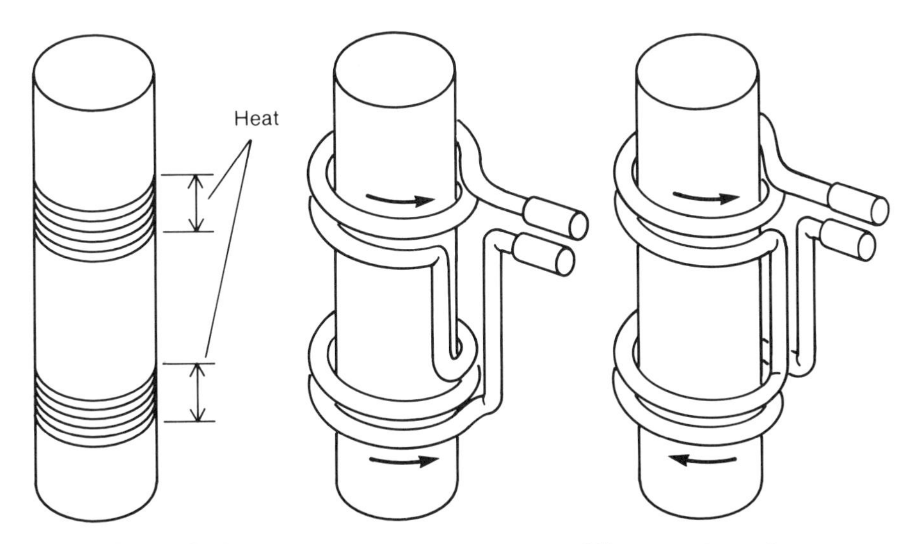

Fig. 8.18. Control of heating patterns in two different regions of a workpiece by winding the turns in opposite directions (from F. W. Curtis, *High Frequency Induction Heating*, McGraw-Hill, New York, 1950)

mediate fields will cancel, and the fields that remain will be restricted. It should be noted that, as shown in Fig. 8.18, lead placement is critical. Having the return inductor spaced far from the coil leads would add unneeded losses to the system. Another example of a counterwound coil is shown in Fig. 8.19; the coil in Fig. 8.19(b) is the counterwound version of the one in Fig. 8.19(a). This type of coil can be used effectively in an application in which the rim of a container is to be heated while the center remains relatively cool.

Another technique that can be utilized in the above circumstances involves the construction of a shorted turn or "robber" placed between the active coil turns. In this case, the shorted loop acts as an easy alternative path for concentration of the excess flux, absorbing the stray field. It is therefore sometimes called a flux diverter. As for the active coil turns, the robber must be water cooled to dissipate its own heat. Figure 8.20 shows a typical installation using robbers to prevent heating of specific areas of a camshaft during hardening. A typical construction is shown in Fig. 8.21.

The use of shorted coil turns to prevent stray-field heating is also used effectively on very large coils where the end flux field might heat structural frames. Figure 8.22 shows a coil for graphitizing of carbon. It is approximately 96.5 cm (38 in.) in diameter and contains a 122-cm- (48-in.-) long active heat zone. There are, in addition, four shorted turns at each end of the coil which act as robbers and thus restrict the stray field.

Flux robbers or flux diverters can also be used in fabricating test coils when it is desired to determine the optimum number of turns empirically. In these situations, a few additional turns are provided that can be added or removed

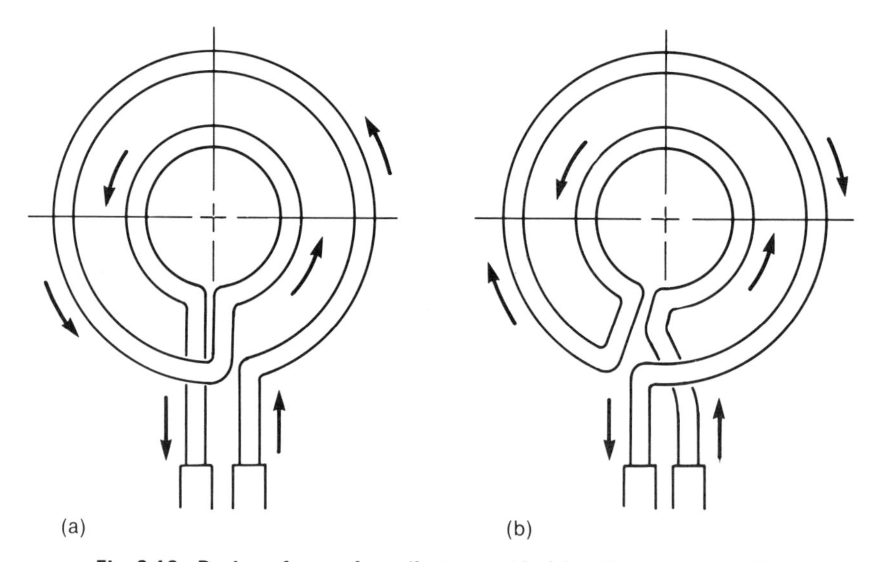

Fig. 8.19. Design of pancake coils to provide (a) uniform, or over-all, heating or (b) peripheral heating only (from F. W. Curtis, *High Frequency Induction Heating*, McGraw-Hill, New York, 1950)

Fig. 8.20. Water-cooled flux "robbers" inserted between adjacent coils in a multizone camshaft-hardening machine (source: American Induction Heating Corp.)

as required. These can be shorted with a copper strap or temporarily brazed while tests are made, and removed pending the outcome of the heating trials.

"Balancing" of Multiplace Coils

Interaction between coils is a factor to be considered when constructing multiplace coils. Whether of multiturn or single-turn construction, the adjacent coil flux paths can interact unless the center-to-center distance between adjacent coils is at least 1½ times the coil diameter. The design technique used in addressing this problem is referred to as coil "balancing." As an example, Fig. 8.23(a) shows a six-place single-turn coil whose construction is interactive due to poor turn spacing. In Fig. 8.23(b), the spacing is increased properly. Because there is a resistive path between the coils, and because the current tends to flow on the internal face of the coil, the parts should be heated evenly. However, because currents tend to take the shortest path, they will not necessarily enter the area between turns. Thus, the flux and heating will not be uniform. One means of "forcing" current into these locations is through the use of saw cuts that direct the path of the current as shown in Fig. 8.23(c). It should also be noted that to provide uniform heating at the end position, additional saw cuts, simulating the lead locations on the interior coil positions, should be made on the two outer locations.

Once coil balance has been achieved, water cooling can be accomplished by brazing or soldering formed tubing to the plate coil, as shown in Fig. 8.24.

Plate coils need not be used solely for heating of parts with similar diameters and masses. It is desirable in some instances to heat a number of parts of different sizes and shapes simultaneously. The problem then becomes one of establishing a balance so that all parts come to the same temperature at the

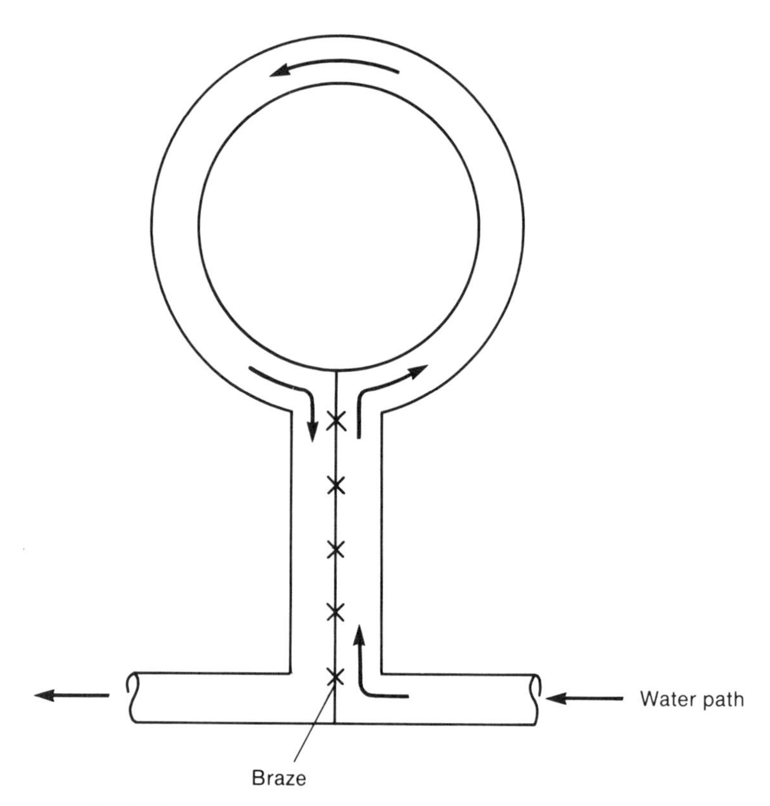

Fig. 8.21. Typical construction of a water-cooled flux robber

Fig. 8.22. Graphitizing of carbon using an induction coil with shorted end turns at top and bottom to restrict stray fields (source: Sohio Carborundum, Structural Ceramics Div.)

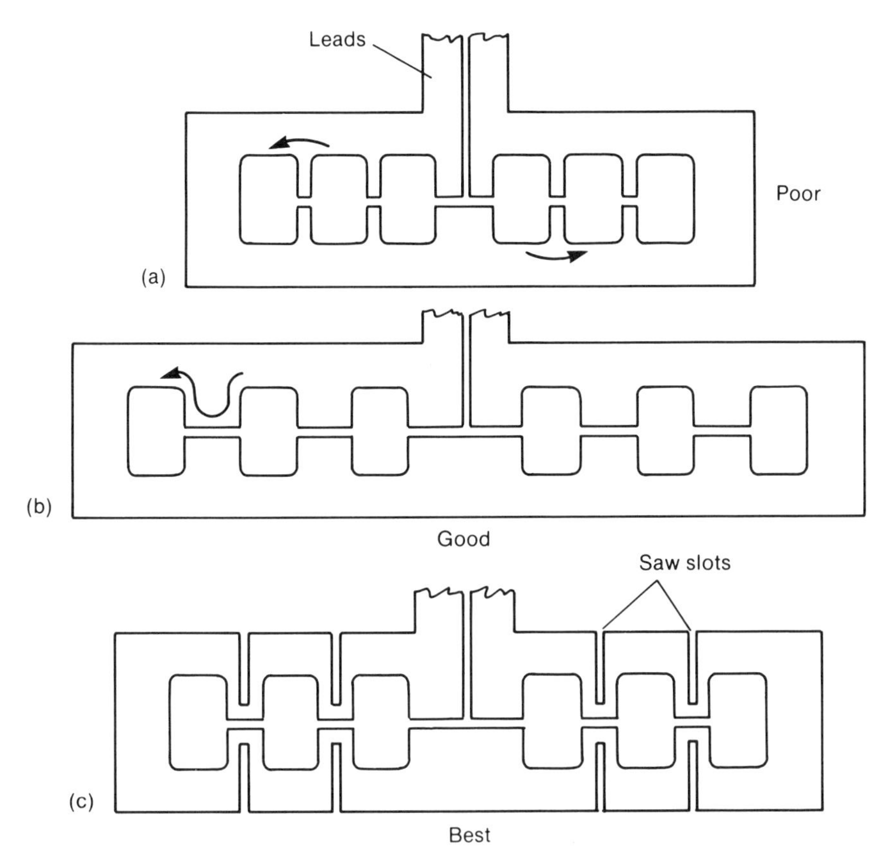

Fig. 8.23. Design of multiplace inductors (from F. W. Curtis, *High Frequency Induction Heating*, McGraw-Hill, New York, 1950)

same time. An example of how this is accomplished is shown in Fig. 8.25. Once the coil is fabricated, balance is achieved by opening the coil coupling in those areas that heat the fastest, thus dropping the heat level in these locations until the pattern is uniform.

Multiplace coils may be of both plate and tubing types. Plate coils provide a rugged assembly for production operations, particularly brazing in which rapid part handling and the corrosive action of a flux can deteriorate coil materials. When part height-to-diameter ratios are high, multiturn coils must be utilized in multiplace arrangements. Construction may be on a feeder bus, as in Fig. 8.26, where individual coils may be easily replaced. The constraint of center-to-center distance being a minimum of 1½ times the coil diameter must still be observed.

SPECIALTY COILS

As mentioned above, coil designs are based on the heating-pattern requirements of the application, the frequency, and the power-density requirements.

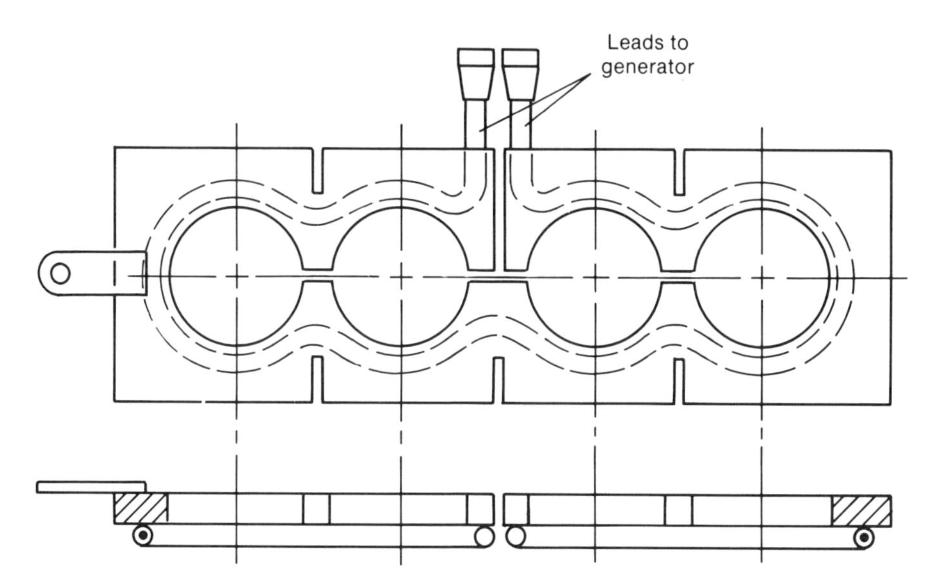

Fig. 8.24. Schematic illustration of the energizing and cooling of a multiplace plate inductor (from F. W. Curtis, *High Frequency Induction Heating*, McGraw-Hill, New York, 1950)

In addition, the material-handling techniques to be used for production also determine to a large extent the coil to be used. If a part is to be inserted in a coil, moved on a conveyor, or pushed end to end, or if the coil/heat station combination is to move onto the part, the coil design must take the appropriate handling requirements into consideration. Accordingly, a variety of specialty coil designs have evolved for specific applications.

Master Work Coils and Coil Inserts

When production requirements necessitate small batches (as in job-shop applications) and a single-turn coil can be used, master work coils provide a simple, rapid means of changing coil diameters or shapes to match a variety of parts. In its basic form, a master work coil consists of copper tubing that provides both an electrical connection to the power supply and a water-cooled contact surface for connection to a coil insert.* A typical design, shown in Fig. 8.27, consists of a copper tube that is bent into the form of a single-turn coil and soldered to a copper band which conforms to the slope of the coil insert and is recessed. Holes in the inserts which match tapped holes in the master coil securely clamp the inserts to the master coil, providing good transfer of electrical energy and heat removal. Inserts are machined from copper with a thickness that matches the required heating pattern and should be

*N. B. Stevens and P. R. Capalongo, "Inductor for High-Frequency Induction Heating," U.S. Patent 2,456,091, December 14, 1948.

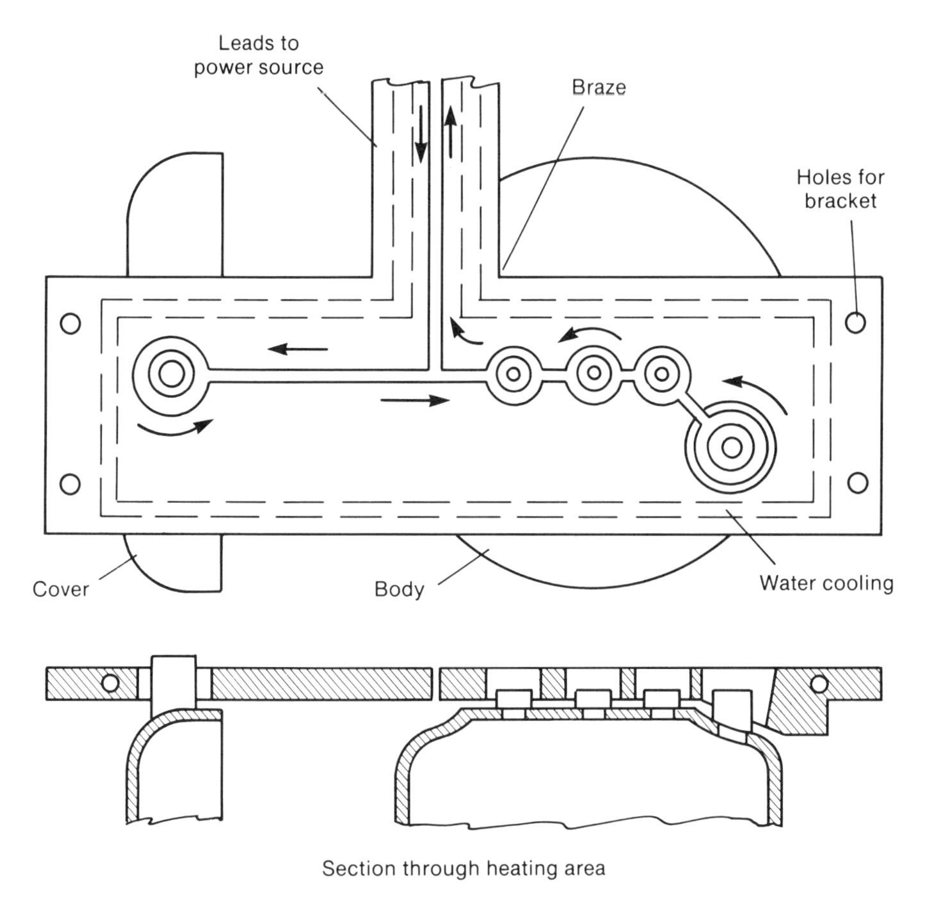

Fig. 8.25. Design of a single-turn, multiplace inductor for simultaneous brazing of different-size couplings in a single operation (from F. W. Curtis, *High Frequency Induction Heating*, McGraw-Hill, New York, 1950)

somewhat greater in thickness than the depth of the recess for easy removal. Special coil shapes are easily configured. It is important to note that, because of the less-than-optimal cooling technique, coil inserts are particularly well adapted to processes requiring short heating times or those in which they are also cooled by the quenching medium.

In machining of coil inserts, care must be taken to relieve sharp corners, unless it is desired to have a deeper heating pattern in these locations. Figure 8.28 shows the effect of sharp corners on a closely coupled part. Flux from both inductor sides couples to the corner, which, due to a lack of mass, tends to overheat relative to the rest of the pattern. Decoupling of the coil from these locations provides the desired pattern but tends to reduce over-all efficiency, thus slowing the heating rate and resulting in a deeper case. Relieving or decoupling of only the corners is a better alternative, particularly when a solid inductor is used, and the relief can be machined as required.

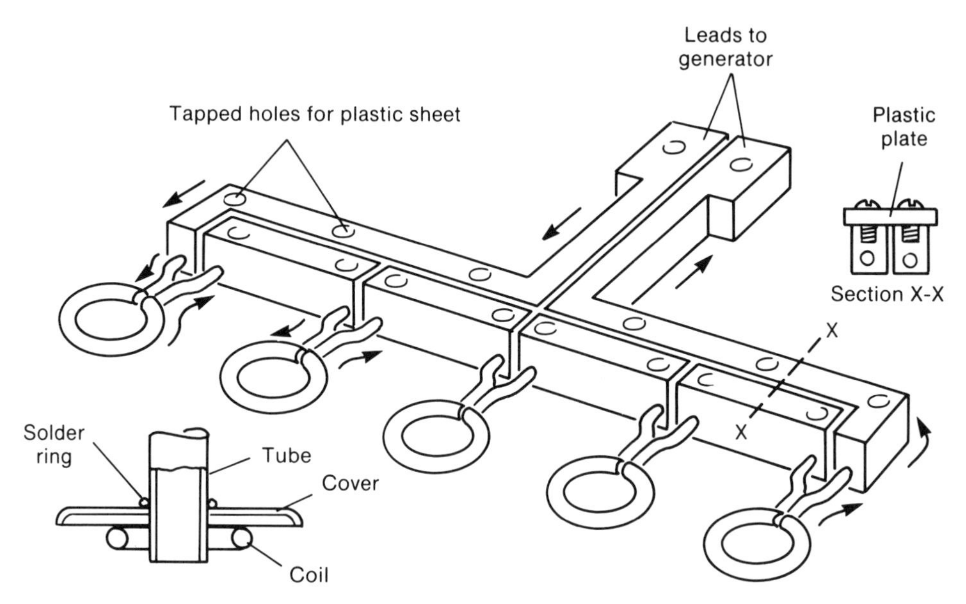

Fig. 8.26. Single-turn, multiplace inductor with individual coils of copper tubing (from F. W. Curtis, *High Frequency Induction Heating*, McGraw-Hill, New York, 1950)

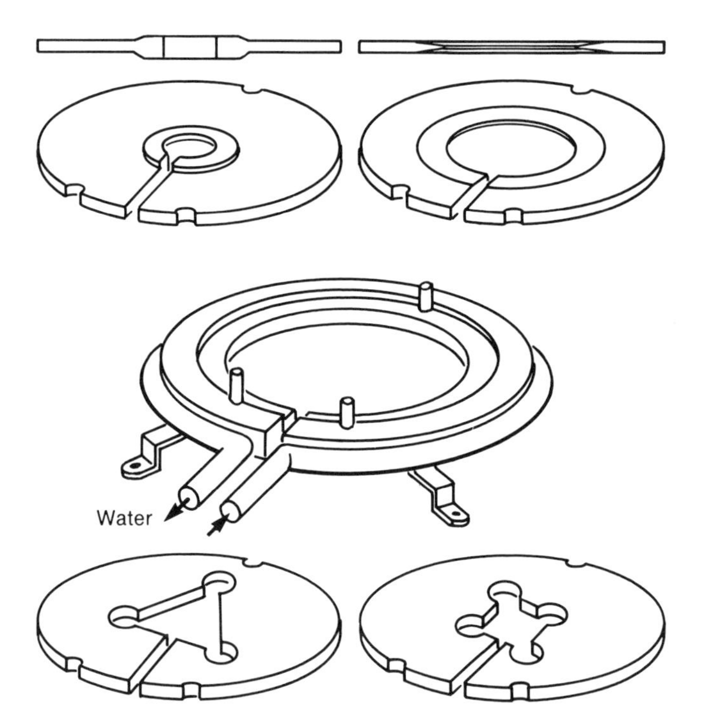

Fig. 8.27. Schematic illustration showing the design of a master coil with changeable inserts (from M. G. Lozinskii, *Industrial Applications of Induction Heating*, Pergamon Press, London, 1969)

Coils for Induction Scanners

Coils for progressive hardening (scanning) are built using two techniques. The simpler of the two employs a simple single-turn or multiturn coil with a separate quench ring which can be mounted on the scanner (Fig. 8.29a). For larger production runs, on the other hand, a double-chamber coil which incorporates both coil cooling and quenching capabilities is often the preferred choice. The scanning inductor shown in Fig. 8.29(b) is typical of the latter type of design. Cooling water flows through the upper, or inductor, cham-

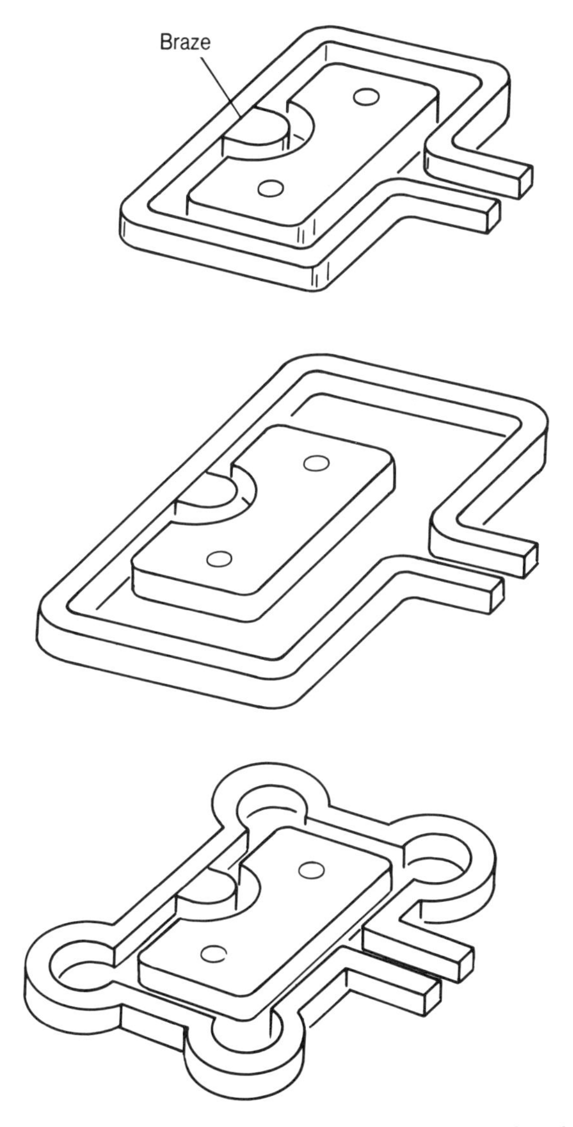

Fig. 8.28. Inductor with a relief designed for the hardening of the lateral surface of a template (from M. G. Lozinskii, *Industrial Applications of Induction Heating*, Pergamon Press, London, 1969)

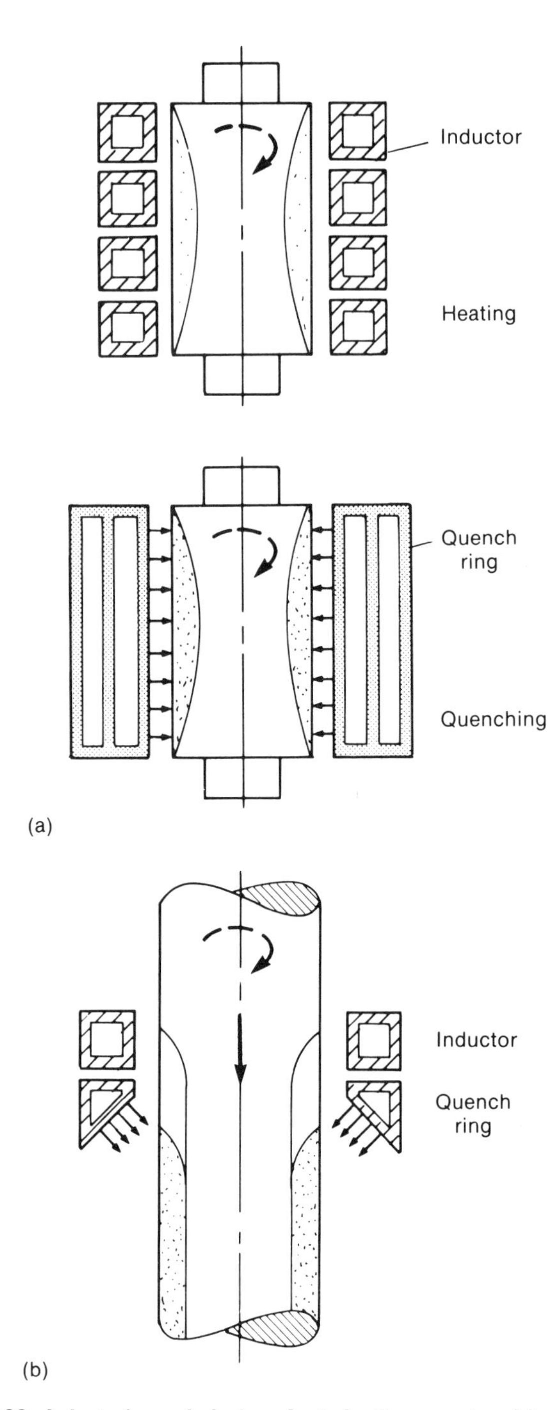

Fig. 8.29. Inductor/quench designs for induction scanning: (a) separate coil and quench and (b) two-chamber, integral coil and quench (from F. H. Reinke and W. H. Gowan, *Heat Treatment of Metals*, Vol 5, No. 2, 1978, p 39)

ber to keep the copper resistivity low. The quenchant is sprayed from perforations in the beveled face onto the workpiece as it exists from the inductor. The beveled face normally is at an angle of 30° to the vertical so that there is some soaking time between the end of induction heating and the quenching operation. This delay time helps to increase uniformity. Proper choice of the spray direction also reduces the amount of fluid runback on the shaft, which could cause variation in bar temperature and result in uneven hardness. Well-directed quench spray holes are required inasmuch as "barber poling" can occur due to erratic or misdirected quenchant that precools the part ahead of the main quench stream.

Split Coils

Split coils are generally utilized as a last resort for applications in which it is difficult to provide a high enough power density to the area to be heated without very close coupling and where part insertion or removal would then become impossible. Typical examples of such situations include hardening of journals and shoulders in crankshafts. In these cases, the split-coil design would also include the ability to quench through the face of the inductor, as shown in Fig. 8.30. Typical methods of hinging split inductors are shown in Fig. 8.31.

It should be noted that with a split inductor good surface-to-surface contact must be made between the faces of the hinged and fixed portions of the coil. Generally these surfaces are faced with silver or special alloy contacts that are matched to provide good surface contact. Clamps are used to ensure closure during heating. Often, high currents at high frequency pass through this interface, and the life of the contact is generally limited due to both wear and arcing.

Coolant for the coil chamber of a split inductor is carried by flexible hoses that bypass the hinge so that excessive heating does not occur in the movable section during the cycle. The quench chamber is fed by a separate hose arrangement. The face of the quench chamber is closest to the work during heating, and therefore carries most of the current. Accordingly, it must be sufficiently thick to preclude either melting or distortion during the heating cycle.

With split coils it is also frequently necessary to provide some means of locating the part in the coil to maintain the proper coupling distance. Ceramic pins or buttons are frequently secured to the face of the inductor. These pins contact the part during the heating cycle and establish rigid relative positioning between part and coil. However, they are subject to thermal shock during the heating and quenching cycles and suffer mechanical abuse as well. Therefore, they should be designed for simple replacement as required. Figure 8.32 depicts an arrangement for the use of either ceramic or metal pins that compensates for these problems. Here, the ceramic pin is approximately 0.64 cm

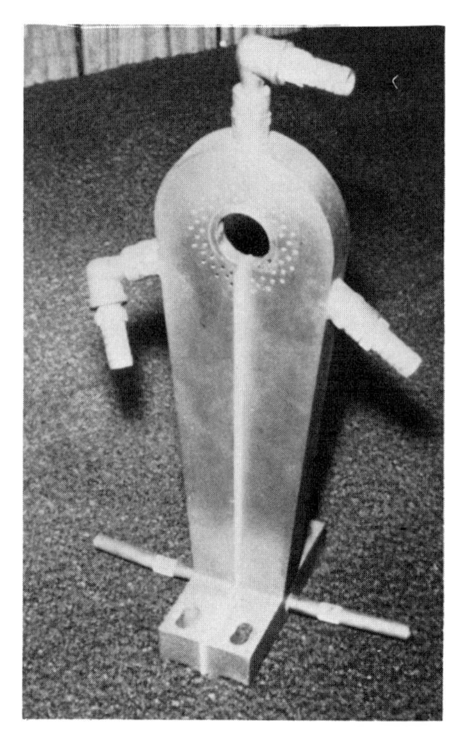

Fig. 8.30. Double-chamber coil for use in induction scanning; perforations are for quench spray from the quench chamber (source: American Induction Heating Corp.)

(0.25 in.) in diameter and 1.3 cm (0.5 in.) long with a 0.69-cm (0.27-in.) head diameter. The rubber packing absorbs the clamping stress. A threaded tube passes through the chamber, and a screw presses the pins against the shaft. In Fig. 8.32(b), a 0.32-cm (0.125-in.) nichrome pin is used with a ceramic tube as an insulator. Being in compression, the tube undergoes comparatively high loads without breaking. The metal pin provides longer life in these conditions than the ceramic pin.

Concentrator Coils

When transformers are not available or when extremely high power densities are required, it is sometimes necessary to utilize an induction coil as the secondary of a current transformer. That is to say, the turns of a primary coil are used to induce high currents in a copper ("concentrator") secondary that is connected to, or serves itself as, the coil which heats the workpiece. For instance, when a complex assembly must be held in a fixture for brazing, it is sometimes simpler to have each fixture include its own coil and inductively couple the energy to the coil.

Figure 8.33 shows an example in which a multiturn pancake coil is connected to the power supply and in turn inductively couples its energy to the concentrator secondary or work coil. The work coil is an integral part of the

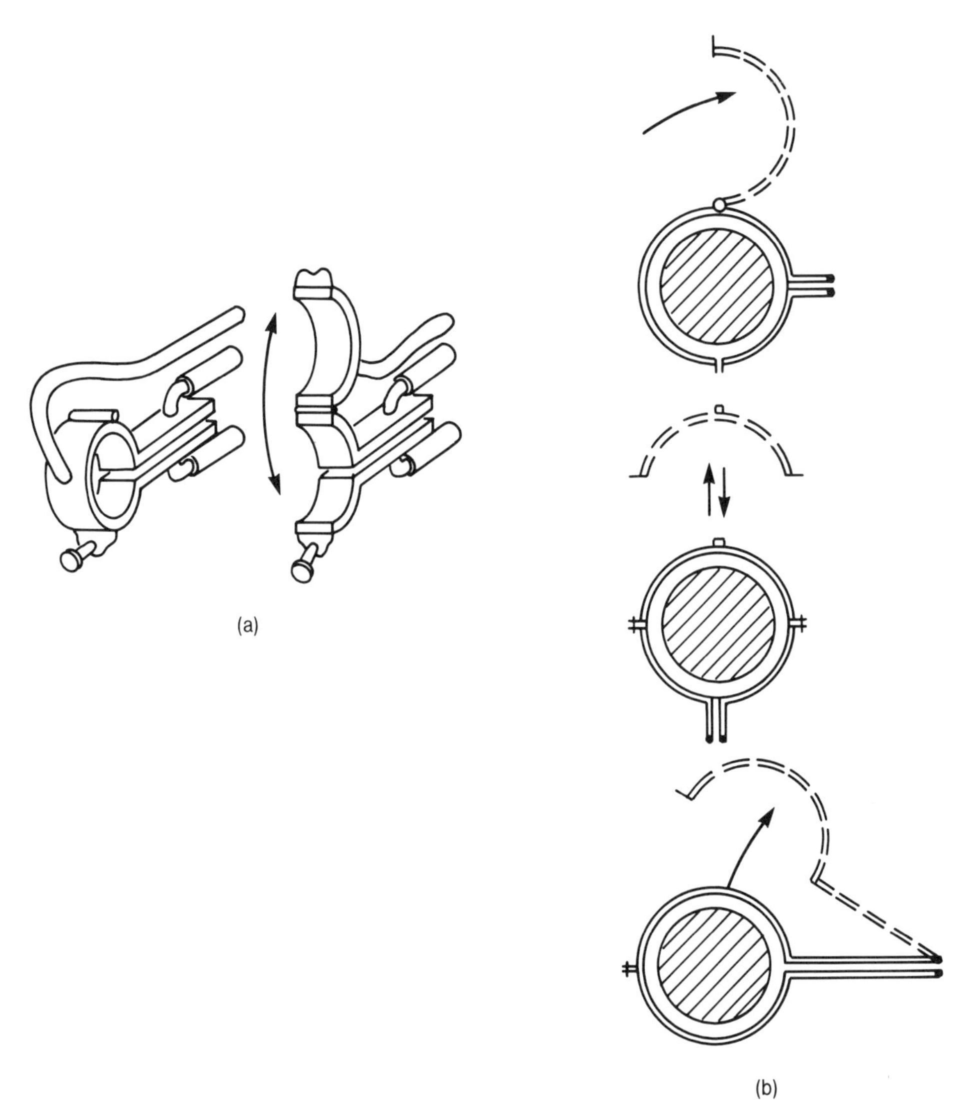

Fig. 8.31. Diagram (a) and schematic illustration (b) of a split inductor used for heating crankshaft journals (from M. G. Lozinskii, *Industrial Applications of Induction Heating*, Pergamon Press, London, 1969)

secondary, and the current is forced to travel through the active portion of the coil to complete its return path. In operation, the secondary/coil combination is mounted to the fixture and the secondary passes *beneath* the primary pancake coil as the combination moves on a conveyor. Heat is generated in the workpiece as the secondary passes through the flux field of the primary. In Fig. 8.34, the transformer secondary moves *through* the primary. The secondary, inside the major flux field of the primary, couples energy to the coil even while it is in motion. This is an ideal way to energize the coil while providing reciprocating motion.

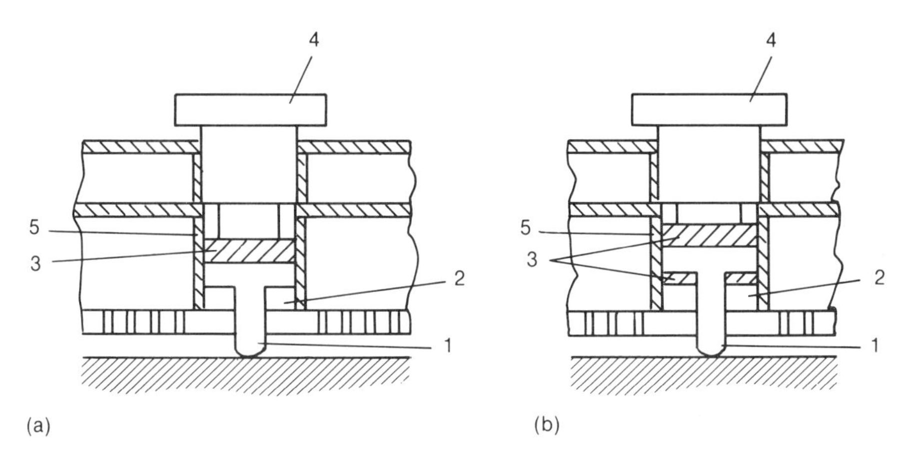

(a) 1-ceramic pin; 2,3-rubber packing; 4-screw; 5-threaded tube. (b) 1-nichrome pin; 2-ceramic tube; 3-rubber packing; 4-screw; 5-threaded tube.

Fig. 8.32. Design of metal and ceramic pins for fixing the position of a split inductor on a crankshaft journal (from M. G. Lozinskii, *Industrial Applications of Induction Heating*, Pergamon Press, London, 1969)

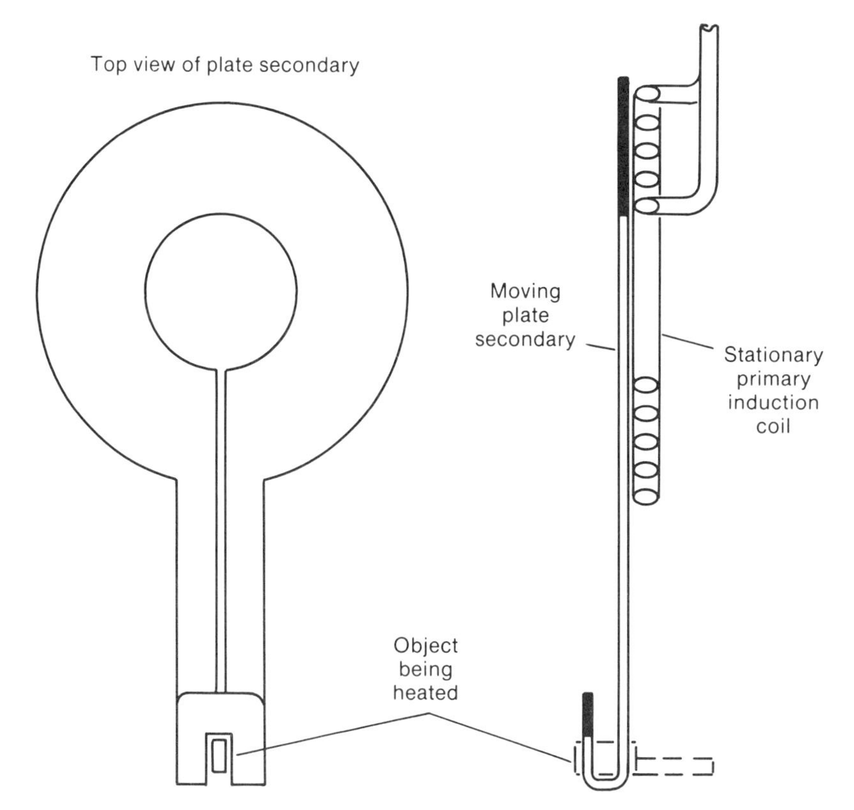

Fig. 8.33. Schematic illustration of inductively coupled primary and secondary coils (source: Lepel Corp.)

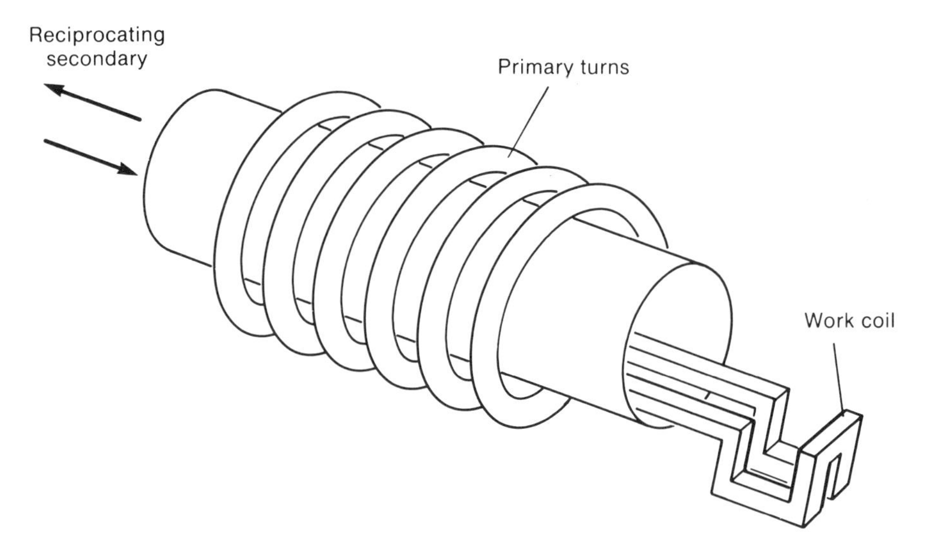

Fig. 8.34. Schematic illustration of a movable secondary concentrator inductor

In some cases, the current path in a concentrator or transformer secondary is constricted to maximize power density. Often, a pancake inductor is coupled to a solid plate or sheet secondary. The current is then forced to travel on the inner path of the secondary, which now becomes the work coil (Fig. 8.35). Occasionally, the first turn on the grounded side of the inductor is brazed to the concentrator. This arrangement then becomes in essence an autotransformer, grounding the secondary and providing some cooling via contact with the water-cooled primary.

Yet another form of current concentrator is depicted in Fig. 8.36. The exterior turns wrapped around the cylindrical portion of the concentrator force the total current to be concentrated on the inner bore or slot, thus transforming a low-heat-density setup into a high-heat-density one in the "active region." Figure 8.37 shows a typical concentrator coil of this type that is used

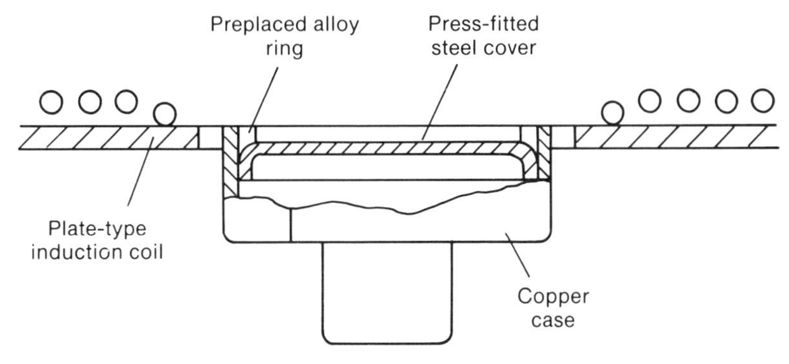

Fig. 8.35. Schematic illustration of a concentrator-type induction coil used for induction soldering (source: Lepel Corp.)

to melt high-temperature alloys off the ends of bars in vacuum to make special powders.

Butterfly Coils

One of the most difficult heating challenges is the creation of an even heating pattern at the end of a bar or shaft. Patterns developed with a pancake inductor produce a dead spot at the center due to field cancellation in this area.

The butterfly coil (Fig. 8.38), so named because of its appearance, utilizes two specially formed pancake coils. The current paths of the adjacent sides are aligned so that they are additive. The "wings" of the butterfly may be bent up to decouple their fields from the shaft, or, if heat is required in this location, they may be coupled with the shaft itself. In winding this coil, it is important that all center turns be wound in the same direction so that they are additive. Further, only these turns should couple directly with the part to produce the desired pattern; part rotation is required to provide uniform heating on the end of the part.

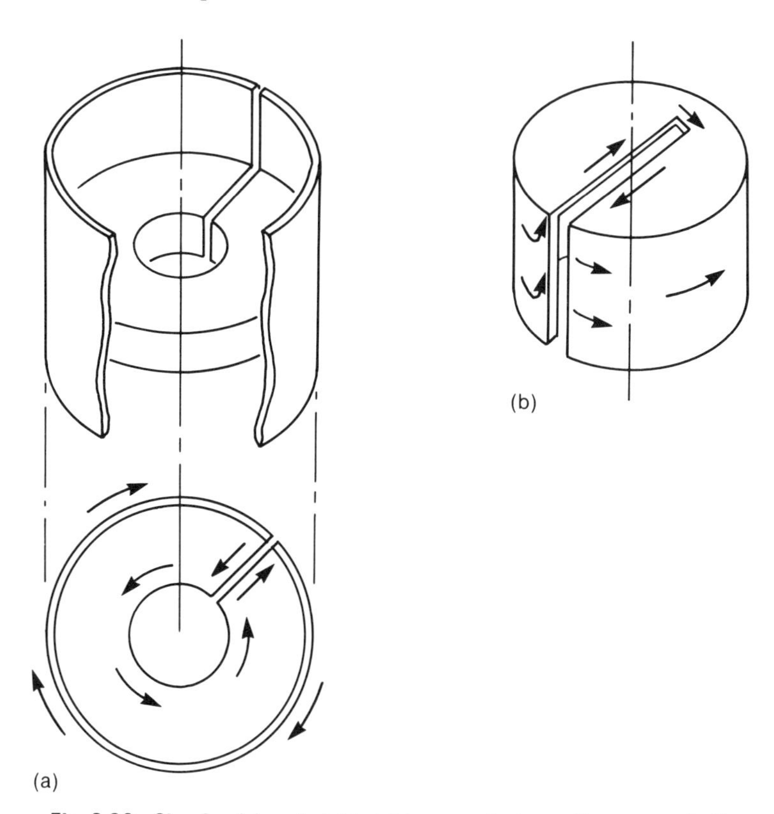

Fig. 8.36. Circular (a) and slotted (b) concentrator coils; arrows indicate direction of current flow (from E. May, *Industrial High Frequency Electric Power*, Wiley, New York, 1950)

Split-Return Inductors

If a narrow band of heat is required and heating must be accomplished from one surface only as in weld-seam annealing, the split-return inductor offers distinct advantages (Fig. 8.39). With this design, the center runner of the work coil carries twice the current of each of the return legs. The pattern on the workpiece, being a mirror image of the coil, produces four times as much heat under the center leg as in each of the return loops. With proper balancing, the high-heat path can then be extremely narrow while the heat produced in each of the return legs is insufficient to affect the remainder of the part.

Tapped Coils

Induction coils can be provided with taps to allow for differences in heated length. A typical application is a forging coil for heating "off the end" of a bar, in which provision must be made to adjust the length being heated (Fig. 8.40). Taps are brazed to the work coil at locations where a water-cooled strap can be moved from tap to tap. The active portion of the coil is then between the power-supply connection and the tap. Water cooling, however, should be maintained through all portions of the coil, both active and inactive.

Fig. 8.37. Concentrator coil for melting the end of a superalloy bar used in production of metal powder in a vacuum (source: Lindberg Cycle-Dyne, Inc.)

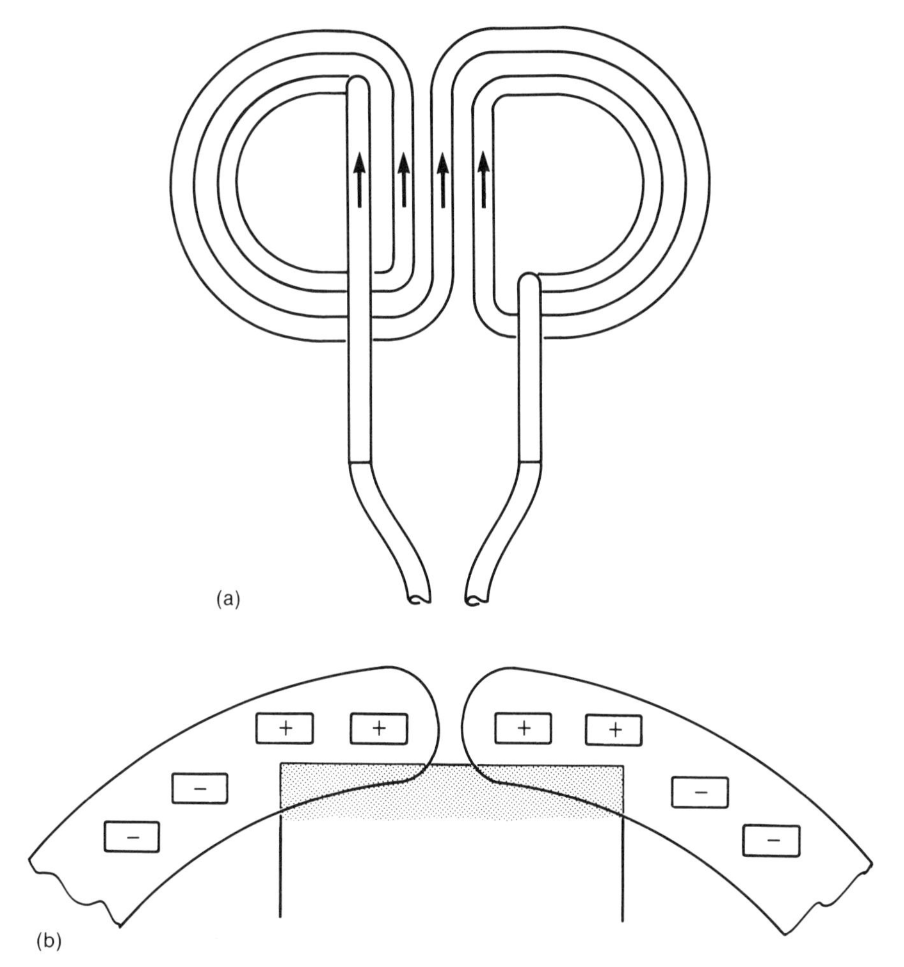

Fig. 8.38. Schematic illustration of a butterfly coil: (a) coil construction (arrows indicate reinforcing type of current flow in coil) and (b) coupling between the turns of the coil and the end of a bar to produce a uniform heating pattern

Transverse-Flux Coils

In heating of parts that have a long longitudinal axis and a thin cross section, a circular coil wrapped around the workpiece produces a heating pattern (Fig. 8.41) that, due to coupling distances, is effective only at the edges. In transverse-flux heating, however, the coil is designed to set up a flux field which is perpendicular to the sheet or similar part. In this way, the path of the eddy currents is changed so that it is parallel to the major axis of the work. For example, in the manufacture of items such as hacksaw blades, the steel moves between the turns of the coil and the eddy-current path is a circular one across the flat of the blade. For heating of wide sheet materials, specially designed transverse-flux inductors have become available in recent years also.

These inductors usually contain laminated iron cores with a series of windings along the length of the strip which induces a circulating current that flows across the width of the strip and returns on itself along the edges under adjacent pole faces. In this type of arrangement, coil sections are placed on both sides of the strip to force the magnetic flux to pass transversely through the

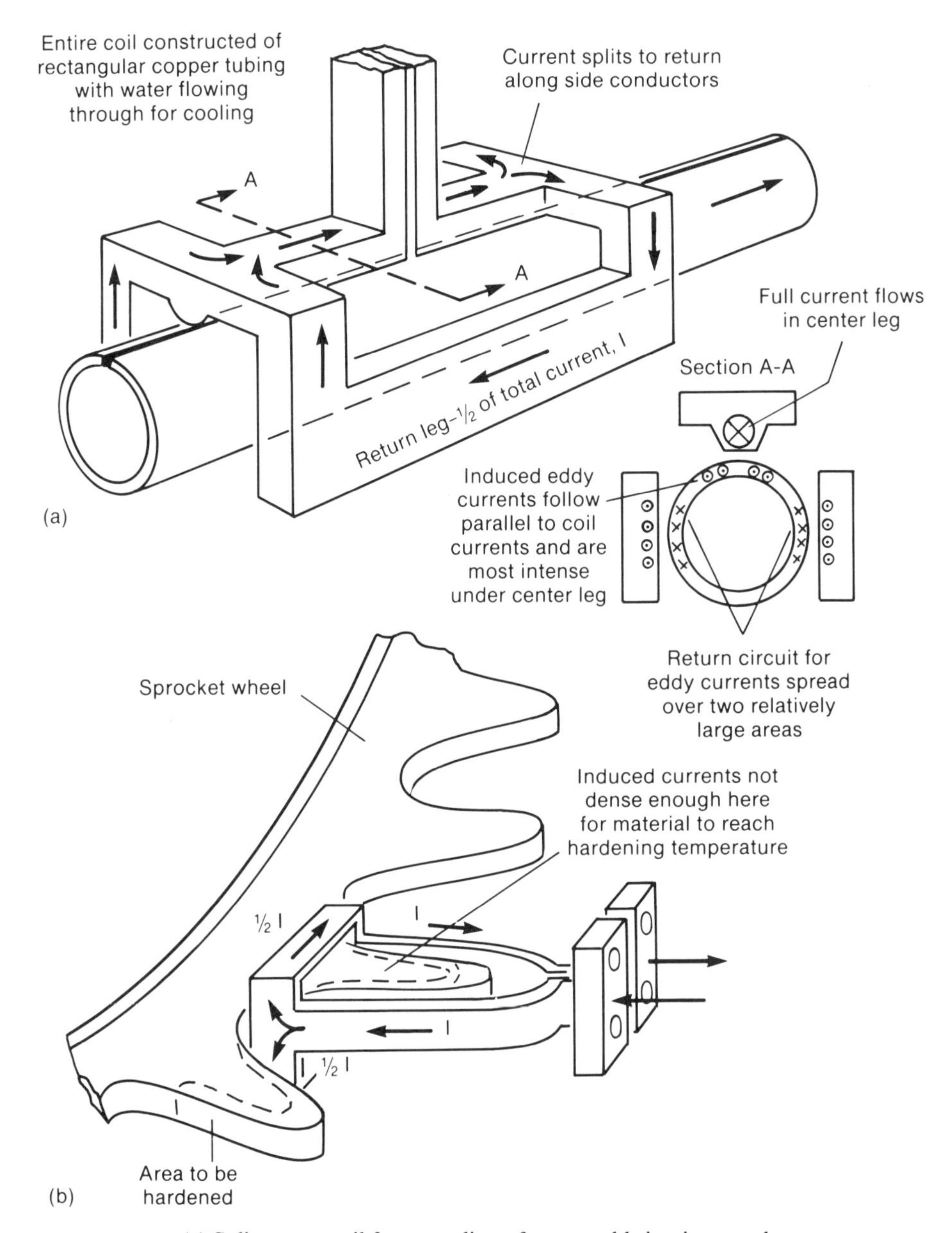

(a) Split-return coil for annealing of seam welds in pipe or tube.
(b) Split-return inductor for hardening of surfaces of large sprocket teeth one tooth at a time (welding fixture not shown).

Fig. 8.39. Two types of split-return coils (from C. A. Tudbury, *Basics of Induction Heating*, Vol 1, John F. Rider, Inc., New York, 1960)

strip as opposed to the typical longitudinal flux pattern developed with a coil which encircles the strip.

Series/Parallel Coil Construction

Often, it is desirable to have long work coils or coils with many turns connected in series. The current through each turn is identical, and it is simple to balance the heating pattern. Furthermore, if an imbalance occurs in any one turn (e.g., it is suddenly loaded more heavily), the other turns are affected equally. Thus, if a slowdown occurs in one location of a conveyor, the power along the whole line can be adjusted proportionally. In some situations, however, as the number of coil turns and resultant length increase, the voltage must also be increased to drive the current through the coil turns. This means that on long coils, the voltage between the workpiece and the coil can be extremely high, and arcing can occur. This may be a problem, for example, in heating of continuous strip, in which speeds are high and coils are therefore very long. In these situations, it becomes advantageous to use parallel coils, thereby reducing the applied voltage.

With parallel coils, the voltage required is limited to that needed to drive current through a single inductor. However, coil resistance and inductance in parallel inductors are such that extreme care must be taken to prevent all the current from going through the inductor closest to the power supply.

Fig. 8.40. Tapped forging coils for heating off the end of a bar (source: American Induction Heating Corp.)

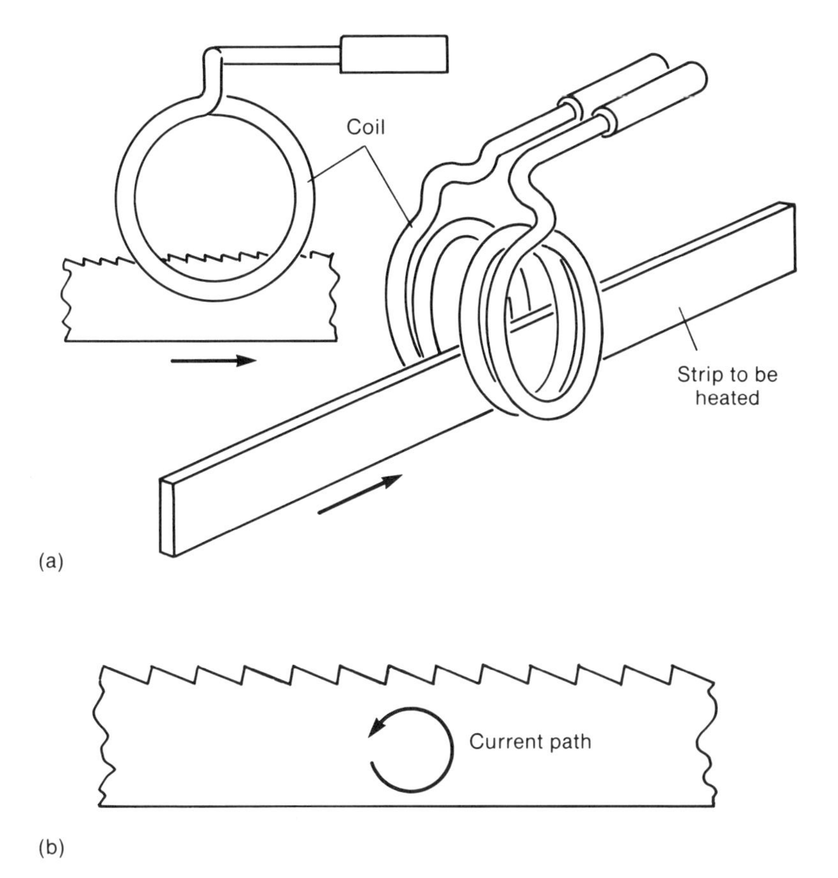

Fig. 8.41. Illustration (a) of one type of transverse coil for heating a thin section; sketch in (b) indicates the current path in the workpiece (from F. W. Curtis, *High Frequency Induction Heating*, McGraw-Hill, New York, 1950)

The best method for parallel coil construction makes use of a wide bus arrangement, as shown in Fig. 8.42. With identical coils mounted to the bus, this method permits rapid, simple replacement of individual sections as required, and fabrication of the individual sections is easily accomplished. A second technique, which ensures an even current distribution, comprises the interwinding of several coils as shown in Fig. 8.43. Here, an even current distribution is ensured in that each coil receives its current from the same point in the bus.

Tuning Stubs (Trombones)

When coils are in parallel, it is often difficult to balance the heating rates of the individual turns, because their inductances are not necessarily the same. In addition, they may be coupled to dissimilar loads of different mass. It then becomes necessary to adjust the individual coils using a tuning stub or "trombone" so that the parts, or different regions of the same part, achieve the same temperature at the same time. A typical application is a low-voltage, multiturn

Fig. 8.42. Photograph of parallel coils on a bus used to reduce coil voltage in heating of moving wire (source: Lindberg Cycle-Dyne, Inc.)

parallel coil (Fig. 8.44) of the type used in the electronics industry for heating a disk-type graphite susceptor. While it might be simpler to utilize a pancake inductor in which the coupling of the inner turns is changed in order to control the heating pattern, a parallel construction is frequently chosen to maintain a lower coil voltage, because the operation is performed in a hydrogen-filled bell jar. Because each of the turns is in essence an individual inductor, a series inductance (i.e., tuning stub) is placed in each parallel leg. This inductance is provided with a shorting bar which, by virtue of the fact that it is movable over the series inductor length, provides an adjustable inductance. The various series inductances may be used to drop the actual usable voltage on each turn, and thus can be used to adjust or "characterize"

Fig. 8.43. Photograph of parallel coils interwound to reduce coil voltage (source: Lindberg Cycle-Dyne, Inc.)

Fig. 8.44. Photograph illustrating trombone adjustment on parallel coils as used in epitaxial deposition (source: Lindberg Cycle-Dyne, Inc.)

the heating pattern of each turn individually in order to achieve balanced heating.

An adjustable inductance or trombone is also used in those instances where a coil is overcoupled. In such a case, the coil tries to couple more energy to the part than the machine can supply, thus overloading the system. Even though this is not an efficient method, rather than rebuild the coil, a trombone is inserted between the output station terminals and the coil. This trombone is then adjusted to limit coil voltage to a region within the capability of the machine. A commercial trombone used for this purpose is shown in Fig. 8.45.

With coils machined from flat stock (e.g., a multiplace, single-turn inductor), drilling a hole in the slot lead and then gradually increasing its diameter performs the same function as a trombone.

Conveyor/Channel Coils

Often when power densities are low and heating cycles not extremely short, parts can be processed by use of a turntable or conveyor in a continuous or indexing mode. The coil must then be designed to permit easy entry and exit of the part. The simplest conveyor or channel coil used in these situations is a modification of the hairpin inductor (Fig. 8.46). With the indexing tech-

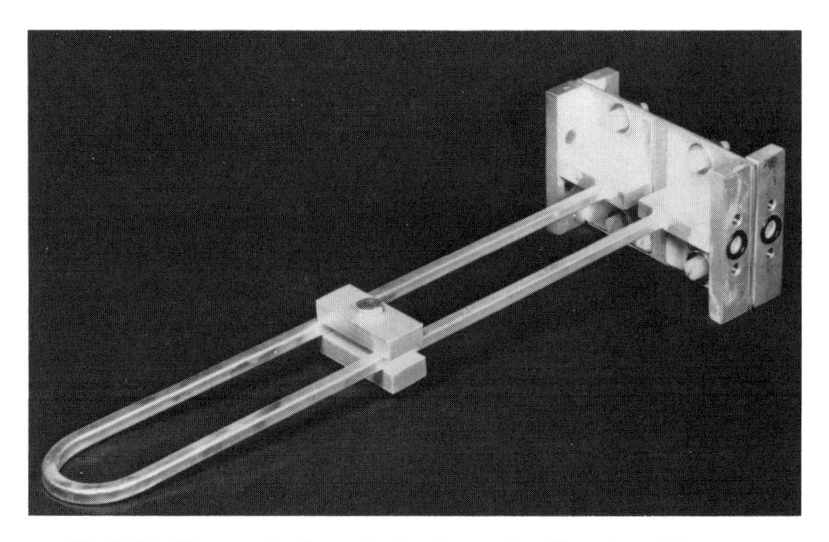

Fig. 8.45. Photograph of a typical tuning stub or "trombone" (source: Lindberg Cycle-Dyne, Inc.)

nique, in which the part is at rest in the coil during the heating cycle, the ends of the hairpin can be decoupled to prevent overheating of the ends. These raised portions or bridges also facilitate passage of the part through the coil. When a wide heating zone is to be produced on the part, coupling over a greater area can be effected through the addition of a liner to the coil turn (Fig. 8.47), or more ampere turns can also be produced with a multiturn channel inductor (Fig. 8.48). Channel-coil liners may also be configured to produce specialized heating patterns where greater heat densities are required in specific areas (Fig. 8.49).

During design of heating operations utilizing channel coils, there is a "fill

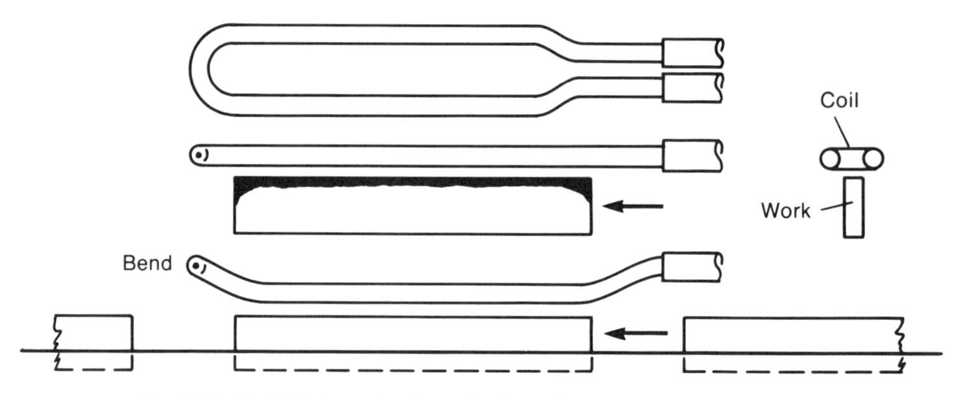

Fig. 8.46. Typical channel coil used to heat the edges of discrete lengths of rectangular bar stock; end of coil is decoupled by bending to prevent overheating of ends (from F. W. Curtis, *High Frequency Induction Heating*, McGraw-Hill, New York, 1950)

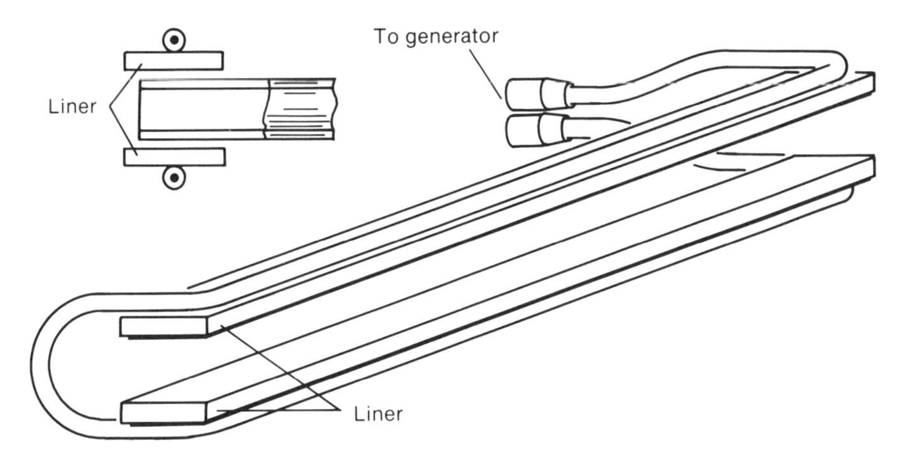

Fig. 8.47. Use of a liner on a single-turn channel coil to provide a wider heating pattern on the workpiece (from F. W. Curtis, *High Frequency Induction Heating*, McGraw-Hill, New York, 1950)

factor" which must be considered from an efficiency standpoint. The unused portion of the coil appears as lead losses. Therefore, parts must be as close as possible to each other, without touching, to utilize the full capabilities of the inductor. Another important consideration in the use of a channel coil is the fact that those areas of the workpiece closest to the coil receive the greatest portion of the flux and therefore heat the fastest (Fig. 8.50). If conduction through the part is slow, the part should be rotated while passing through the coil. Sufficient time (in an indexing conveyor or turntable) or speed variation

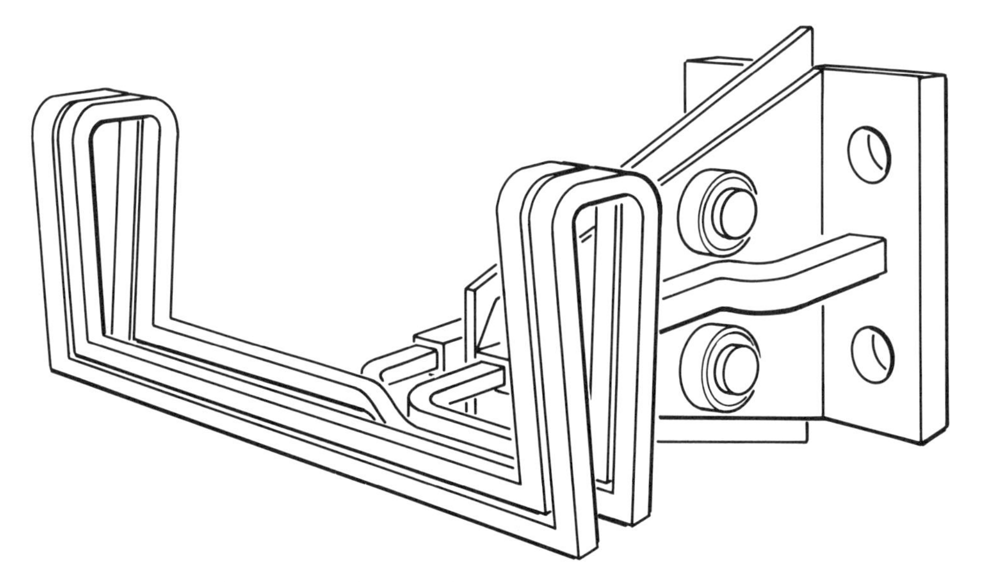

Fig. 8.48. Multiturn channel coil used to increase the ampere turns coupled to an induction heated workpiece (source: Lindberg Cycle-Dyne Inc.)

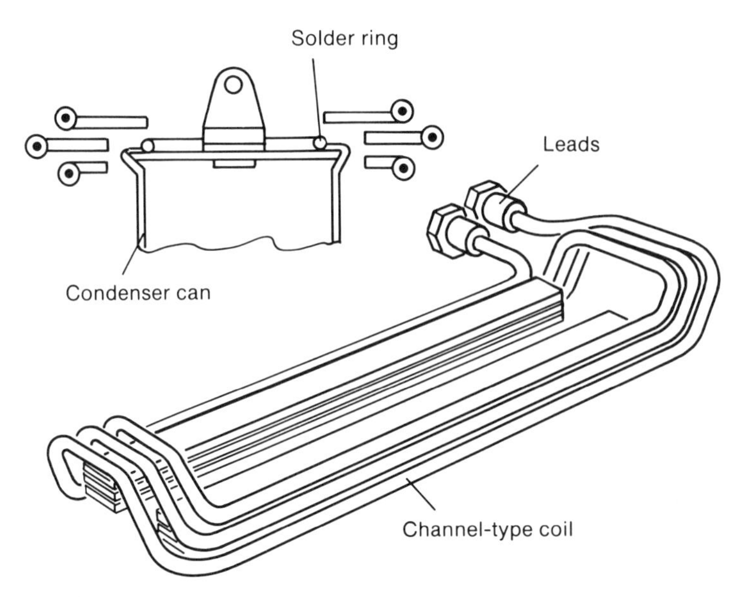

Fig. 8.49. Multiturn channel coil with a liner added to control the heating pattern (from F. W. Curtis, *High Frequency Induction Heating*, McGraw-Hill, New York, 1950)

(in a continuous-motion device) must be provided to allow heat uniformity to occur in part areas farthest from the coil turns.

COIL FABRICATION

Because of its low resistivity, fully annealed, high-conductivity copper is most commonly used in the fabrication of induction heating coils. The copper is typically in a tubular form (with a minimum outer diameter of 0.32 cm or 0.125 in.) to allow for water cooling. Material of this kind is available in a wide range of cross sections (round, square, and rectangular) and sizes.

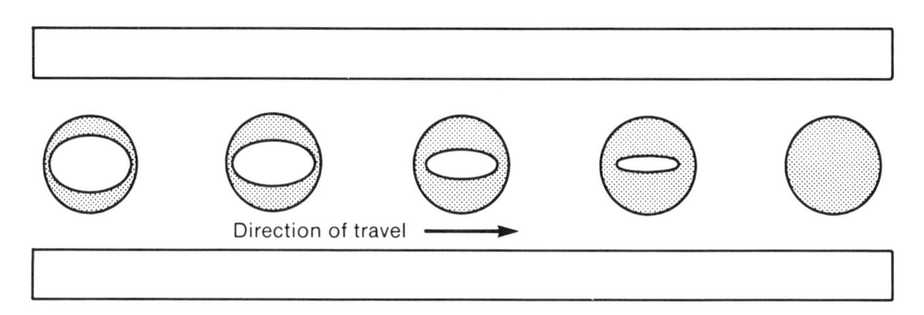

Fig. 8.50. Development of the heating pattern in parts moved through a channel coil

Selection of Tubing

In addition to the I^2R loss due to its own resistivity, the coil surrounds the load and absorbs additional heat through radiation and convection from the heated surface. Therefore, it is essential that the tubing selected for the work coil have a sufficient cooling path to remove this heat. Otherwise, the resistivity of the copper will increase due to the temperature increase, thus creating greater coil losses. In some instances, such as large coils, it may be necessary to break up the individual water paths in a coil to prevent overheating and possible coil failure. Figure 8.51 shows a coil used for a line-frequency application which has a number of manifolded water paths to prevent steam formation.

Another factor in the selection of tubing for induction coils relates to the fact that the current in the work coils is traveling at a specific reference depth which depends on the power-supply frequency and the resistivity of the copper. Accordingly, the wall thickness of the coil tubing should be selected to reference-depth limits similar to those used for induction heating of copper. Suggested wall thicknesses for various frequencies are shown in Table 8.2. However, copper availability must be considered, and often wall thicknesses less than twice the reference depth are used with only a nominal loss in overall coil efficiency. In coils for low-frequency operation, particularly those used at 60 Hz, the large depth of penetration requires tubing with a particularly heavy wall. Because the current flows on the ID face of the coil (i.e., adjacent to the workpiece), coils for such applications are generally fashioned of

Fig. 8.51. Photograph of a low-frequency induction coil with manifolded water taps/cooling paths (source: American Induction Heating Corp.)

Table 8.2. Selection of copper tubing for induction coils

Frequency	Theoretical wall thickness (= 2 × reference depth)(a), mm (in.)	Typical wall thickness available, mm (in.)	Minimum tube diameter(b), mm (in.)
60 Hz	16.8 (0.662)	14.0 (0.550)	42.0 (1.655)
180 Hz	9.70 (0.382)	8.13 (0.320)	24.3 (0.955)
540 Hz	5.59 (0.220)	4.67 (0.184)	14.0 (0.550)
1 kHz	4.11 (0.162)	3.43 (0.135)	10.3 (0.405)
3 kHz	2.39 (0.094)	1.98 (0.078)	5.97 (0.235)
10 kHz	1.32 (0.052)	1.07 (0.042)	3.30 (0.130)
450 kHz	0.15 (0.006)	0.89 (0.035)	0.38 (0.015)
1 MHz	0.08 (0.003)	0.89 (0.035)	0.19 (0.0075)

(a) Resistivity of copper assumed to be 1.67×10^{-6} Ω·cm (0.66×10^{-6} Ω·in.). (b) Tube ID requirements for adequate cooling-water flow should also be considered.

heavy, edge-wound copper with a brazed cooling path at the outer edge to reduce cost. Specially drawn tubing (Fig. 8.52) is also available with an offset extruded cooling path.

Square copper tubing is also commercially available and is frequently used in coil fabrication. It offers a considerable advantage in that it couples more flux to the part per turn than round tubing (Fig. 8.53). Moreover, it is more easily fabricated in that it will not collapse as readily on bending. It is also easily mitered (Fig. 8.54) to create sharp, close bends as required. If only round tubing is available, it can be flattened in a vise or other simple device (Fig. 8.55) to adjust the resultant thickness dimension. This flattening can be done with minimal decrease in dimension of the water-flow path.

Coil Forming

In fabrication of copper coils, it must be noted that the copper work hardens with increasing deformation. Thus, most fabricators anneal the tub-

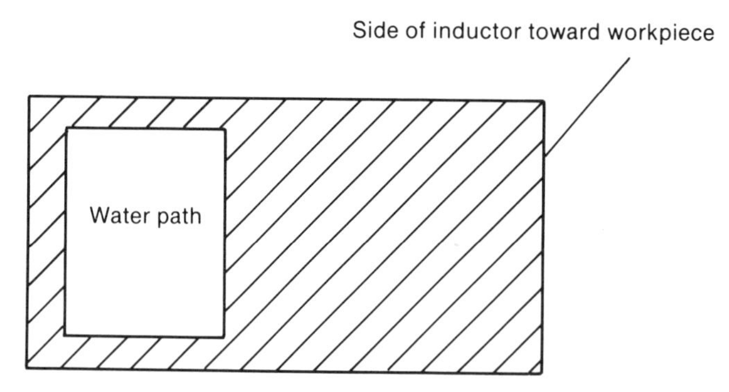

Fig. 8.52. Heavy-wall tubing used in construction of induction coils

ing every few bends to relieve this condition by heating the tubing until it is bright red, then cooling it rapidly in water. These intermediate anneals prevent fracture of the tubing during fabrication.

In some forming operations, it may be desirable to fill the coil with sand or salt to preclude collapse of the tubing. In addition, there are several low-temperature alloys (with melting points below 100 °C or (212 °F) that are normally used to perform this same function. When the coil is completed, it is immersed in boiling water. The alloy then flows out freely and can be reused at another time. With any of these techniques, once filled, the tubing acts as a solid rod during forming and can be simply cleared on completion.

Joining of Coils to Power-Supply Leads

Joining of an induction coil to a power supply frequently involves brazing of copper tubing or connecting small- to large-diameter tubing. In brazing of copper tubing, it is generally preferable to use a low-silver alloy such as phos-copper. These alloys do not require the use of flux when making copper-to-copper joints, and in addition are sluggish when molten. They are thus useful

Fig. 8.53. Comparative heating patterns produced by using round vs. square tubing for a solenoid induction coil (from M. G. Lozinskii, *Industrial Applications of Induction Heating*, Pergamon Press, London, 1969)

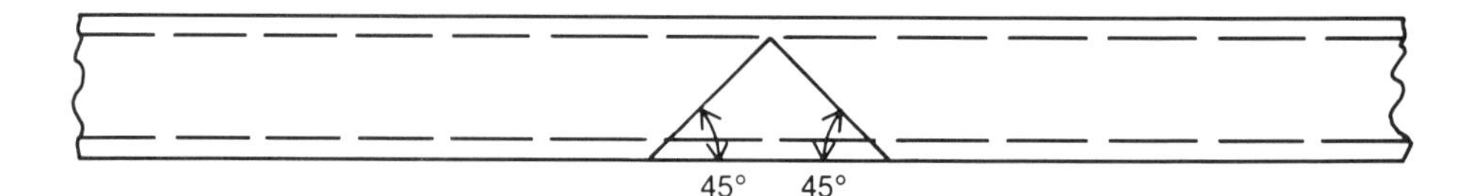

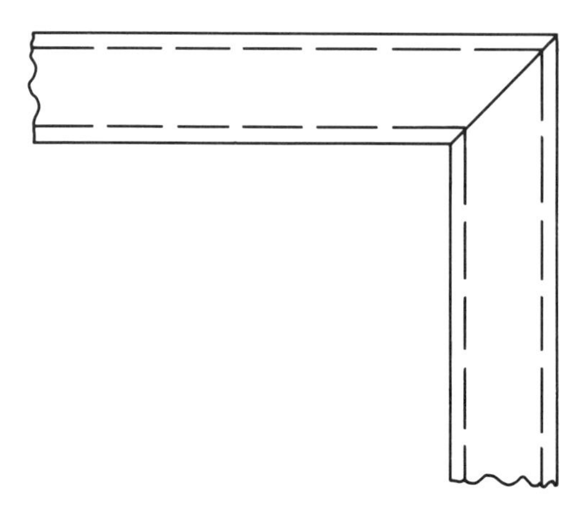

Fig. 8.54. Mitering of square tubing to provide right-angle bends in an induction coil

in filling pinhole leaks, but will not fill the water path in brazing of small-diameter tubing.

Most power-supply terminals utilize rather large tubing. Thus, it is frequently necessary to join small-diameter tubing to larger tubing at some point. Figure 8.56 shows a simple technique for accomplishing this. The larger tubing is crimped after inserting the smaller-diameter stock, and the joint is closed

Fig. 8.55. Device used to flatten tubing during manufacture of induction coils (source: Lepel Corp.)

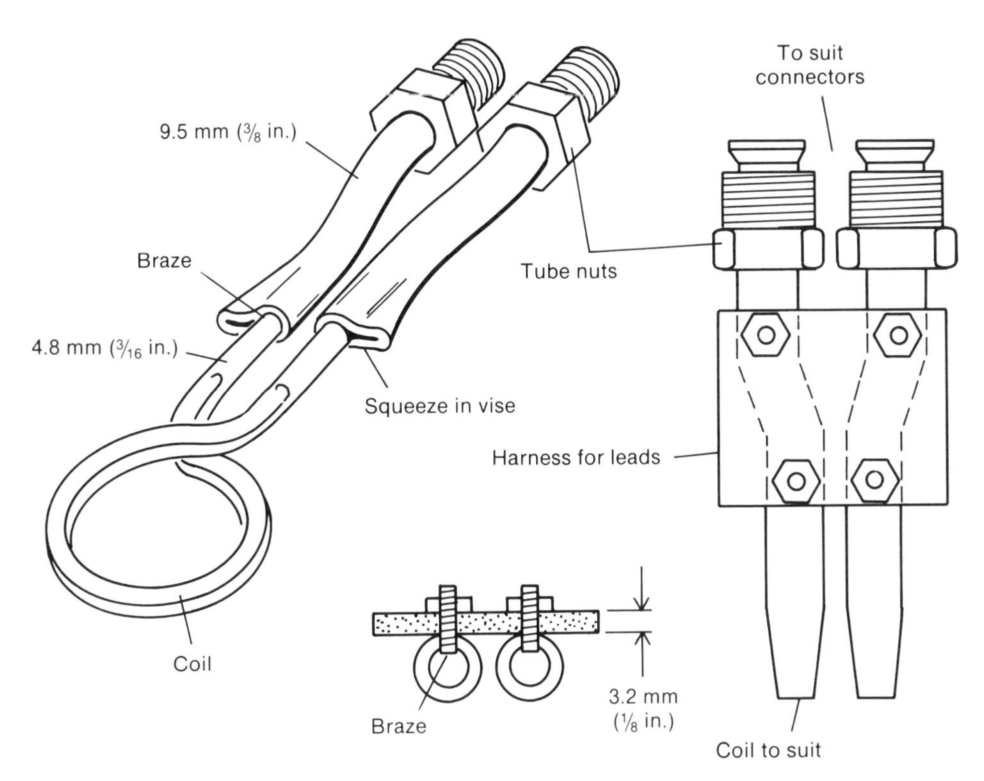

Fig. 8.56. Illustration of method for joining small- to large-diameter copper tubing for induction coils (source: F. W. Curtis, *High Frequency Induction Heating*, McGraw-Hill, New York, 1950)

by brazing. The illustration also depicts the common use of a flare or compression fitting when making coil connections; these are generally limited to low-power, high-frequency (RF) coils. Although useful when coils are changed frequently, such fittings are subject to water leakage under constant changing. In addition, the twisting imparted to the coil during attachment tends to distort the coil configuration.

Most low-loss fishtail leads (discussed below) utilize solid block connectors that are cross-drilled for water flow and are mated with counterbored holes and "O" ring seals. In these cases, the coil is often held to the output terminals by brass or phosphor-bronze screws. This type of construction permits coil changes while preventing bending of the coil during installation (Fig. 8.57).

Bracing of Coils

Because electric currents flow in both the workpiece and the coil, magnetomotive forces between the two are developed. The magnitudes of the forces depend on the magnitudes of the currents. If sufficiently large, the forces may cause the part to move in the coil. If the part has a large mass, however, the coil will tend to move relative to the workpiece. The turns may

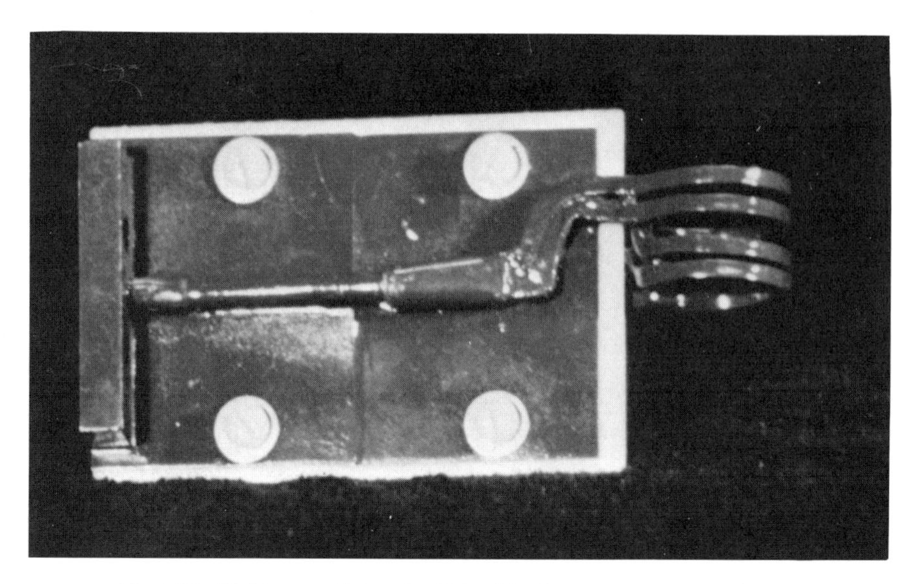

Fig. 8.57. Multiturn coil showing fishtail-lead construction (source: Lindberg Cycle-Dyne, Inc.)

also tend to move relative to each other. It is important, therefore, that the coil turns be suitably braced to prevent movement and possible turn-to-turn shorting. Furthermore, coil motion relative to the part must be prevented to avoid undesirable changes in the heating pattern.

Much of the acoustic noise generated during low-frequency operations also occurs due to coil vibration, much as a speaker coil and magnet structure work in an audio system. Bracing and physical loading of the coil to restrict its movement will aid in reducing this condition. On very large, high-current coils, the magnetomotive force exerted can be extremely large, and if proper bracing is not provided, the coil may gradually work harden and finally fail. With such coils, care should thus be taken to place braces from end board to end board so that the coils are in compressive loading, thereby minimizing these effects.

Typical bracing techniques are illustrated in Fig. 8.58. In Fig. 8.58(a), brass studs are brazed to every other turn. These studs are then secured to insulator posts to hold them in a fixed relation to each other. Nuts on each side of the stud at the insulator allow adjustment for characterization of the heating pattern. In Fig. 8.58(b), the insulation has been contoured to hold the turns relative to each other after the end turns are secured with studs. Figure 8.59 shows construction of a 3-kHz coil used in the forging industry. The insulator boards are notched and hold the turns in rigid location.

The insulation used for bracing applications must meet the criteria for the coil design. In addition to the installation being capable of withstanding the heat radiated from the workpiece, its electrical capabilities must permit it to

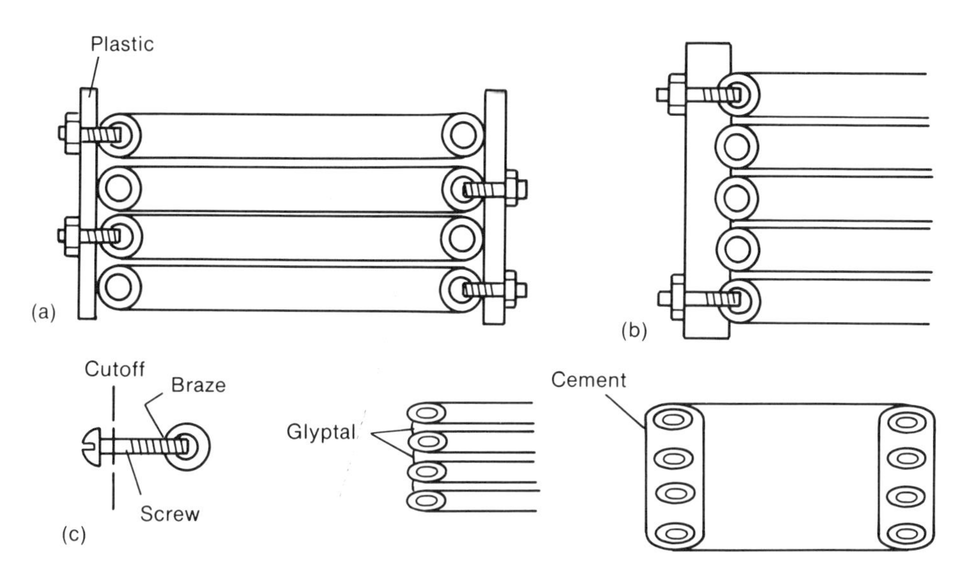

Fig. 8.58. Typical techniques for bracing of induction-coil turns (from F. W. Curtis, *High Frequency Induction Heating*, McGraw-Hill, New York, 1950)

withstand the voltage between the mounting studs or the turn-to-turn voltages of the coil. This is of particular concern when using high-voltage RF coils where up to 12,000 V may be impressed across the total coil. It may be necessary in these instances to provide slots between the stud locations in the insulator boards to increase the creepage path between the studs. It may also be necessary to increase the heat-resistant characteristics of the insulation by facing the area exposed to the heated surface with a sheet of high-temperature insulation.

For purposes of rigidity, cleanness, and protection, it is sometimes desirable to encapsulate work coils in a plastic or refractory material. The same kind of care with respect to voltage and temperature characteristics must be taken with these materials as with insulating boards. For low-temperature induction heating applications, epoxy encapsulation of the coil is quite common. For heating of steel billets, coils are usually cast in a refractory cement to prevent scale from the part from falling between the turns (Fig. 8.60). In coating of coils with refractory materials, care must be taken to match the pH of the refractory to that of the material being heated; for example, an acidic refractory is required for the ferrous scale which drops off during high-temperature heating of steels. On larger, low-voltage installations, such as those encountered in line-frequency heating, the individual turns are typically wrapped with fiberglass tape and then varnished. This permits close spacing of coil turns. The coil in Fig. 8.61 is a 60-Hz forging coil being fabricated of offset tubing having one heavy wall which is faced toward the billet. Note the multiple water exits which will provide the parallel water paths.

Fig. 8.59. Photograph of a billet-heating induction coil that is braced using contoured boards to separate the coil turns (source: American Induction Heating Corp.)

POWER-SUPPLY LEADS

Design Considerations

All coils represent an inductance to the tank circuit. However, in practice, the working portion of the coil may in fact be only a small portion of the inductance presented to the tank. Between the output terminals of the generator or heat station and the heating portion of the work coil, there may be a considerable distance of output lead. This distance can be minimized through use of a remote heat station. In any case, however, some finite distance exists between the heat-station terminations and the actual coil. Design and construction of these work-coil leads can be a major factor in determining job feasibility.

The effect of lead construction on system performance can be best understood with respect to the tank circuit of which it is a part (for example, see Fig. 8.62). The coil/load inductance is represented by L_2. Each lead connecting the tank capacitor to the coil has its own inductance (L_1, L_3). If the voltage in the tank, E_T, is impressed across the total of these inductances, then some voltage drop appears across each. The full voltage will thus never appear across the work coil. Nevertheless, if the inductance of the coil (L_2) is at least approximately ten times the total inductance of the leads (L_1 plus L_3), a maximum of 10% of the total voltage will be lost in the leads. Any loss less than this can be considered nominal.

Fig. 8.60. Photograph of a billet-heating induction coil that has been mounted in a cast refractory for purposes of bracing (source: American Induction Heating Corp.)

Coils for low-frequency applications often have many turns, a large cross-sectional area, and thus fairly high inductance. Hence, the comparative lead inductance is small. As the frequency increases, coils often become smaller in size, and their inductance and inductive reactance decrease. As the distance between the heat station and coil increases, therefore, these lead inductances can become critical.

Several coil designs which illustrate the effect of lead design are shown in Fig. 8.63 and 8.64. In Fig. 8.63(a), a coil with leads far apart is depicted. The space between the leads presents an inductance almost equal to that of the coil. Thus, a major portion of the voltage will not appear in the work-

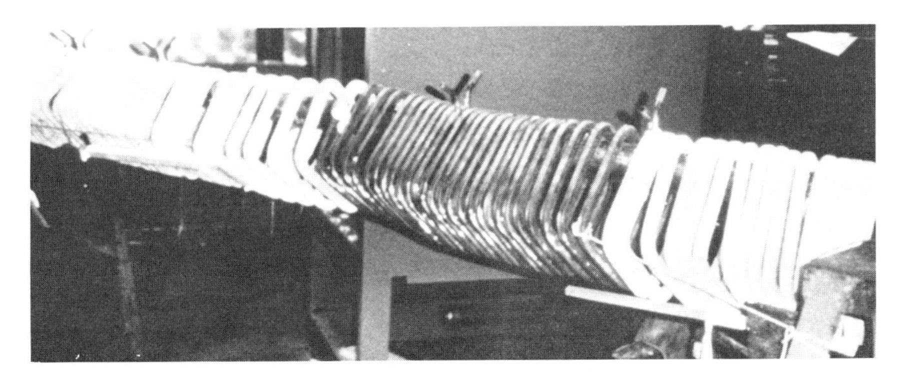

Fig. 8.61. Photograph of edge-wound, offset coil stock being prepared for winding of a 60-Hz induction coil for heating of forging billets (source: American Induction Heating Corp.)

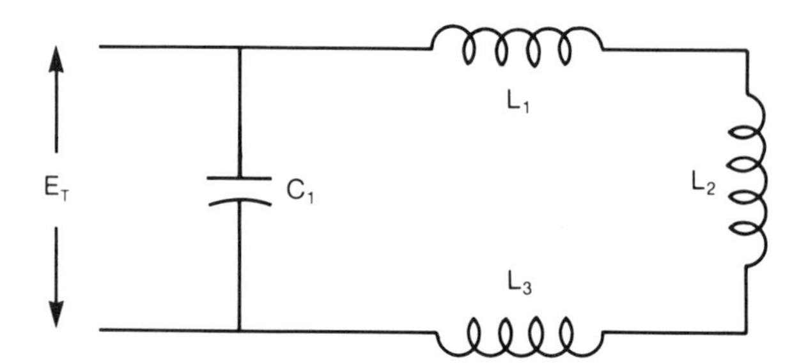

Fig. 8.62. Schematic circuit diagram indicating the inductance of the coil leads and induction coil itself: L_1,L_3–lead inductances; L_2–induction-coil inductance; C_1–tank capacitance; E_T–tank voltage

ing area. A better design (Fig. 8.63b) minimizes this gap and thus improves heating efficiency. Figure 8.64 also shows single-turn, multiplace coils with an extremely poor and an improved lead design.

Another factor to consider is the interaction of the leads with nearby metal structures. Because all leads have some inductance, they can act as work coils. Thus, a conductor placed within their field will be heated. Leads placed adjacent to metal structures will tend to heat them. In addition to unwanted heat, this loss reduces the power available to the load. It is important that lead-to-lead separation be minimized and proximity to metallic structural members be considered. Whenever possible, duct housings, trays, or conduits must be of low-resistivity or insulating materials, such as aluminum or plastic.

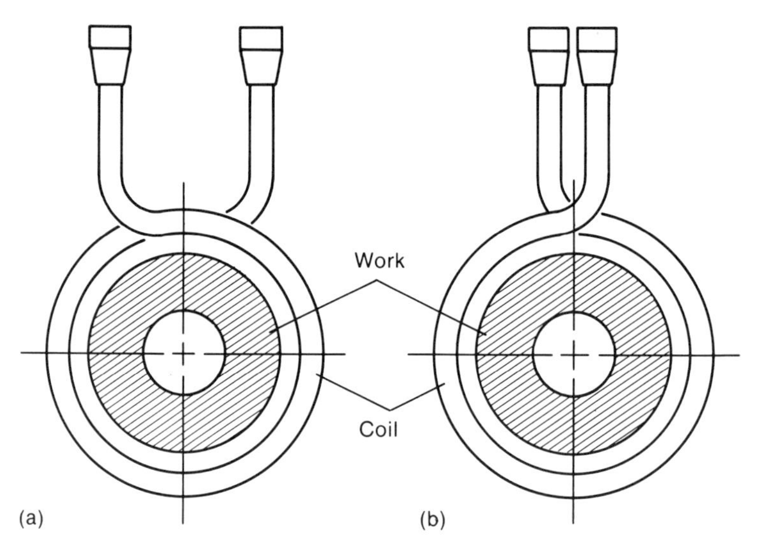

Fig. 8.63. Effect of coil-lead spacing on lead inductance; closer spacing, as in (b), reduces lead inductance and thus power losses (from F. W. Curtis, *High Frequency Induction Heating*, McGraw-Hill, New York, 1950)

Typical Lead Design

Induction heating lead designs typically make use of water-cooled copper plates or tubes.

When coil voltages are low (≤800 V), a low-inductance structure known as a fishtail is often utilized. A fishtail (Fig. 8.65) consists of a pair of parallel copper plates which are water cooled to maintain low resistivity. They are placed with their wide bus faces parallel and are either separated physically with air as an insulator or held together by nylon bolts and nuts with teflon or a similar material acting as a spacer. Extending from the heat station to a point as close as possible to the operating area of the work coil, they present minimum inductance and provide maximum power at the coil. Depending on conditions and construction, efficient runs of approximately 5 m (15 ft) are practical. The thickness of the copper plates should be consistent with the frequency, as noted in Table 8.2, and cooling-water paths and sizes must be consistent with the power being transmitted as well. The copper plates should increase in width with generator power and the distance of the run. Moreover, they should be spaced as close together as possible with only enough space for proper insulation to prevent arcing.

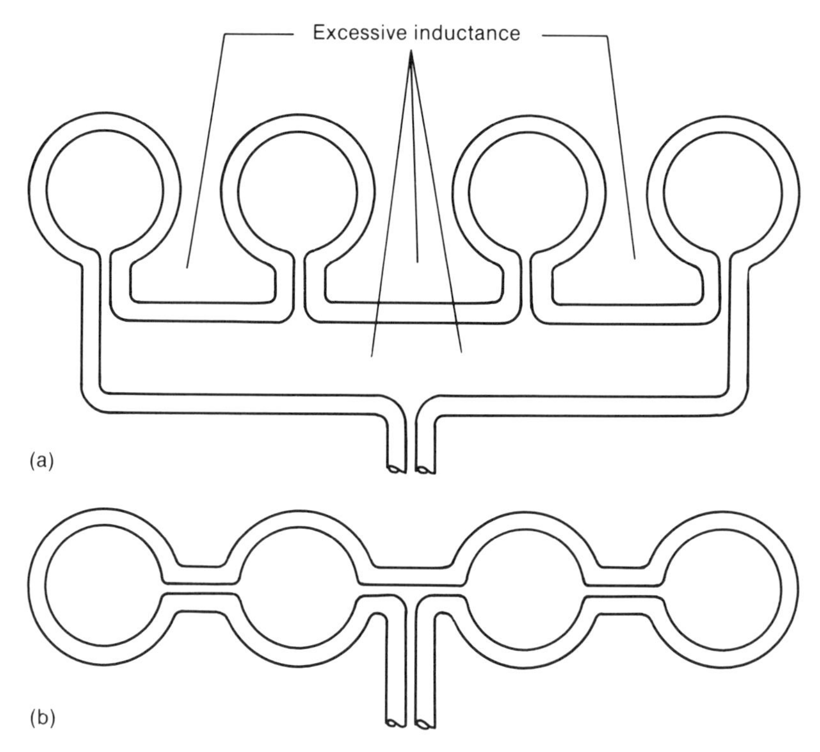

Fig. 8.64. Lead construction for multiplace inductors; lead design in (b) is preferable because of lower lead inductance (from F. W. Curtis, *High Frequency Induction Heating*, McGraw-Hill, New York, 1950)

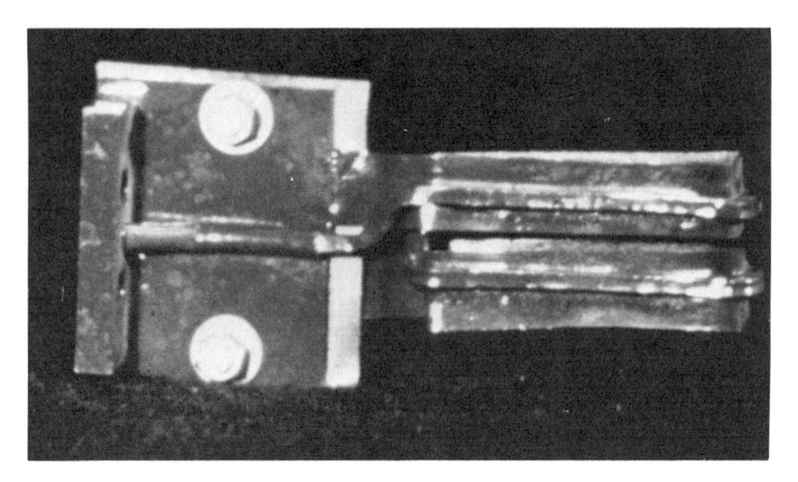

Fig. 8.65. Photograph of a radio-frequency two-turn coil with a liner, showing fishtail-lead construction (source: Lindberg Cycle-Dyne, Inc.)

As the coil inductance increases (e.g., as the number of turns or the coil diameter increases), lead length becomes less critical, and plain copper tubing leads then become more practical. However, larger coils also require higher terminal voltages. These leads must also be kept as close as possible to each other while maintaining sufficient spacing to prevent arcing. However, good practice still dictates that coil leads be kept to a minimum length and that copper tubing sizes be used that are consistent with frequency, current, and cooling requirements.

Rigid leads, whether tubing or bus, built to the above guidelines are inherently more effective than flexible, water-cooled cable. In some cases, however, it is absolutely necessary to use flexible connections. There are several variations in flexible leads, but it must be kept in mind that the inductive lead losses in flexible cables are usually much greater than those for rigid connections. The most common flexible lead is generally used in applications similar to tilt-type induction melting furnaces and consists of a water-cooled, spiral-wound inner conductor (similar to BX cable, but made of copper) with an outer insulating covering. These leads are used in pairs with one for each lead connection. Not only must they be sized for current and frequency, but the insulation must be capable of handling the voltage rating of the system. To keep lead inductance to a minimum, four flexible cables should be used where possible (with similar polarities opposite), and the flexible leads should be tied together with insulating straps.

Coaxial leads are also available and may be rigid or flexible. They consist of an inner conductor and an outer sheath or housing which is also used as the return conductor. This outer sheath is generally at ground potential. In addition to providing an extremely low-inductance lead, the outer ground acts to eliminate possible strong radiation or inductive coupling to adjacent structures.

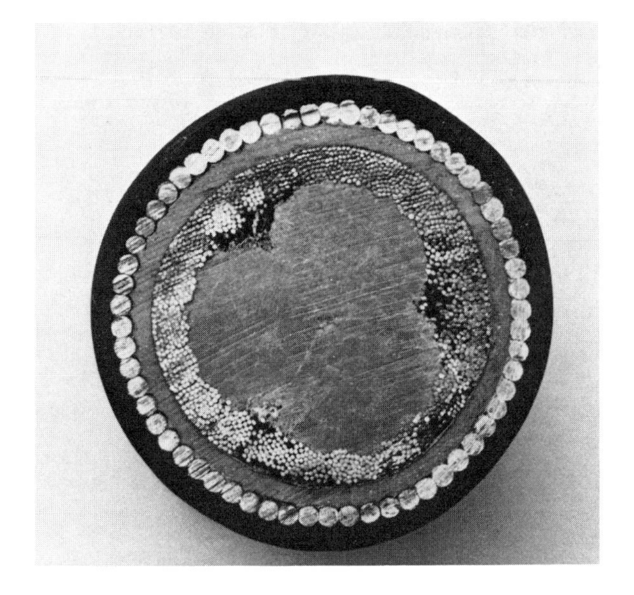

(a)

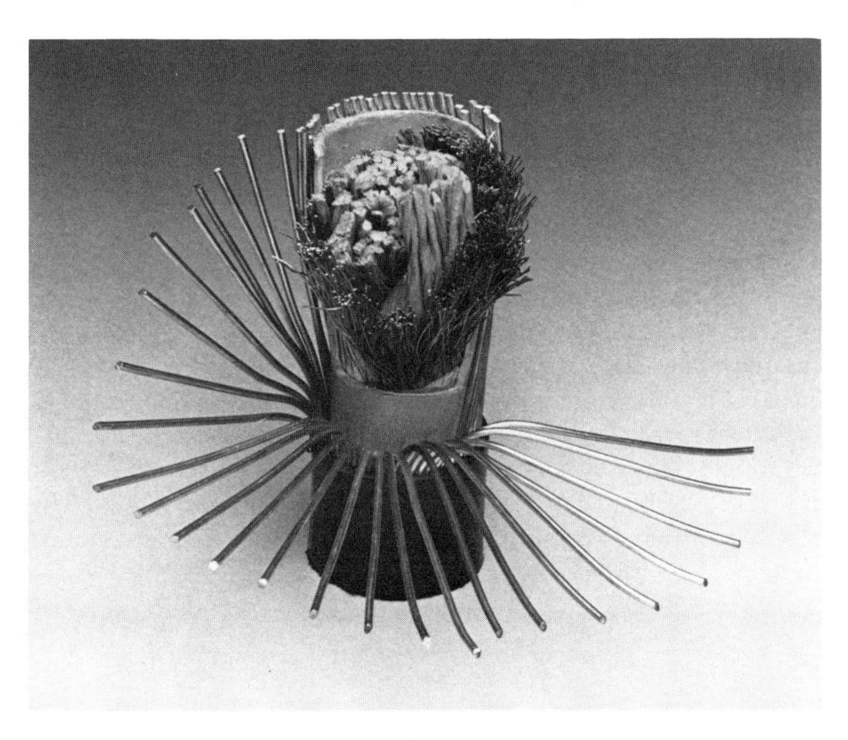

(b)

Fig. 8.66. Semiflexible coaxial cable used to transmit power between medium- or low-frequency power supplies and heat stations: (a) cross section of coaxial cable and (b) cable "broken out" for assembly (source: American Induction Heating Corp.)

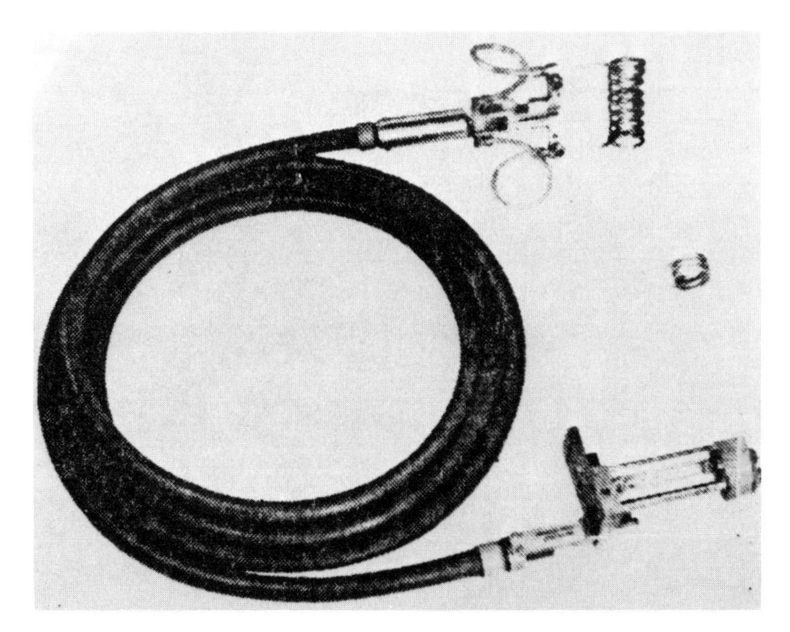

Fig. 8.67. Water-cooled flexible coaxial cable used in radio-frequency induction heating to connect the inductor to the power supply (source: L. C. Miller Co.)

Rigid coaxial lead is generally quite expensive and is usually limited to those applications where it is imperative to transmit high power at high frequency over some distance. A more recent development is the semiflexible solid coaxial cable (Fig. 8.66). It is used to connect medium- and low-frequency generators to remote heat stations. Composed of an outer layer of heavy copper conductor and an inner core of many small-gage wires, it can be pulled like regular cable and requires no conduit.

A third type of coaxial cable is the water-cooled type generally used at radio frequencies (Fig. 8.67). It consists of a low-inductance, braided inner conductor which runs through a water-cooled tube and an outer return braid which is also water cooled. This construction is generally utilized with medium-to-high-inductance coils because its construction does not greatly minimize inductance but does provide flexibility. This last type of lead is most common when the operator must physically move the coil from part to part as in bottle sealing.

Chapter 9

Flux Concentrators, Shields, and Susceptors

To a large extent, the induction coil and its coupling to the workpiece determine the precise heating pattern that is developed. However, it is often desirable to modify this pattern in order to produce a special heating distribution or to increase energy efficiency. At other times, the high heating rates characteristic of induction are needed for processing of nonconductors. Three broad methods of accomplishing such objectives are described in this chapter. They make use of devices known as flux concentrators, shields, and susceptors. Flux concentrators and shields are used to modify the field of magnetic induction (thereby shaping the heating pattern) or to prevent auxiliary equipment or certain portions of a workpiece from being heated, respectively. Susceptors are materials which are readily heated by induction and which subsequently are used to heat electrically nonconductive materials through radiation or conduction heat-transfer processes.

FLUX CONCENTRATORS

When placed in an induction field, magnetic materials tend to gather the lines of flux. Such materials are said to have high permeability. On the other hand, nonmagnetic materials (e.g., copper, aluminum, etc.) do not exhibit this property when placed in a magnetic field; thus, their permeability is equivalent to that of air. In induction heating design, the *relative* magnetic permeability, or the permeability relative to that of air, is of importance. The relative magnetic permeability of air is assigned the value of one. By contrast, magnetic materials have relative magnetic permeabilities from approximately 100 to 1000, depending on the strength of the magnetic field in which they are placed. Above the Curie temperature, however, they lose their ferromagnetic properties, and their relative magnetic permeability drops to unity.

Flux concentrators are magnetic materials which are utilized to gather the flux field set up during induction heating and thus to modify the resultant heating pattern. The means by which this is accomplished is illustrated in Fig. 9.1. Here, it can be seen that a flux concentrator in the form of a permeable iron core causes distortion of the flux field surrounding the induction coil. The permeable material presents an easier path for the flux lines; thus, they concentrate in the permeable material. By concentrating the field and providing a better path, stray flux can be used to advantage. Because the voltage generated in an induction heated workpiece is proportional to $\Delta\phi/\Delta t$, or the time rate of change of the number of lines of flux passing through the part, a greater number of flux lines leads to a higher voltage and therefore larger eddy currents in the workpiece.

Materials for Flux Concentrators

Materials for flux concentrators are of two basic types: (1) packages of laminated silicon steel punchings used at frequencies below 10 kHz, and (2)

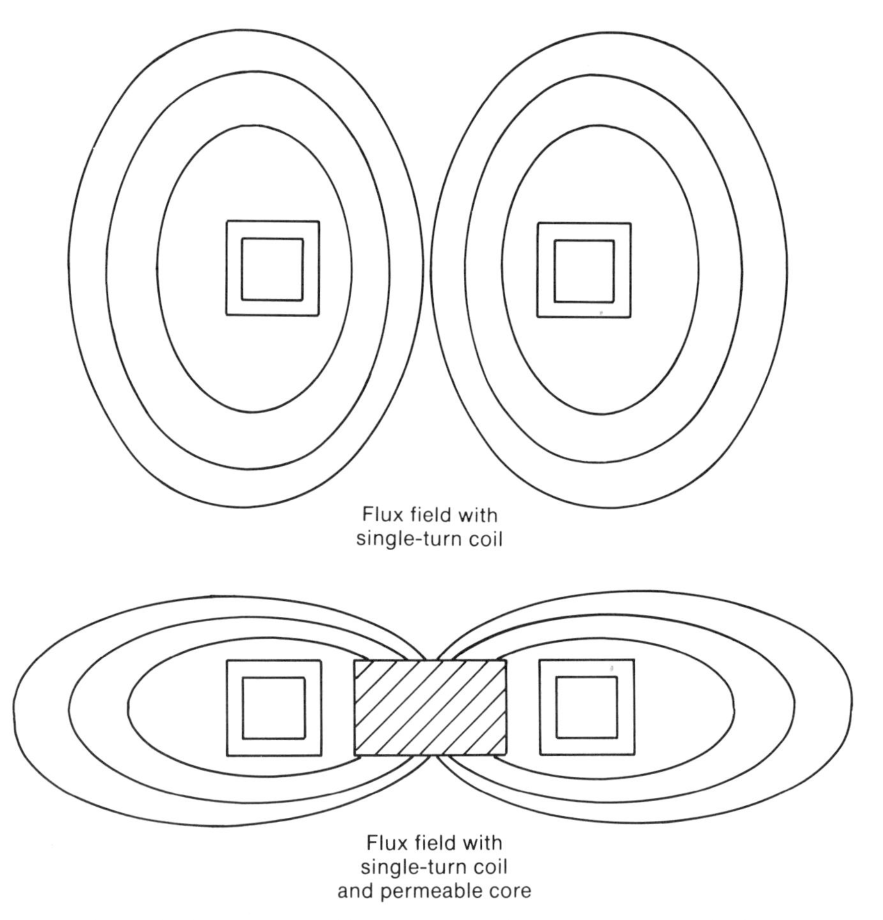

Fig. 9.1. Concentrating effect of a permeable core on a magnetic flux field

ferrites or powdered iron combinations for higher frequencies, including those in the radio-frequency (RF) range.

The laminations for low- and medium-frequency concentrators are usually of the same grain-oriented silicon steel used for high efficiency in power transformers and electric motors. Such steels would tend to heat dramatically in the magnetic field were it not for the fact that they are in the form of a laminated structure. By this means, the eddy-current paths are minimized to keep losses low. The thickness of the individual laminations should be held to a minimum. Generally, below 3 kHz, the laminations should be on the order of 0.38 mm (0.015 in.) each. At 10 kHz, a thickness of 0.20 mm (0.008 in.) is used. The laminations have a highly oxidized or phosphated surface which acts as insulation between the layers, thereby electrically isolating each for maximum efficiency.

At RF frequencies, powder metallurgy (PM) materials and ceramics are generally used because the particle size of the permeable material must be very small—i.e., on the order of 40 μm (1.57 mils) each. Since these must be isolated from each other, they are usually sintered compacts of magnetic powders which may consist of ferrites or similar components. By design, they have a narrow hysteresis loop for low loss, low cohesive force, and a high flux density at low field strength. Most of these materials are brittle and require grinding or special machining to fit the application. However, some new materials which are both flexible and machinable have recently been developed (Fig. 9.2).

Fig. 9.2. Flux concentrators made from a machinable form of ferrite that are used for induction heating applications (source: R. S. Ruffini and Associates, Inc.)

Because all flux-concentrator materials conduct flux at high densities, there is some loss generated as heat within their structures. Generally, the laminations in the coil area are cooled by virtue of the fact that they are in contact with the coil. In high-power-density situations, it is important therefore that the coil water-cooling system be designed to handle and remove this additional heat.

With the ferrites used at RF frequencies, the problem becomes somewhat different. Although the magnetic particles heat, they are insulated from each other by a ceramic binder which also provides good thermal insulation. Consequently, although the outer surface of the flux concentrator may be cooled adequately, the inner particles may continue to increase in temperature. Accordingly, when flux density is high and high duty cycles are employed, the inner portions of these materials may attain temperatures above their capabilities and fail in service. They should therefore be designed only for systems in which they can be readily replaced.

Application of Flux Concentrators

Flux concentrators, whether laminations or ferrites, should be located directly in or on the coil. For example, placement in the center of a pancake coil collapses the over-all field to provide a higher density at the coil surface (Fig. 9.3). In the same manner, insertion of a concentrator in a helical coil collapses the end flux outside the coil center. These lines then provide a higher-density field closer to the coil surface.

Size and location of concentrators can effectively control the heating pattern. Figure 9.4 shows the effect of varying the width of a laminated concentrator on the length of a heating pattern. With RF systems, the same type of control can be achieved using ferrite flux concentrators. In addition, ferrites can sometimes be loaded into the core of an output-impedance-matching transformer to increase its efficiency. However, water cooling of the core material in such a high-density field is difficult, and, unless the duty cycle is low, ferrite life will be very limited.

SHIELDS

In many cases, the flux field surrounding a conductor may produce heat where it is not desired. For instance, the field at the end of a coil (which is effective at distances up to 1½ times the coil diameter) may heat the parts of an adjacent loading mechanism. As another example, stray flux may heat the center boss on a gear blank while the teeth are being hardened. Similarly, when separate and independent coils are placed adjacent to each other, as in a long forging line, the end-effect fields of the coils might cause interaction. It then becomes necessary to provide a means of cancelling or shielding these fields from the affected parts or from each other. Devices known as shields are used for this purpose.

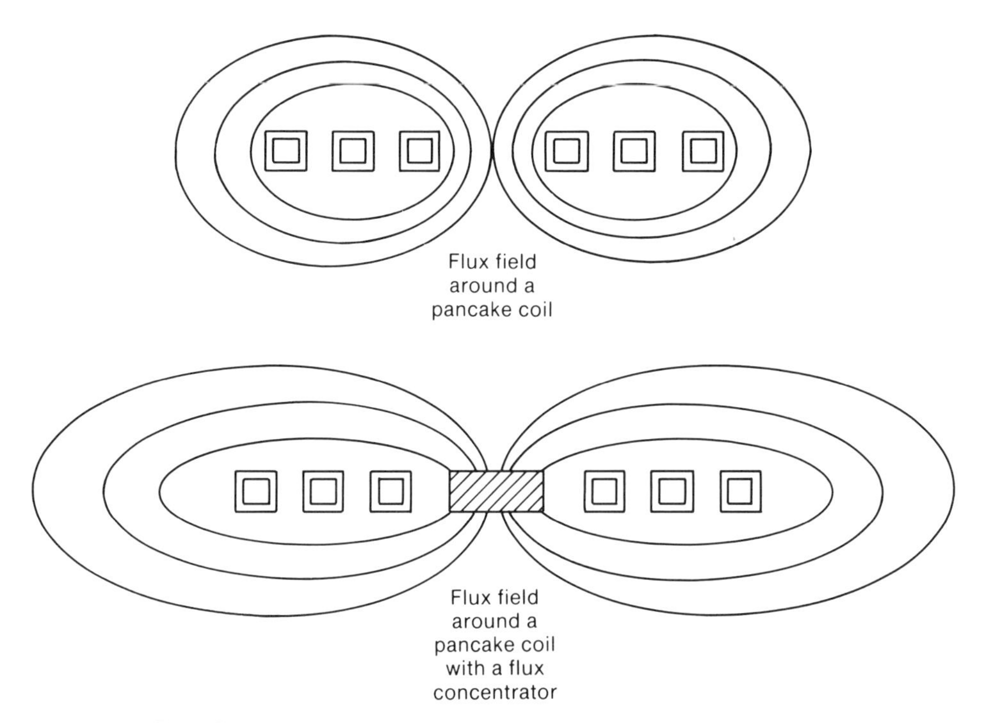

Fig. 9.3. Use of a flux concentrator in the center of a pancake coil to increase the amount of magnetic flux at this point and thus the overall heating efficiency

Shield Design

Shields are flat or formed pieces of sheet metal whose thickness is not less than four times the depth of penetration of the shield material at the particular frequency used in the induction heating application. Because of its thickness, the shield will absorb the total field to which it is exposed.

As discussed in Chapter 2, the depth of penetration is proportional to the square root of the electrical resistivity. Therefore, highly conductive materials (e.g., copper and aluminum) are used as shield materials even at low induction heating frequencies. At 10 kHz, copper sheet which is 2.5 mm (0.10 in.) thick is sufficient to act as a shield. At 450 kHz, copper as thin as 0.61 mm (0.024 in.) can be used reliably.

The power absorption into a shield from the electromagnetic field is a function of its permeability and resistivity. Copper, with a permeability of $\mu = 1$, will absorb approximately 30 to 50 times less power than mild steel in the same field. However, some power is always absorbed, and water cooling of the shield may be necessary under some circumstances.

Typical Applications of Shields

Much like flux robbers (Chapter 8), most shields form simple eddy-current loops or minor modifications thereof. Typical of these designs is the end plate shown in Fig. 9.5. It is used on a forging-line induction coil to prevent the fields of adjacent inductors from affecting each other. The plate is fabricated of nonmagnetic stainless steel. Design of the current path involves splitting of the closed loop and provision of a larger circuit with annular slots, thus increasing the resistance and decreasing the respective current. In this way, shield heating is greatly minimized.

Shields may also be used where a high-power-density field used for hardening might change the metallurgical structure of adjacent areas on a part. For example, a copper shield is often used to protect gear teeth while a boss is being tempered (Fig. 9.6). Similarly, in Fig. 9.7, a collar is used to protect the shoulder on a shaft while the part is hardened.

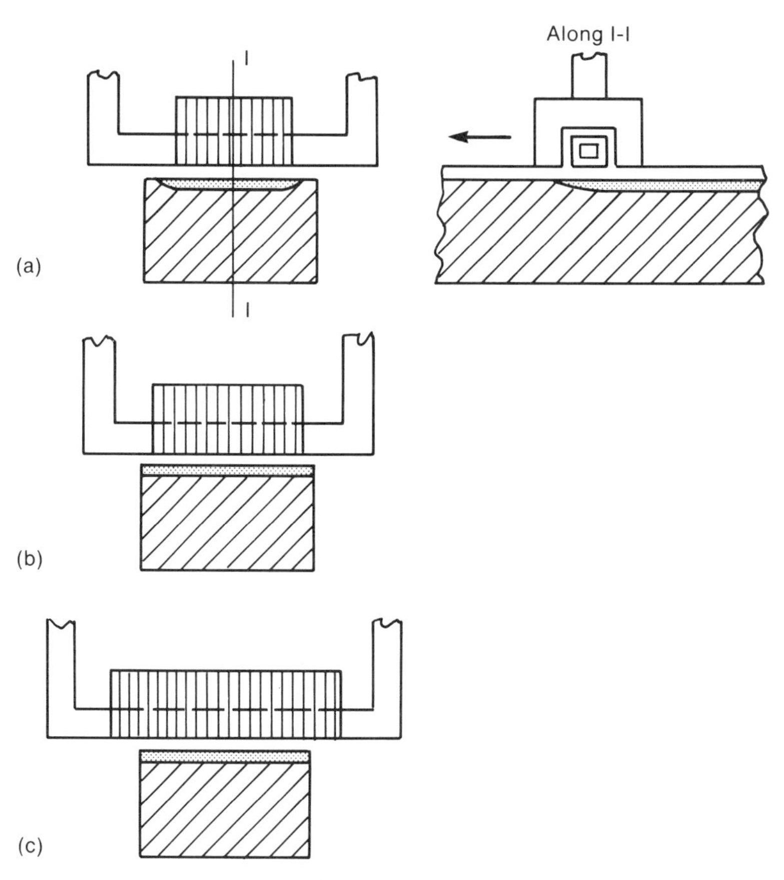

The short pattern in (a) can be widened, as in (b), by increasing the number of concentrator laminations. An even wider concentrator pattern (c) provides deeper heating at the edges of the part.

Fig. 9.4. Effect of flux-concentrator design on induction heating patterns (from M. G. Lozinskii, *Industrial Applications of Induction Heating*, Pergamon Press, London, 1969)

Shields provided for reducing the end-heating effect on large-diameter coils may also be of the split-current-loop design. In those instances in which the end turns of the coil cannot be short circuited to collapse the field (Chapter 8), a split shield may be separately mounted (Fig. 9.8). Coils and shields of this type are often used in large steel chambers for sintering in vacuum or atmosphere. In these circumstances, the flux field surrounding the coil may also heat the vessel. The same high-permeability laminations that are used for flux concentrators can be used to *prevent* the field from heating the vessel. Spaced at intervals around the coil, and running its full height, the laminated stacks (Fig. 9.9), here referred to as "flux diverters," provide a path that directs the field away from the wall of the vessel.

Shields for RF applications are somewhat different in their requirements. Because radio waves travel in straight lines, shields cannot be split but must fully encompass the current-carrying conductor. Aluminum or copper shields over transmission lines or high-impedance coils must be fabricated with either interlocking surfaces at right angles (Fig. 9.10) or tightly mating surfaces using conductive gaskets held together with many screws. In either case, the electrical contact must be very good.

SUSCEPTORS

A susceptor is a material which is heated as a result of its presence in the induction field, then passes its heat to the workpiece by conduction, convec-

Fig. 9.5. End plate on an induction coil used to heat forging preforms that acts as a shield to prevent interaction of the magnetic field with an adjacent coil (source: American Induction Heating Corp.)

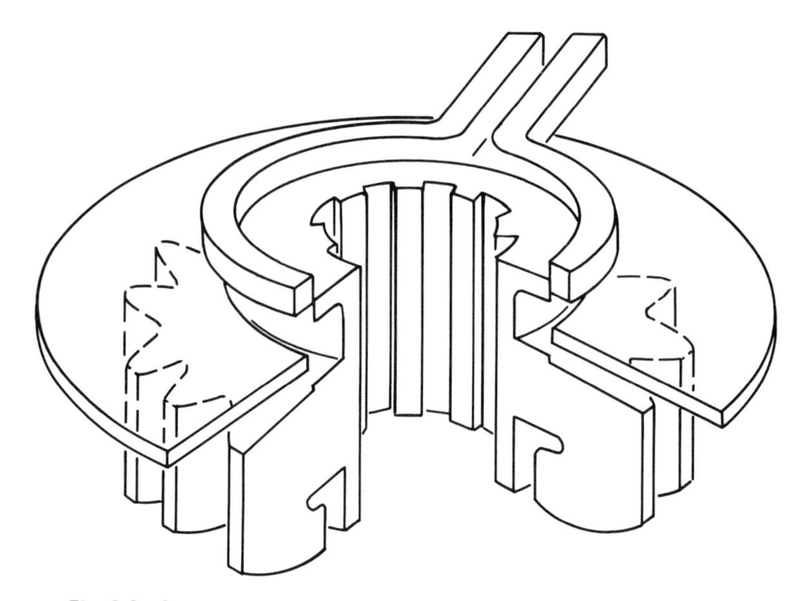

Fig. 9.6. Shield used to protect the hardened surface of a gear while the shoulder on the gear is being tempered (from M. G. Lozinskii, *Industrial Applications of Induction Heating*, Pergamon Press, London, 1969)

tion, or radiation. It is used where the workpiece cannot be heated to the desired temperature directly by induction. Because it is solely a source of heat and is not part of the process itself, a susceptor must exhibit certain properties:

1. It must be readily heated by induction to the process temperature.
2. It must be inert with regard to the process.
3. It must be readily formable or machinable.

Typical uses for susceptors at high temperatures occur in the semiconductor and fiber-optic industries where silicon, germanium, gallium arsenide, zirconia, etc. are heated to high temperatures. These materials have extremely high resistivities, almost to the point of being insulators at frequencies below 8 MHz. At extremely high temperatures and high frequencies, these materials can be induction heated, but they first must be brought to these temperatures.

Susceptor Materials

The most common susceptor material in induction heating applications is graphite. It has been used successfully as a susceptor material at temperatures up to 3000 °C (5430 °F). Having a fairly high resistivity, graphite presents an ideal load to the power supply. The material is readily machined (Fig. 9.11), and can be purchased in the form of a barrel, a cylinder, a boat, or other shape that is suitable for a specific application. Graphite also comes in a number of grades. The specific type for each application must be determined in

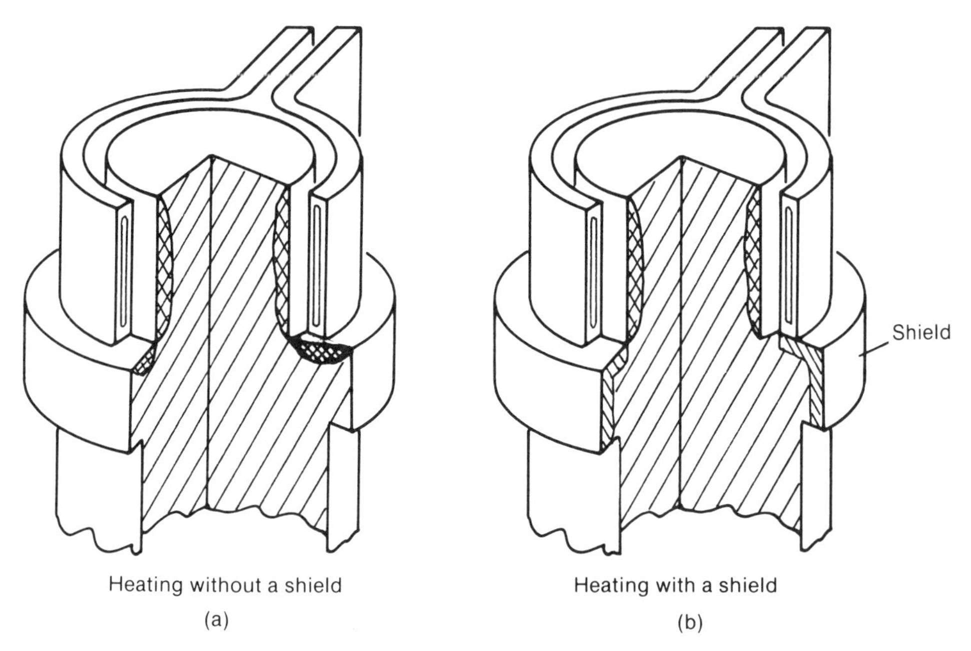

Fig. 9.7. Shield used to prevent heating of a shoulder on a shaft while the remainder of the shaft surface is being hardened (from M. G. Lozinskii, *Industrial Applications of Induction Heating*, Pergamon Press, London, 1969)

Fig. 9.8. Split shields used to reduce stray magnetic flux at the ends of a large induction coil; laminations on the vessel wall act as flux diverters (source: American Induction Heating Corp.)

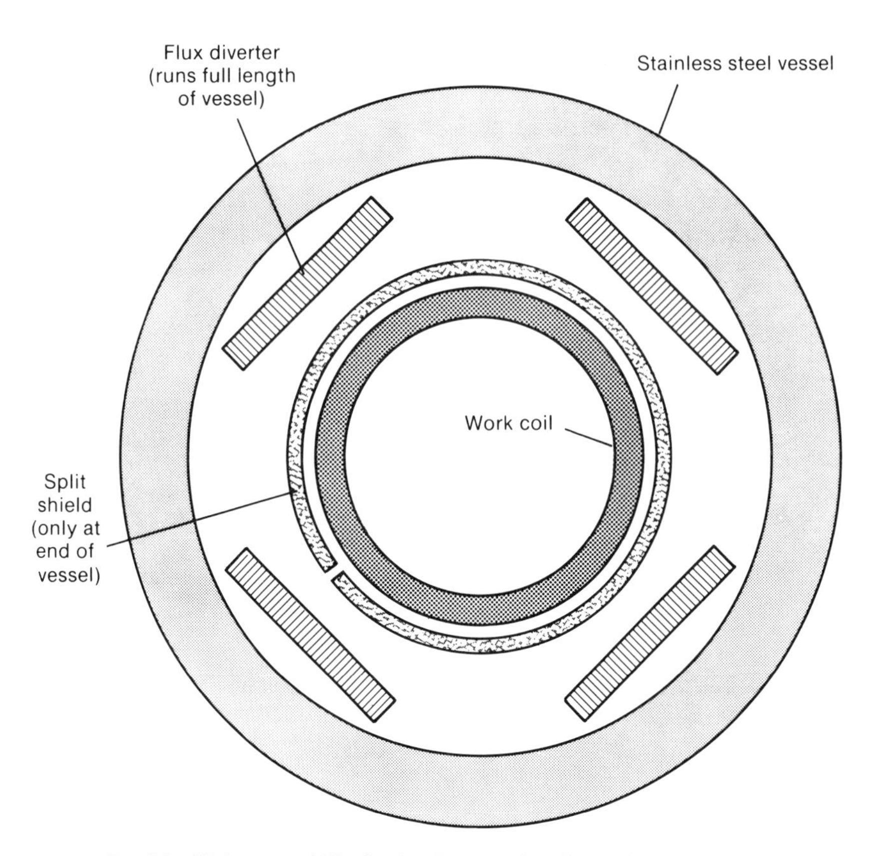

Fig. 9.9. High-permeability laminations used to divert the magnetic flux field and thus prevent stray heating within a stainless steel vessel

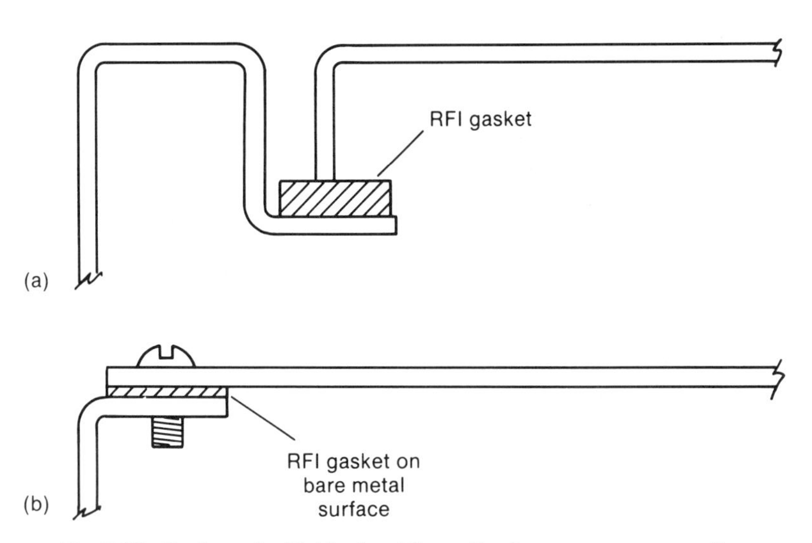

Fig. 9.10. Design of shields for RF applications to prevent radio-frequency interference (RFI). Joints or seams that interlock (a) are preferable to lap joints (b)

coordination with the supplier. In addition, there is a form of material called pyrolytic graphite, which, depending on the plane at which it is machined, acts as an insulator rather than a susceptor. Because it is a form of graphite, it will withstand the operating temperatures required of other graphite susceptors. In some semiconductor operations, a carbon-graphite susceptor is contained in a matching receptor (or cover) of pyrolytic graphite. This receptor acts as insulation to reduce the losses from the susceptor to the atmosphere.

Graphite oxidizes readily. If it is operated in an oxygen environment, it will deteriorate. For this reason, it is generally used in applications where air is not present, as in a vacuum or where a protective gas is used either as part of the process or specifically to protect the susceptor. In semiconductor processing, in which the susceptor must be acid etched occasionally to remove secondary materials, the graphite is protected by a silicon carbide coating.

In some operations, such as sintering of iron-base PM products, in which carbon can contaminate the workpiece, molybdenum is a common susceptor material. Although considerably more expensive than graphite, it has excellent electrical and mechanical properties for induction heating applications. It is also inert to most processes. Molybdenum is generally used in a hydrogen atmosphere and, in larger systems, can be used for sintering high-temperature materials such as tungsten or molybdenum.

Typical Applications of Susceptors

As mentioned above, susceptors may transfer the heat generated in them to the workpiece by processes such as conduction and radiation. When this heat transfer occurs by conduction, the susceptor and the workpiece may both form a part of the final product. For example, when steel tanks are coated with glass, the body of the vessel is heated by induction. This in turn passes its heat to the glass frit, which then flows to provide the final glass coating.

Fig. 9.11. Range of sizes and shapes of graphite susceptors used in induction heating applications (source: GTE Products Corp.)

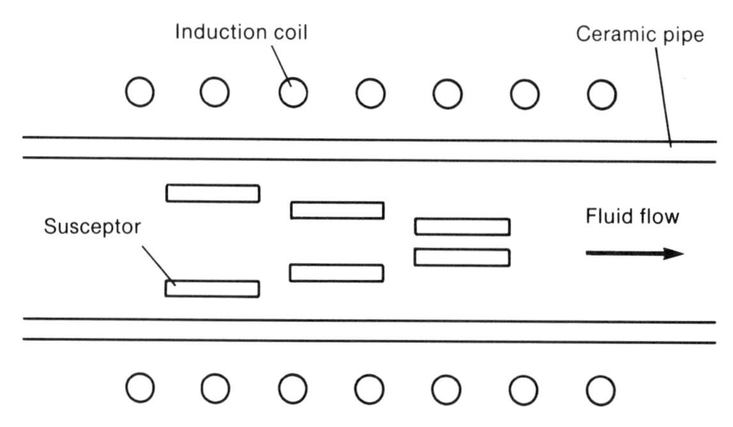

Fig. 9.12. Illustration of the application of susceptors for heating electrically nonconductive fluids flowing through ceramic or glass pipes. The susceptors are heated by induction, and the fluid by conduction of heat from the susceptors

In chemical processing, conductive piping can be used as a susceptor to transfer heat directly to the process fluid. Generally these are low-frequency systems in which the pipe is heated with line-frequency coils wrapped in a low-temperature insulation. Air cooling may be sufficient for the inductor in these cases. In some instances, it is necessary to heat fluids passing through pipes of nonconducting materials such as glass or ceramics. A susceptor of suitable material, coated to prevent interaction with the process, may be suspended in the fluid path (Fig. 9.12). This in turn is heated by an inductor on the outside of the pipe, transferring its heat to the process material by conduction.

Susceptors need not have high resistivity if they meet the other requirements of the process. Accordingly, aluminum foil is the basic susceptor material in packaging where it is used to transfer its heat to a plastic coating such as polyethylene for making seals.

Although the susceptors discussed above essentially transfer heat by direct conduction to the workpiece, some applications use radiation to provide a noncontact method of heating for special processes. In one instance, high temperatures are used in the continuous production of carbon fibers. The material itself cannot be heated directly by induction due to its small diameter. The filament is passed through a graphite tube, which is induction heated. The filament then, in effect, is passing through a high-temperature muffle, and is heated by radiation from the interior walls of the tube.

A final application involves the manufacture of optical fibers from quartz using a zirconia susceptor. The dependence of the resistivity of zirconia on temperature is such that it cannot be induction heated efficiently below 870 °C (1600 °F). Hence, it is radiation heated to this temperature by an auxiliary graphite susceptor, after which it is heated directly using a radio-frequency (3-MHz) induction power supply.

Chapter 10

Materials Handling

Because of its speed and ease of control, induction heating is a process which is readily automated. It can be integrated with other processing steps such as forming, quenching, and joining. Hence, the past several decades have seen the development and marketing of a large number of completely automated heating/handling/control systems by induction equipment manufacturers. Power supplies, coil design, and controls for these systems have been covered in previous chapters. This chapter deals with materials handling and automation. Basic considerations such as generic system designs, fixture materials, and special electrical problems to be avoided are summarized first. This discussion is followed by a description of typical materials-handling concepts and illustrations of specific applications.

BASIC CONSIDERATIONS IN MATERIALS HANDLING*

Part Movement Through Induction Coil

The design of the coil and workpiece often suggests methods of material handling. For example, it is easy to visualize continuous motion of parts through, into, or under solenoid, pancake, or channel coils (Fig. 10.1). Similarly, indexing arrangements involving either movement of a discrete part into a fixed coil or movement of the induction coil about a stationary part may be used (Fig. 10.2). Stationary coils frequently allow shorter electrical leads from the power supply, accommodate coaxial leads, and are therefore preferred. However, part geometry or other design considerations may necessitate coil motion (e.g., the arrangement in Fig. 10.2b). The movements that are required in either type of design (part moving or coil moving) are generally short. They are normally implemented using simple hydraulic or pneumatic systems actuated electrically. The hydraulic type is thought to be more reli-

*H. U. Erston and J. F. Libsch, *Lepel Review*, Vol 1, No. 16, p 1.

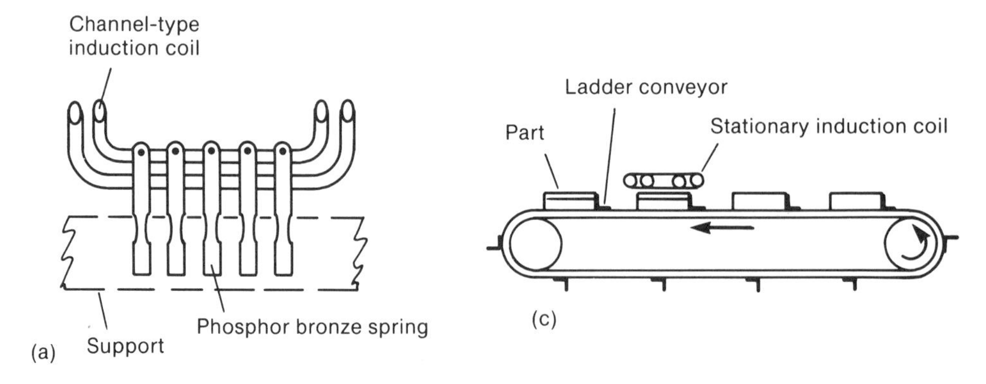

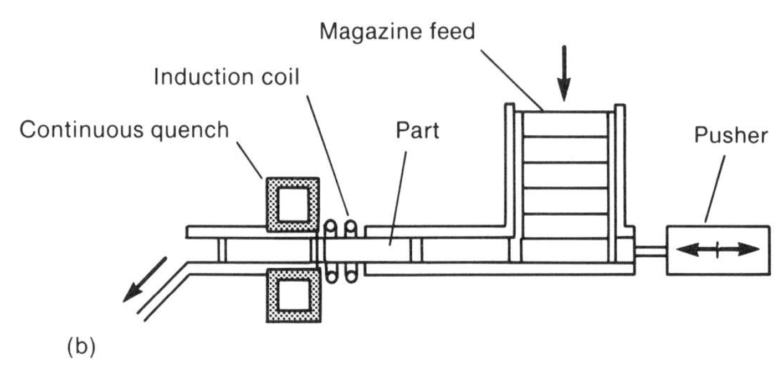

(a) Continuous movement through a channel-type coil. (b) Periodic movement through a solenoid coil. (c) Continuous movement under a pancake coil.

Fig. 10.1. Material-handling arrangements involving a fixed coil and moving parts (from H. U. Erston and J. F. Libsch, *Lepel Review*, Vol 1, No. 16, p 1)

able because the quality and quantity of the air supply used to power pneumatic drives may not be consistent.

Frequently, automatic handling involves another important requirement, namely the delivery of the part in the correct position at the correct instant in time. Among the delivery techniques are automatic conveyors with ingenious pickups, magazine feeds, and rotary tables. Tooling for a specific application can often be constructed about a standard unit using conventional movements such as:

- Geneva driver. This type of indexing machine, electrically or air operated, is most useful for heavy workpieces requiring large-diameter rotary stations.
- Barrel-cam driver. This is a high-speed, motor-driven indexing device that has nearly zero blacklash. It has better accuracy and acceleration characteristics than the Geneva driver. If the required heating time is to be longer than the dwell time of the constantly turning indexing cam, an electromagnetic clutch should be used in conjunction with the programmer.
- Air-operated ratcheting-type indexing turntable. Positive stopping of this kind of device can be improved by equipping it with a hydraulic damper.

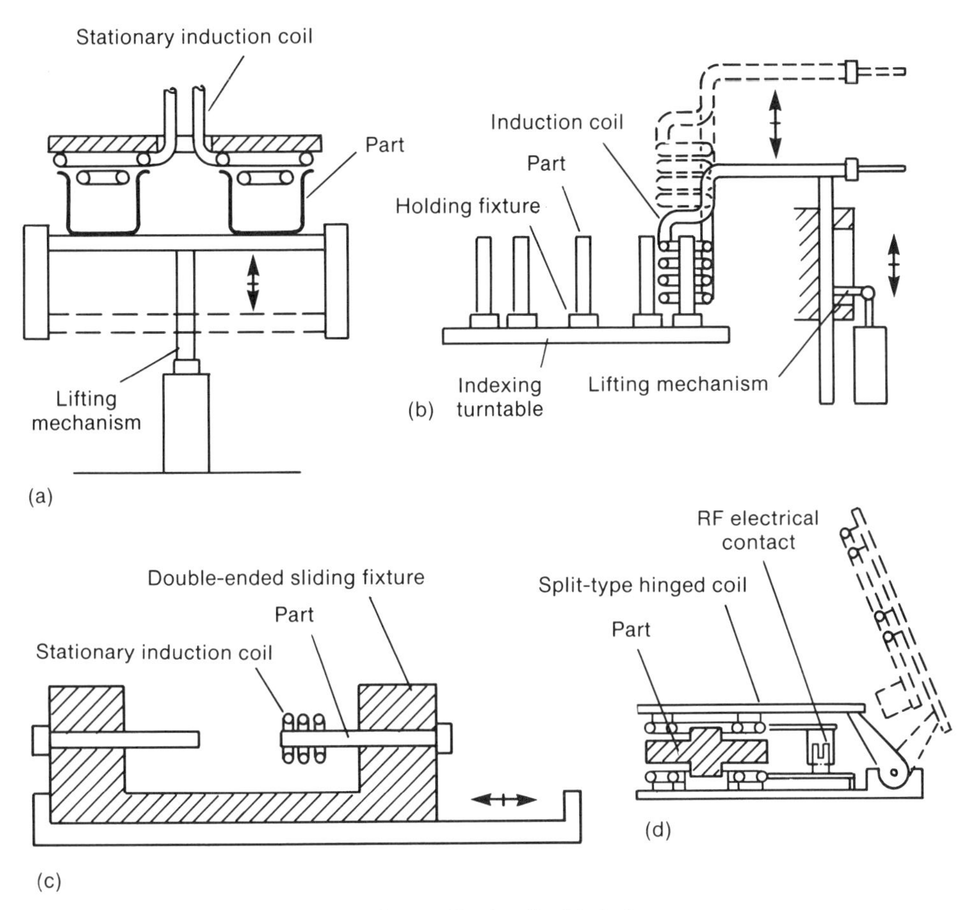

(a) Part movement into a fixed coil. (b) Coil movement over parts on an indexing turntable. (c) Double-ended sliding fixture. (d) Split-type hinged coil.

Fig. 10.2. Indexing arrangements for handling-fixture design (from H. U. Erston and J. F. Libsch, *Lepel Review*, Vol 1, No. 16, p 1)

Materials for Handling Fixtures

Automation of induction heating processes usually involves continuous feeding of parts into or through the induction coil and auxiliary equipment. Typically, this involves handling of individual pieces in some sort of fixture. Because the electromagnetic field extends beyond the coil, fixtures made of electrically conducting materials (e.g., steel) may be heated also, leading to a costly loss of energy. For this reason, fixture materials such as aluminum, which heats less readily and which is easy to cool, and nonconducting materials are frequently used. Some typical selections for these types of applications are listed in Table 10.1.

In some induction heating applications, such as end heating of bars, parts can be held in place far from the coil, thereby permitting irons and steels to be used in fixturing. On the other hand, in operations such as soldering and brazing, bonding of the fixture to the workpiece as well as the corrosive and cumulative effects of fluxes must be considered. Aluminum, titanium, and some nonmetallic materials have been found useful in these instances. In addition, when the fixture must precisely locate the components

Table 10.1. Structural materials for induction heating fixtures (from H. U. Erston and J. F. Libsch, *Lepel Review*, Vol 1, No. 16, p 1)

Material	Characteristics and comments
Nonmetals	
Diamonite(a)	Aluminum oxide; very hard, dimensionally stable; standard and special shapes; resists high temperatures
Epoxy Fibre Glass FF91(b)	Coil-mounting supports; good electrically; also useful for soft soldering to about 230 °C (450 °F)
Transite II(c)	Work-table tops; heat resistant; avoid for coil supports (electrically poor)
Mycalex, Supramica(d)	Various grades; useful for high-temperature coil supports; special shapes available
Silicone rubber (RTV)(e)	Molds accurately; flexible; useful for soft soldering to 315 °C (600 °F)
Fired Lava(f)	Easily machinable before firing; good heat resistance; less strength than Diamonite
Nonmagnetic Metals(g)	
Aluminum alloys	Fixture base plates, work tables; useful for soft soldering (does not bond)
Brass (free machining)	Supporting screws, locators adjacent to coils, sinks, etc; corrosion-resistant pins; quench and recirculating tanks
Titanium	Useful for positioning of parts to be silver brazed; does not stick
Nichrome	Excellent high-temperature strength; oxidation resistant; used for locating and holding of parts
Inconel alloys	Useful for locating parts, radiant (susceptor) heating of thin materials
Magnetic Metals(g)	
Low-carbon steel	Structural members; cabinets; work tables
Alloy steel (hardenable)	Moving parts of fixtures subject to wear
Stainless steel (hardenable)	Moving parts of fixtures subject to wear and corrosion

(a) Diamonite Products Div., U.S. Ceramic Tile Co. (b) Formica Div., American Cyanamid Co. (c) John-Manville Co. (d) Mycalex Corp. of America. (e) Dow-Corning Co. (f) American Lava Corp. (g) Materials listed are available from a number of manufacturers.

of the assembly to be joined, careful attention should be given to the thermal expansion coefficients of the workpiece and fixture.

The induced heating of tooling near the coil, which may not necessarily be a part of the fixturing, should also be taken into account. Inadvertent coupling between the field of the coil and a conductive loop in the tooling is a common error. It can be avoided by breaking the conductive path with insulation at a convenient point in the tooling.

Use of Controlled Atmospheres or Vacuum*

Despite the rapid heating associated with induction methods, it is sometimes necessary to utilize a controlled atmosphere, be it inert, reducing, oxidizing, or vacuum. Applications requiring such precautions include vacuum melting of special alloys, brazing of stainless steels, and thermal processing of silicon and germanium in the semiconductor industry.

When the parts to be heated are relatively small and few in number, they may be processed inside a bell-jar fixture. If a reducing atmosphere such as hydrogen, which is explosive, is used, the bottom of the bell jar is left a short distance above the work table. Under normal operating conditions, this allows slow escape of excess gas. Furthermore, if the gas is fed to the top of the fixture, it may be lifted to remove a finished workpiece and to put a new one in its place without loss of atmosphere, particularly when the atmosphere is composed of a light gas such as helium or hydrogen.

Because induction heating is frequently localized, other simple methods of providing a controlled atmosphere in the heated area are often satisfactory. For example, a quartz or refractory tube can be used to maintain an atmosphere. The arrangements shown in Fig. 10.3 enable continuous movement of parts to be heated under controlled atmosphere through the induction coil. In Fig. 10.3(a), brass parts are brazed in a forming-gas atmosphere; the components, properly located in ceramic holders on a turntable, pass through the channel coil and are cooled in the atmosphere before exiting from the chamber. An elevator-pushrod arrangement for moving parts continuously through a controlled atmosphere is depicted in Fig. 10.3(b). Lighter-than-air gases such as helium or hydrogen are introduced at one elevator and flow throughout the elevator–ceramic tube system. With suitable inversion of the fixture, heavy gases may also be used.

Sometimes the coil and/or holding fixture itself can be used to introduce or contain the controlled atmosphere, as shown in Fig. 10.4. While these arrangements may not provide perfectly controlled atmospheres, they are sufficient for many applications.

When a controlled atmosphere is used, the amount of gas required is typically very small. However, the cooling effect of the gas should not be over-

---

*Anon., *Lepel Review*, Vol 1, No. 10, p 3.

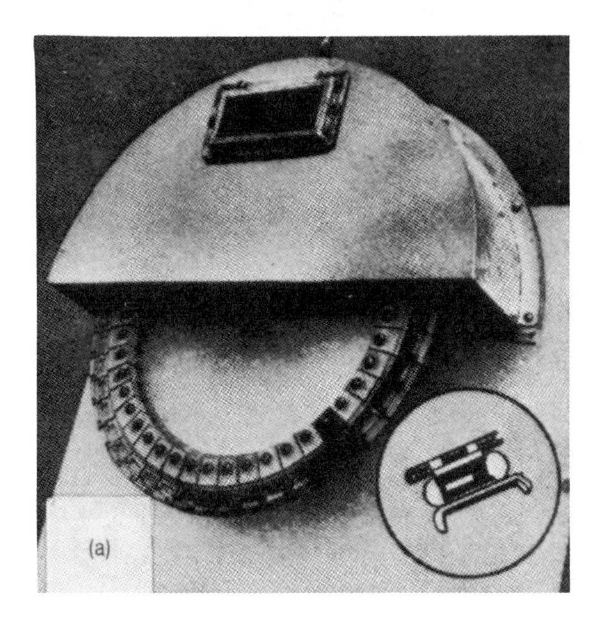

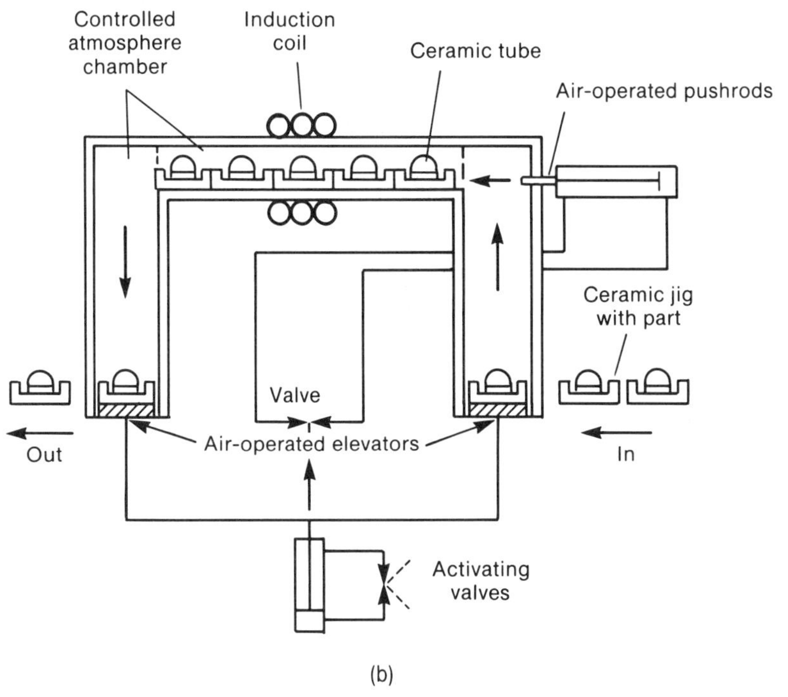

(b)

(a) Brazing of brass watchband clips without flux in a forming-gas atmosphere. (b) Fixture using an elevator-pushrod arrangement.

Fig. 10.3. Methods for continuous movement of parts through a controlled atmosphere during induction heating (from Anon., *Lepel Review*, Vol 1, No. 10, p 1)

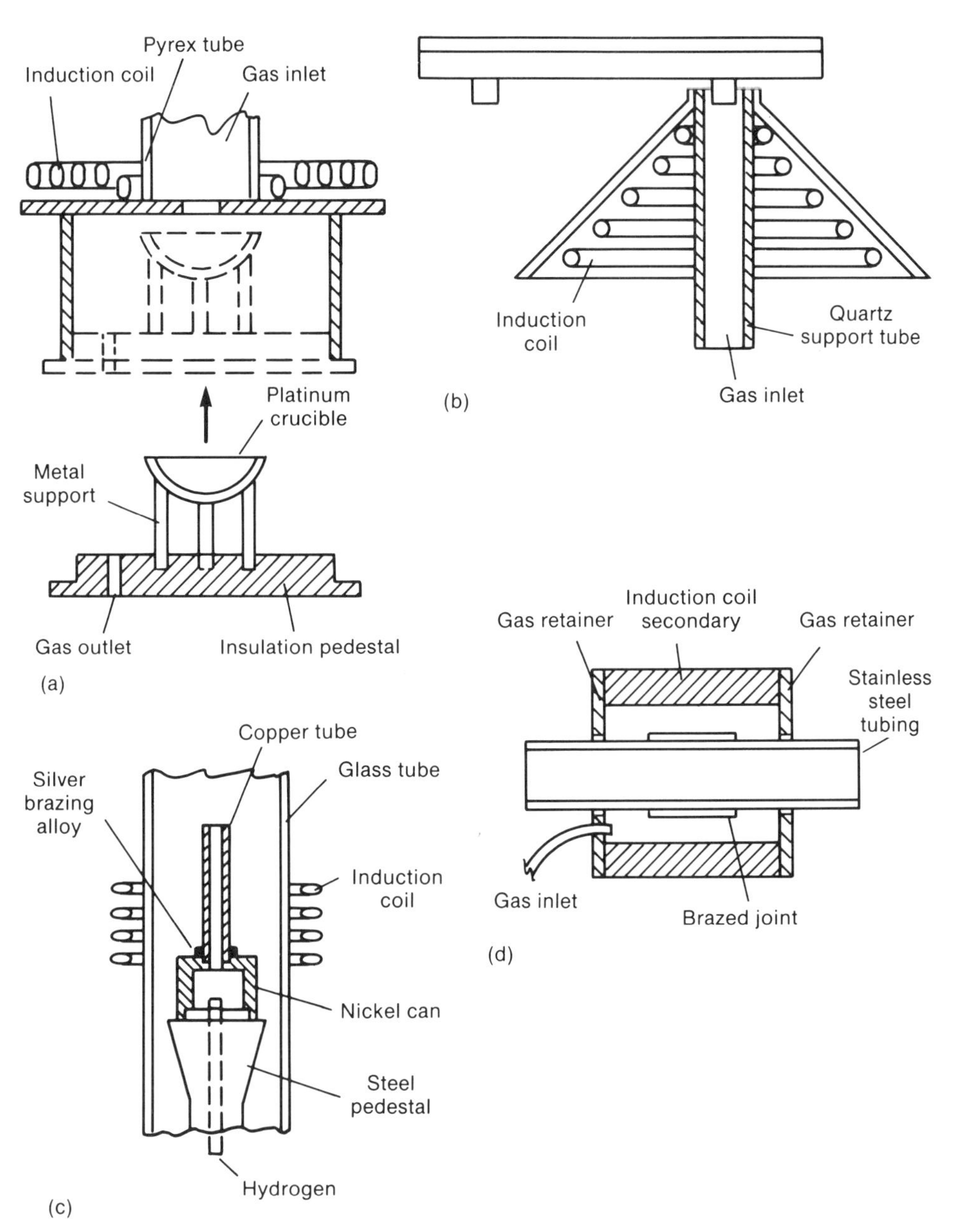

(a) Coil and fixture forming a chamber for an argon atmosphere. (b) Quartz tube–conical coil arrangement for providing a localized protective atmosphere. (c) Atmosphere introduced through a support fixture. (d) Single-turn induction coil and retainer plates providing an atmosphere chamber for brazed joints.

Fig. 10.4. Adaptations of coils and holding fixtures for introduction or containment of controlled atmospheres (from Anon., *Lepel Review*, Vol 1, No. 10, p 1)

looked. Often, this necessitates an increase in power to maintain the same production speed as that employed under an ambient (air) atmosphere.

MATERIALS HANDLING IN INDUCTION BILLET AND BAR HEATING

Handling of stock for induction heating prior to forging, extrusion, and other fabrication processes usually involves equipment for unloading of cut preforms, for feeding of the stock through the induction coil or coils, and finally for discharging of the heated material into the fabrication area. Typical devices for each operation are shown in Fig. 10.5. They include the following:

- Loading: magazines, hopper feeders, bundle unscramblers, and random-loaded hopper feeders
- Feeding: dual opposed chain, hydraulic or pneumatic charge cylinders, walking beams, and drive rolls
- Discharging: extractor forks, power rolls, vee rolls, gravity chutes, and robots.

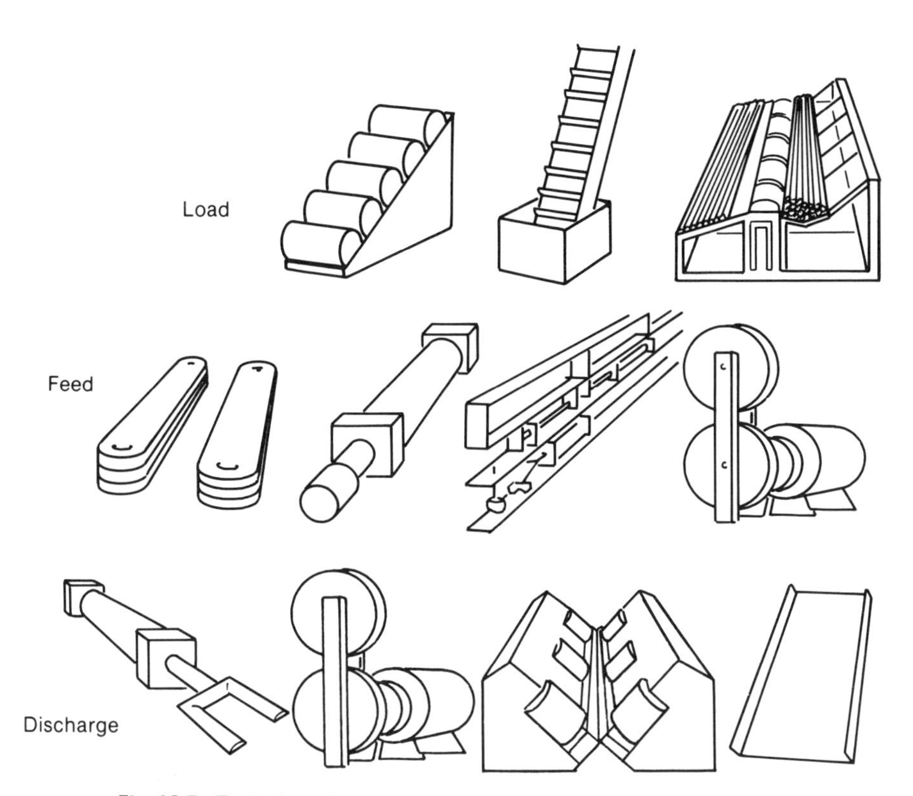

Fig. 10.5. Typical methods of material handling in automated induction heating operations (source: TOCCO Brochure DB-2032-8-80, Ferro Corp.)

In most cases, the over-all handling system must be flexible enough for heating of billets or bars with greatly differing transverse dimensions. In the case of circular billets, for example, it is usually required that one unit be capable of heating stock whose diameter varies by a factor of two to three. At the same time, it must be capable of heating billets of various lengths.

Feed Mechanisms

Loading and discharging equipment is usually selected on the basis of billet or bar geometry. In contrast, a number of different mechanisms can be used to perform the same feeding task. These include continuous, walking-beam, and billet/bar end heaters.

As their name implies, continuous feed devices (Fig. 10.6) are used to heat the entire length of a series of bars or billets. The stock is pushed through a series of coils by variable-speed pinch rolls. Semiautomatic and fully automatic systems are also available. With semiautomatic feeds, bars or billets are pushed from a manually loaded magazine, for instance, by a hydraulic or pneumatic cylinder. They are pushed end-to-end through the induction coils

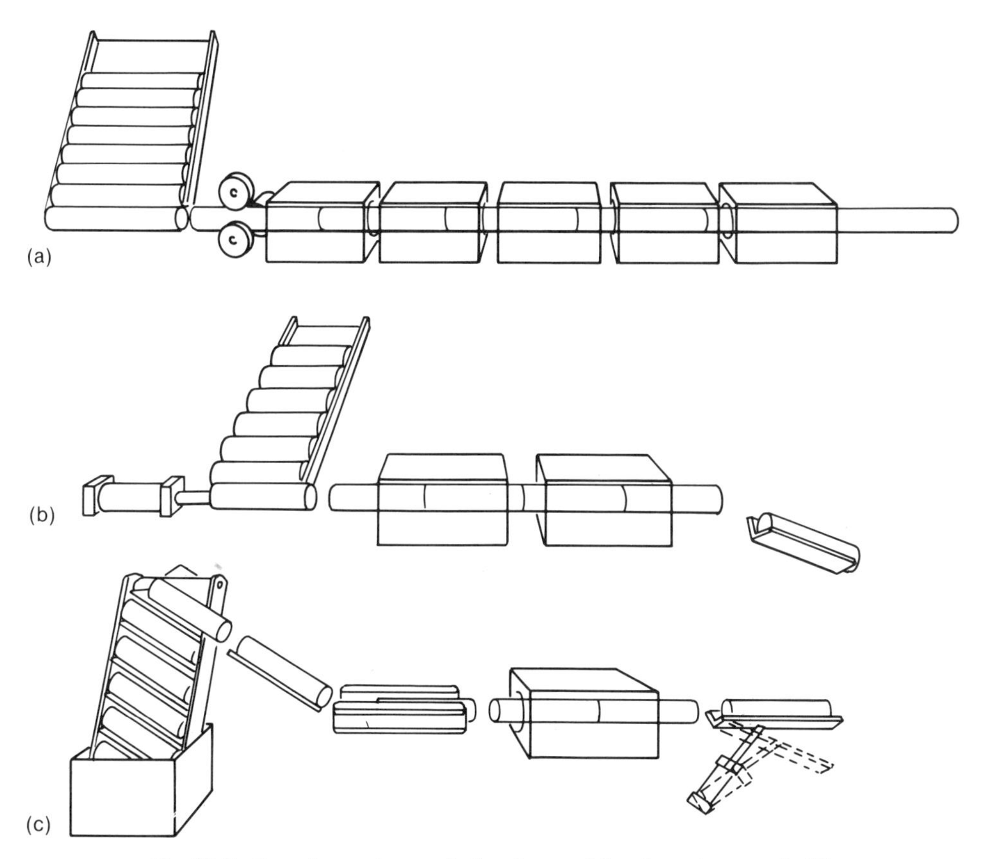

Fig. 10.6. Generic continuous in-line heater (a) and semiautomatic (b) and fully automatic (c) versions of it (source: Inductoheat)

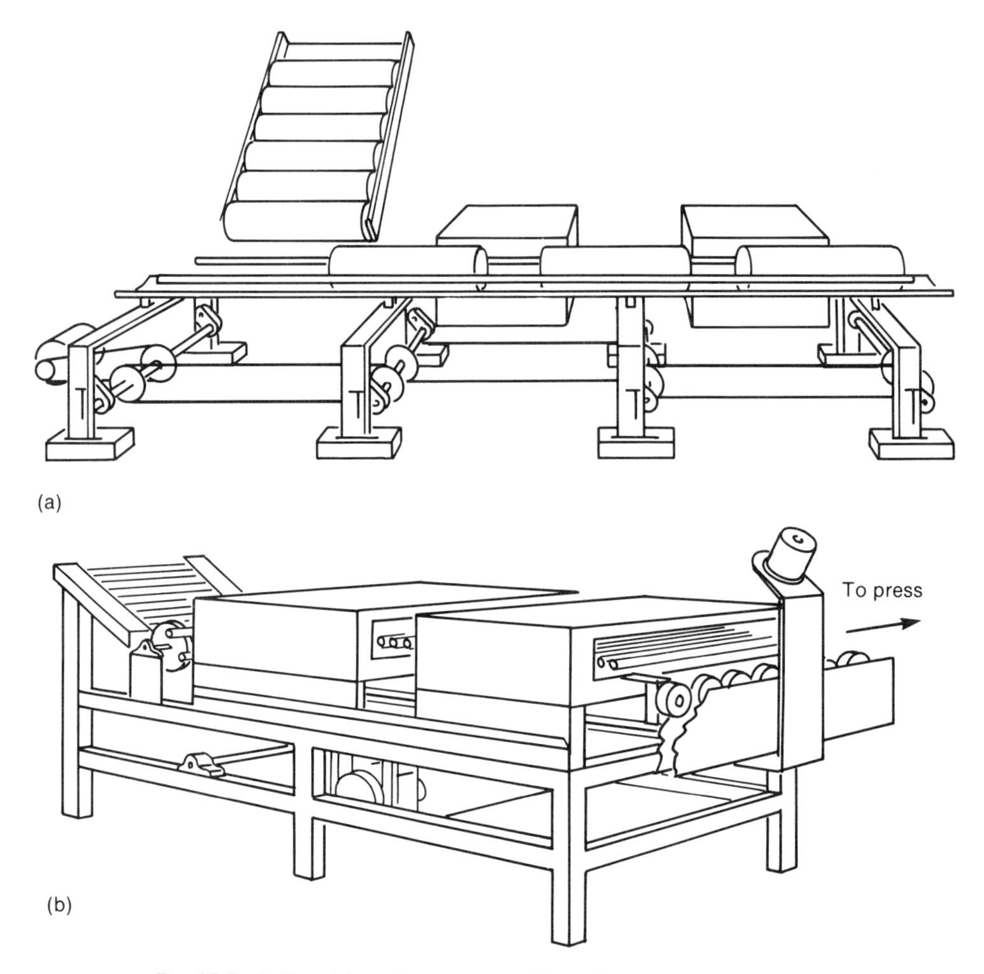

Fig. 10.7. In-line (a) and transverse (b) walking-beam feed devices (sources: Inductoheat, Westinghouse Electric Corp.)

and, upon exiting, move to the forging, extrusion, or other station under the force of gravity. In a fully automatic system, loading and unloading of workpieces is done automatically.

Walking-beam feeding of bars or billets (Fig. 10.7) is accomplished without touching of consecutive pieces. At the loading end, individual workpieces are placed on a loading vee by a hydraulic pusher, magazine, or other means. A second device then loads the billet or bar onto the walking beam, where it is "walked" through the line by a series of lifting, horizontal transfer, and lowering motions into consecutive vees in the guidance system. Each motion is actuated through a dependable mechanical-linkage system. The last billet comes through by itself, eliminating shutdown scrap. Walking-beam feeds are of two basic designs. In an in-line heater (Fig. 10.7a), the bar or billet axis is parallel to the feed direction. With a transverse walking-beam heater (Fig. 10.7b), the material feed is transverse to the axis.

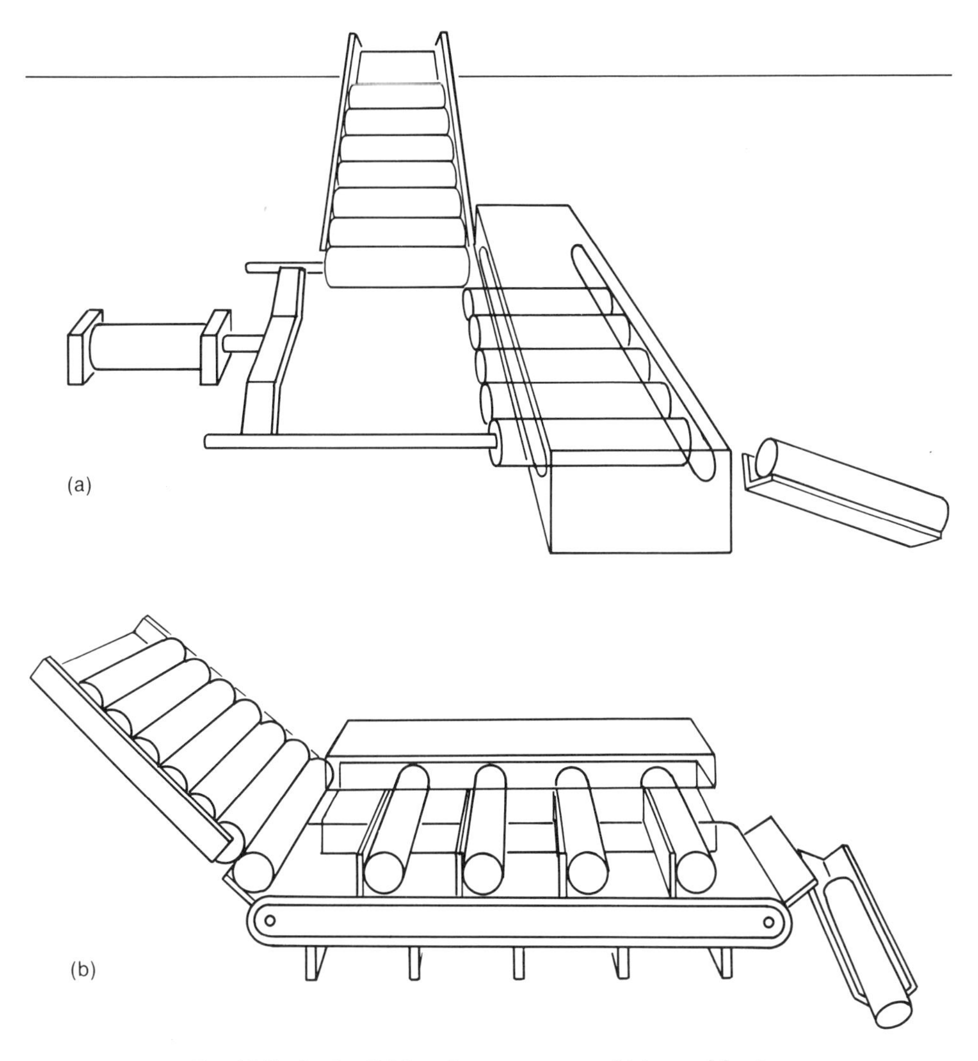

Fig. 10.8. Oval-coil (a) and conveyor-type (b) bar-end heaters

Feed mechanisms of the third type—those used for applications involving heating of bar ends—come in a number of forms. The two most common are the oval-coil feeder and the conveyor-type bar-end heater. Loaded from a magazine, bars in an oval-coil feed machine (Fig. 10.8a) are pushed one at a time into the upper end of an inclined-oval inductor. They roll down as previously loaded bars reach the aim temperature and are pushed out by another device to the forming machine. The zig-zag motion that the bar follows gives rise to an alternative designation for this device—a Z feeder. Conveyor-type bar-end heaters (Fig. 10.8b) make use of a channel induction coil. Magazine-loaded bars are removed by carriers onto a belt. Upon moving through the coil, the end is progressively heated to the required forming temperature.

Feed-Mechanism Selection Criteria

Each of the various feed mechanisms has advantages and disadvantages that must be considered during system design. The continuous systems offer the following advantages:

- The mechanical feed device is very simple whether it is driven hydraulically or by compressed air.
- The rate at which workpieces leave the coil line can be easily changed by regulating a timer.
- The equipment is inexpensive.

On the other hand, continuous lines suffer from the problem of billets sticking together. Furthermore, heating along the axis of the bar or billet may be nonuniform if its length is not a multiple of the active length of the inductor. Caution must also be taken when billets become shorter, lines longer, and peak temperatures higher; under these conditions, the tendency for line buckling increases, and consideration should be given to walking-beam feed mechanisms which totally eliminate problems of buckling and sticking together of parts.

Other advantages of the walking-beam method include reduced maintenance and wear of the coil refractory lining and individual support/guide rails. However, it has two basic drawbacks. First, the inside diameter of the inductor must be greater than that required by other systems with the result that the efficiency decreases. Secondly, the over-all cost of the feed system is considerably greater because of its complexity and because of the extra guides needed along the transport path.

One other design feature that must be considered with regard to feed equipment is related to the feed rolls, guide rails, or supports that may hold the workpieces in place. Feed rolls are used for heating long bars. When guide rails are used (such as for very short pieces), either two or four are usually required (Fig. 10.9), depending on the cross-sectional geometry of the material being heated. These rails are usually constructed of steel tubing which is water cooled, although high-temperature superalloy and nonmetallic (refractory) guides, which are not water cooled, are also in wide use. The number of guides should be kept to a minimum because they tend to absorb part of the thermal energy developed in the system through radiation, conduction, and convection, as well as energy due to induced eddy currents if they are metallic. Keeping the diameter of the guide rails as small as possible helps in minimizing such energy losses.

MATERIALS HANDLING IN INDUCTION HEAT TREATMENT

Handling of materials during induction heat treatment, be it hardening, tempering, or another process, depends to a great extent on part shape. How-

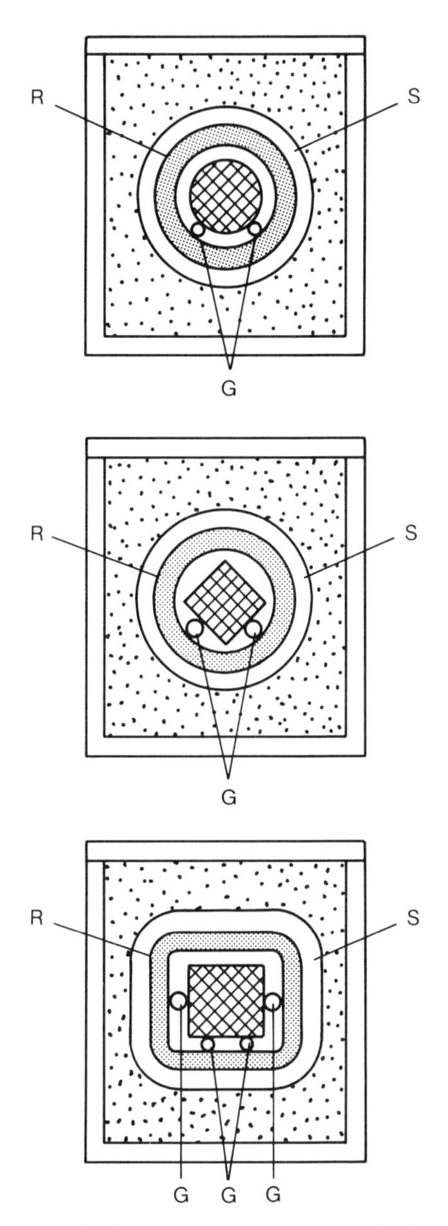

S–Induction coil. R–Refractory insulation. G–Guide rail.

Fig. 10.9. Cross sections of common induction coil–guide rail arrangements used in bar and billet handling (from C. Dipieri, Electrowarme Int. B, Vol 38, No. 1, February, 1980, p 22)

ever, handling devices can be broken into two major categories: those designed for continuous parts of regular cross section, and those for discrete parts which typically preclude continuous in-line feeding. Several typical handling schemes are discussed next and illustrated with examples. It should be borne in mind, however, that the kinds of equipment for induction heat treatment are innumerable. New fixtures are constantly designed to fit new and unique applications.

Continuous Heat Treatment

For parts of regular cross section such as round bars and tubes, materials-handling systems very similar to those described above for induction heating prior to hot working are employed. These systems include loading, feeding, and other equipment. The only major exceptions pertain to added stations needed for quenching or air cooling or to integration of austenitizing and quenching stations, for instance. Several examples discussed below illustrate typical setups of continuous induction heat treatment lines.

Continuous Heat Treatment of Bars and Shafts. Because of high-volume production, the continuous heat treatment of automotive parts such as shafts is readily carried out by induction heating processes. One such line, developed by Westinghouse Electric Corporation,* is shown schematically in Fig. 10.10. It includes an automatic handling system, programmable controls, and fiber-optics sensors. Mechanically, parts are handled by a quadruple-head, skewed-drive roller system, abbreviated QHD, after being delivered to the heat treatment area by a conveyor system. The roller drives, in conjunction with the chuck guides, impart rotational and linear motion to the incoming workpiece. Once a part enters the system, the fiber-optics sensor senses its position and initiates the heating cycle for austenitization. This sensor is also capable of determining if the operation is not proceeding normally (e.g., if the part is not being fed properly) and can automatically shut down the system.

In the hardening cycle of the QHD system, the induction generator frequency is generally either in the radio-frequency range (approximately 500,000 Hz) for shallow cases or about 3000 to 10,000 Hz for deeper cases. In either instance, a temperature controller automatically senses if the part has been heated to too high or to too low a temperature in order to prevent an improperly austenitized piece from passing through the system. Assuming that the part has been heated properly, it then passes through a quench ring which cools it to a temperature of 95 °C (200 °F) to form a martensitic case prior to its movement into the tempering part of the heat treatment line. Again, a fiber-optic system senses the presence of the part and begins the heating cycle using low-frequency current generally around 3000 Hz, because the desired tempering temperature is approximately 400 °C (750 °F), or a temperature at which the steel still has a large magnetic permeability. Once again, the part is automatically heated, quenched, and moved from the heat treatment station, this time onto a conveyor which takes it to the machining area for grinding.

---

*E. Balogh, *Heat Treating*, Vol 14, No. 9, 1982, p 33.

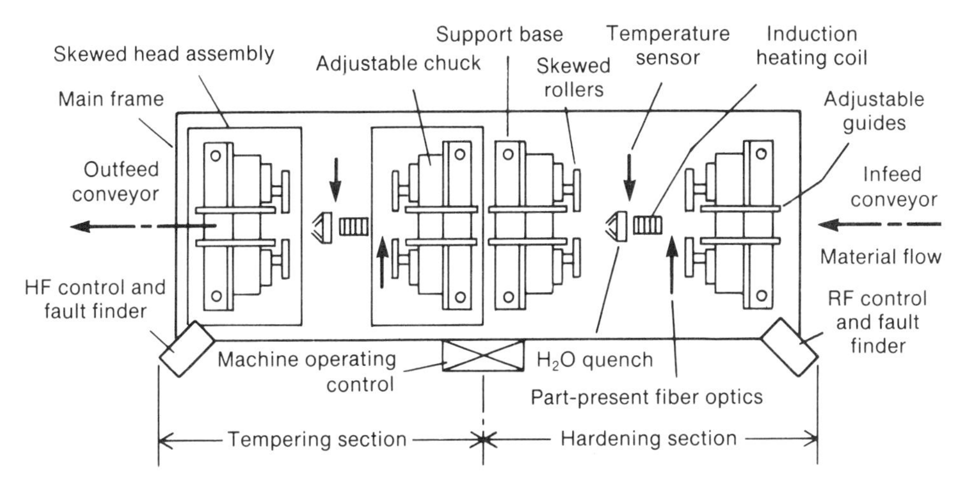

Fig. 10.10. Automated, quadruple-head, skewed-drive roller system used for in-line induction hardening and tempering of automotive parts (from E. Balogh, *Heat Treating*, Vol 14, No. 9, 1982, p 33)

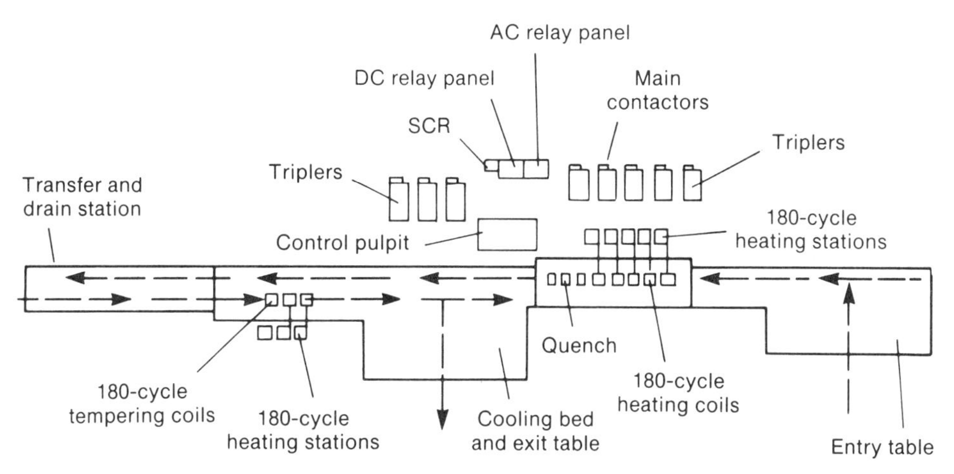

Pipe enters from the right, is austenitized, quenched, drained, and tempered. Following tempering, the pipe is transferred to cooling beds for air cooling.

Fig. 10.11. Schematic illustration of equipment used for in-line induction through hardening and tempering of pipe-mill products (from R. M. Storey, *Metal Progress*, Vol 101, No. 4, April, 1972, p 95)

Continuous Heat Treatment of Line Pipe. Probably the largest application of induction through hardening and tempering is for piping or tubular goods used for oil wells and gas pipelines. In a typical installation, such as that at Lone Star Steel* (Fig. 10.11), processing is conducted on a continuous line on which the steel is austenitized, quenched, tempered, and finally cooled to room temper-

*R. M. Storey, *Metal Progress*, Vol 101, No. 4, April, 1972, p 95.

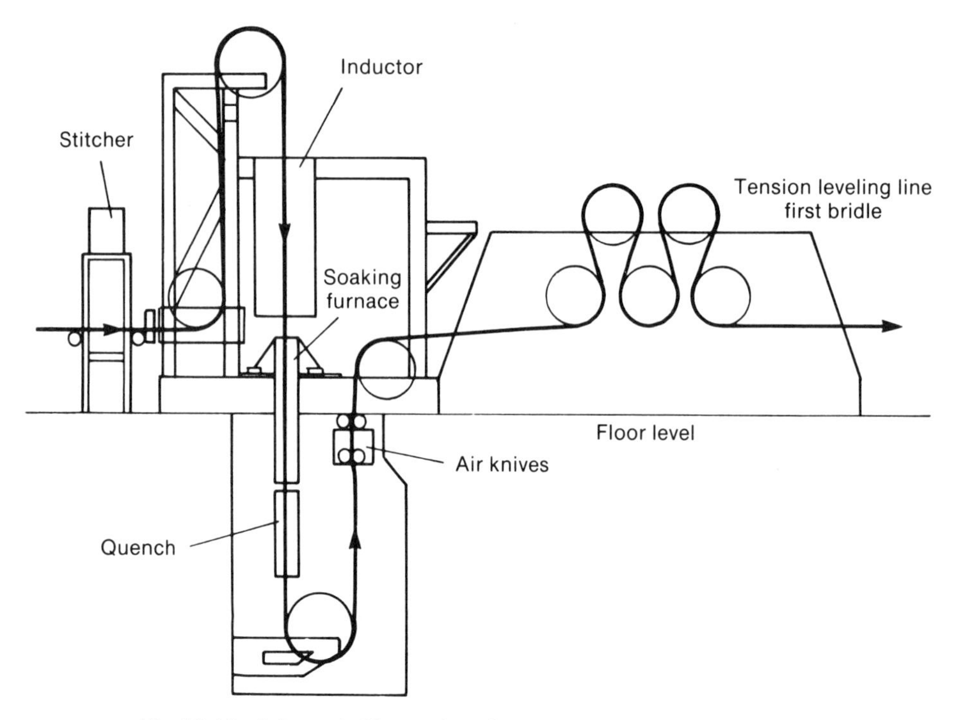

Fig. 10.12. Schematic illustration of a transverse-flux induction heat treatment line used for processing of aluminum alloy sheet (from R. Waggott, *et al.*, *Metals Technology*, Vol 9, December, 1982, p 493)

ature at successive stations. In this system, pipes are loaded onto an entry table and fed onto the conveyor as soon as the heat treatment of the pipe preceding it is completed. As each pipe passes through the austenitizing station (consisting of five coils), it is rotated on skewed rollers to ensure temperature uniformity. Also, because only a small portion of the pipe is heated at one time, distortion is readily controlled.

After austenitizing, the pipe enters the quench ring several feet down the line. Upon drawing the water off, it then moves to the tempering station. Following tempering, the pipe continues along the conveyor to cooling beds. It is rotated during the entire cooling cycle to ensure straightness and lack of ovality.

Continuous Heat Treatment of Sheet Metal. High-speed continuous heat treatment of thin sheet metal is also possible using induction heating techniques. It is most economical with a technique known as transverse-flux induction heating. In this method, eddy currents circulating in the plane of the sheet are set up by a magnetic flux field perpendicular to the sheet. A typical sheet-handling system* for carrying out the processing is shown in Fig. 10.12. Coiled sheet

*R. Waggott, *et al.*, *Metals Technology*, Vol 9, December, 1982, p 493.

stock is fed from a payoff reel through a stitcher, which allows continuous processing of consecutive coils without line shutdown. The sheet is then guided through the inductor at a constant speed by a set of drive rolls. The inductor is mounted on wheels to allow automatic alignment of the sheet within it. The drive rolls also provide front and back tension to help maintain sheet flatness. After being heated, the sheet is quenched and then dried by a set of air knives. Upon exiting the line, the material may be coiled, fed into a leveling line, or routed to a subsequent processing station for coating or some other operation. Because the gage may vary, and to account for small variations in sheet speed, the line is also fitted with thickness and speed monitoring devices whose signals are fed into a controller which adjusts the inductor power to maintain the desired temperature.

Heat Treatment of Discrete Parts

Parts which are irregularly shaped or which are selectively heat treated usually cannot be processed on a continuous basis. In these cases, special fixtures or handling techniques are required. Some of the most common devices include special scanning and single-shot equipment, indexing turntables, rotary tables, and robots, which are briefly discussed below.

Shaft Hardening. Single-shot and scanning equipment is used widely in the automotive industry for hardening of various kinds of shafts either along their entire length or selectivity. In a typical vertical induction scanning device, two to six parts are loaded manually in a group of spindles. Each is then simultaneously heated and quenched by an individual scanner. Preset variables control the scanning speed and the areas to be hardened. The power is also controlled by preset parameters, as are the quench cycle and the need for part rotation. Parts are then unloaded manually. Nowadays, however, robots (Fig. 10.13) are being used with vertical scanners to eliminate the manual operations.

If in-line automatic processing is preferred over manual operation of a vertical multispindle device, shafts can be heat treated one at a time in a horizontal single-shot or scanning device.* In one setup, shafts are fed to the machine by a horizontal conveyor or walking-beam device. Here they are picked up between centers and chucked into a four-station turret. The flange of a typical shaft is clamped in a three-fingered, rotating chuck, while the spline end is supported by a rotating center. A center is also provided at the flanged end, but this only aids in the pickup and initial locating operations; it is retracted as soon as the flange is chucked. The shaft remains clamped throughout the machine cycle until it is unloaded. After the initial loading, the shaft is indexed to the next station, at which a single-shot nonencircling inductor

---

*H. B. Osborne, *Proc. Eleventh IEEE Conference on Electric Process Heating in Industry*, IEEE, New York, 1973, p 59.

moves over the rotating part, heating it to the desired temperature. A guide shoe on the inductor maintains the proper air gap between the inductor and the part to ensure uniform heating and to minimize distortion. The rotating center at the spline end automatically compensates for thermal expansion as the shaft is heated by applying a fixed load. After being heated, the workpiece is indexed to a quenching station for the actual hardening operation. The fourth, and final, station is another quench which removes any residual heat. Following these operations, the shaft is unloaded by a walking beam onto a conveyor which then transfers it to the tempering unit, which may be a similar induction-based unit or a gas furnace.

Fully automatic, horizontal shaft hardening by scanning methods is also popular. Some of these devices allow loading and scan hardening of multiple workpieces without operator assistance. This is made possible by careful design of walking-beam feeding systems and parts chucking in the heat treating equipment.

Indexing Turntables. A popular machine for automated handling and heat treating of small parts is the rotary indexing turntable. Parts are loaded into an elevated hopper that feeds the machine along an inclined steel track. Various fixtures can be designed to locate the parts when they are dropped onto the table as well as to hold them in place. This is particularly important for maintenance of precise distances between the part and an induction coil or an ejector (which may be used to push the parts into a quench tank) at various stages

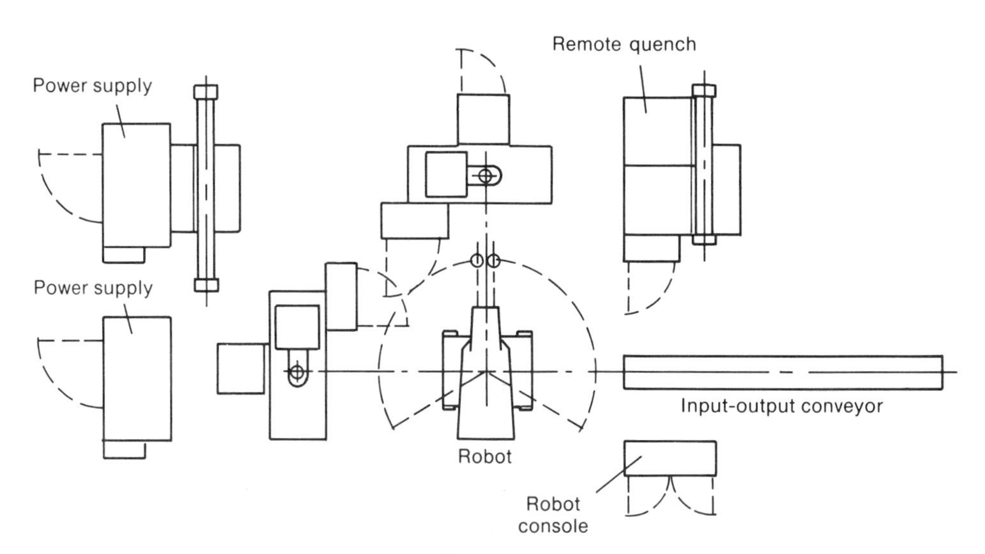

Fig. 10.13. Schematic illustration of an automated heat treatment cycle including two vertical induction scanning units, a quench station, and a part-handling robot (from P. J. Miller, *Heat Treating*, Vol 14, No. 5, May, 1982, p 40)

during the processing sequence. Often the first stage in such a turntable consists of some sort of checking device. It detects improper loading and is electrically interlocked so that indexing can be stopped if a mislocated part, which might damage the induction coil, is present. These machines often have very high outputs—sometimes as high as several thousand parts per hour.

Indexing devices also find use in tooth-by-tooth hardening of gears.

Rotary Tables. Rotary tables, without indexers, also find application in induction heat treating of small parts. In this instance, a number of parts are usually loaded into a groove or nest on the periphery of the table and their ends are heated by a channel-type coil as the table rotates. One application, the hardening of 20-mm armor-piercing projectiles,* is illustrated in Fig. 10.14. The table is water cooled from beneath. The overhead induction coil is flared outward at its lower edge to prevent excessive heating and thus hardening of the lower portion of the projectile.

As the table carries the projectiles from beneath the coil, they are tipped into a tank which provides an agitated water quench. The agitation also causes the parts to be transported along a helical ramp toward a discharge opening onto a second chute. Here, the projectiles are dropped into a perforated barrel revolving in a soluble oil. This likewise has a helical partition which finally elevates and drops the workpieces by gravity into a tray for stress relief in a batch furnace.

Robots. The development of flexible manufacturing systems and automation in general has come about in large measure because of industrial robots. In induction heat treating, robots are now used to transport, load, and unload parts as well as to perform unpleasant tasks such as quenching.

The application of robots can be illustrated with a heat treating process for internal ring gears described by Loth.† Following a washing stage, a robot first loads the gear in an induction preheating stage. After being heated to 345 °C (650 °F), the part is transported by robot to the second, or final, heating/austenitizing station. At the completion of this heating cycle, it is automatically lowered into a quench tank. Next, the gear is transferred by robot to a two-stage washer to remove the quench oil from the part prior to the tempering operation. Again, the robot takes the part to the tempering station, after which it removes it and places it onto a conveyor leading to a machining area.

The use of robots for part handling/manipulation is sure to increase as the "vision" and tactile capabilities of such machines are developed more fully.

---

*Anon., *Lepel Review*, Vol 1, No. 2, p 8.
†R. W. Loth, *Production*, Vol 89, No. 4, April, 1982, p 80.

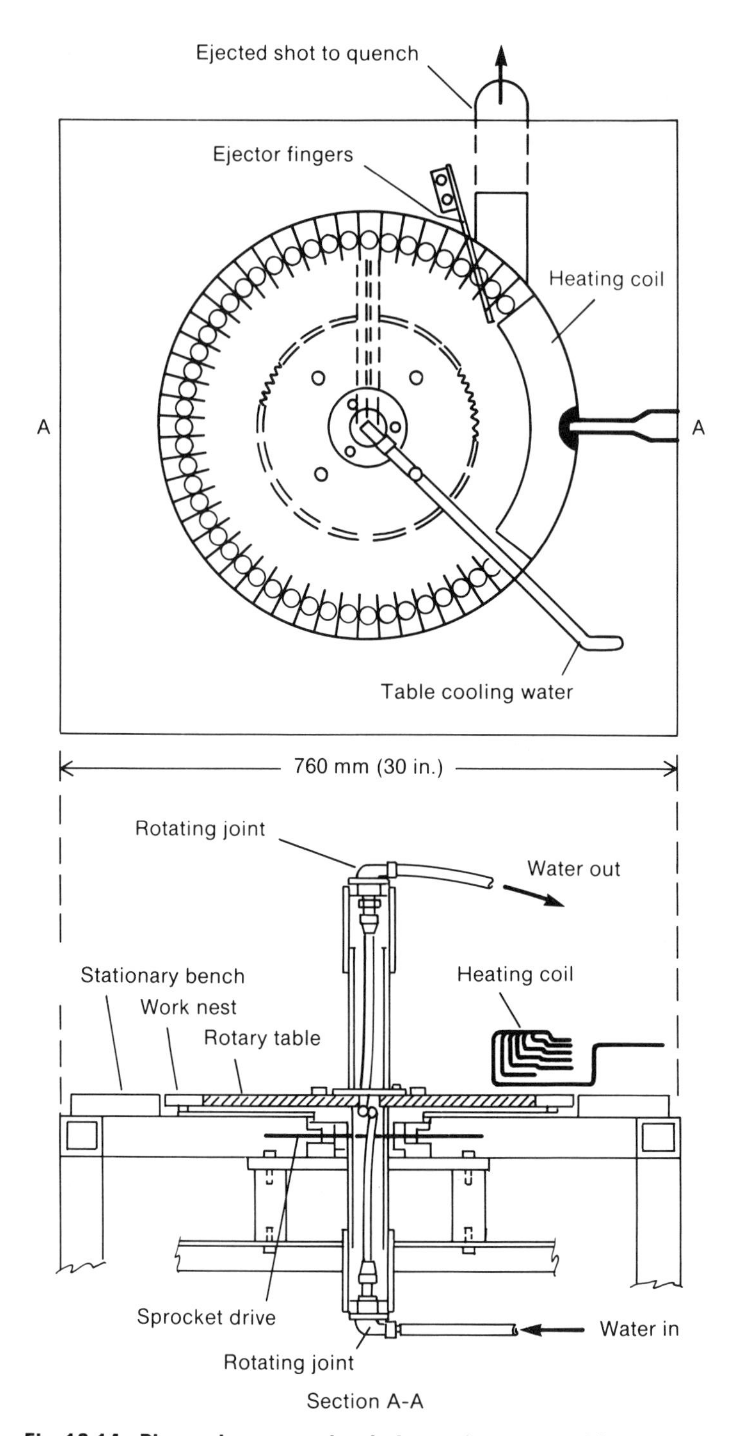

Fig. 10.14. Plan and cross-sectional views of a rotary table arrangement used in induction heat treatment of armor-piercing projectiles (from Anon., *Lepel Review*, Vol 1, No. 2, p 8)

MATERIALS HANDLING IN INDUCTION SOLDERING AND BRAZING

Soldering and brazing, like heat treatment, frequently involve induction heating of discrete parts. Often these parts are very complex in geometry. Thus, materials-handling considerations in these instances consist primarily of design of special jigs to hold one or a series of components in place while they are being heated for the joining operation. Several examples from the *Lepel Review** illustrate the kinds of designs used; in practice, however, each induction brazing or soldering application requires its own specially designed fixture.

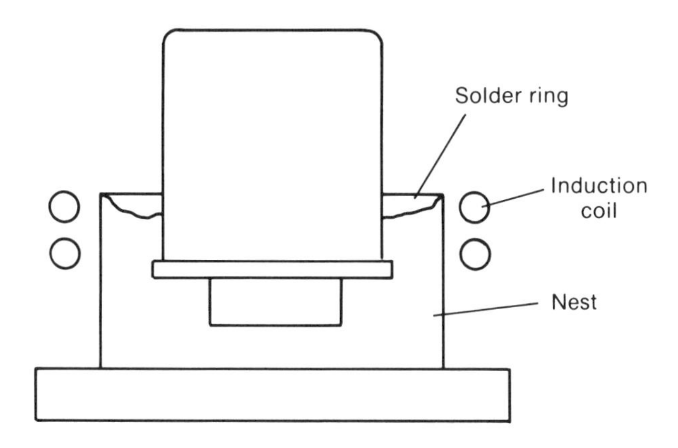

Fig. 10.15. Nest fixture used in induction soldering when the preform is difficult to hold in place; the nest also permits component alignment prior to joining if preassembly is inconvenient (from J. Libsch and P. Capolongo, *Lepel Review*, Vol 1, No. 5, p 1)

Figure 10.15 depicts a ceramic or fiberglass nest used to position solder preforms employed in joining of the components of capacitor cans by induction heating. The small nest is designed to fit into the inductor and to provide proper positioning of the parts to be joined by a close fit with the capacitor-can contour.

Several concepts for carrying out multiple soldering operations sequentially or simultaneously are shown in Fig. 10.16 and 10.17. Figure 10.16 shows a turntable with a conveyor-type coil used to join parts on which solder and flux have been previously placed. The components are moved continuously through the coil at a previously determined speed in order to bring the joint area to soldering temperature. Alternatively, the turntable may be set up with an indexing fixture in cases for which the heat must be localized. With this type

---

*J. Libsch and P. Capolongo, *Lepel Review*, Vol 1, No. 5, p 1.

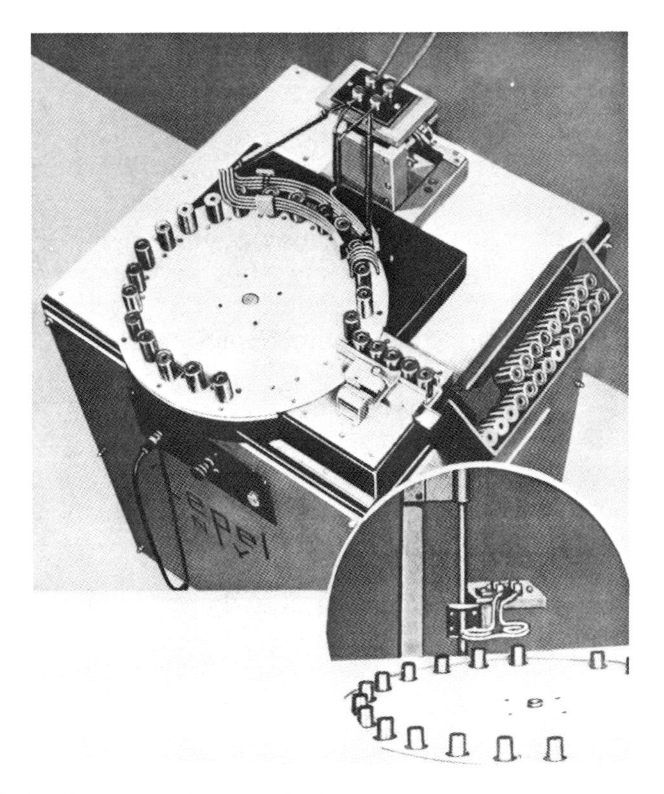

Fig. 10.16. High-production induction soldering fixture utilizing a turntable and conveyor-type coil which allows continuous part movement; the inset shows an indexing arrangment for applications where localized heating is required (from J. Libsch and P. Capolongo, *Lepel Review*, Vol 1, No. 5, p 1)

of arrangement, the coil is automatically raised during indexing and lowered prior to the beginning of the heating cycle.

Figure 10.17 shows a multiple-position fixture for induction soldering of four assemblies at once. After the parts to be soldered have been placed in a jig, the jig is closed. Pressure pins located in the top of the fixture are utilized to force each tube into proper position in the flange when the preplaced solder melts.

A last example, one involving induction brazing, is illustrated in Fig. 10.18. This is a simple, manual arrangement for brazing gold rings to settings with a gold brazing alloy. The ring and setting are held in position by a spring-loaded fixture constructed of an insulating material such as transite. There is also a split nickel alloy liner surrounding the workpiece, but inside the coil. The liner is induction heated to a temperature determined by its thickness; however, it does not prevent the setting assembly from being induction heated as well. Therefore, the ring/setting components are brazed as a result of both induction and radiation heating.

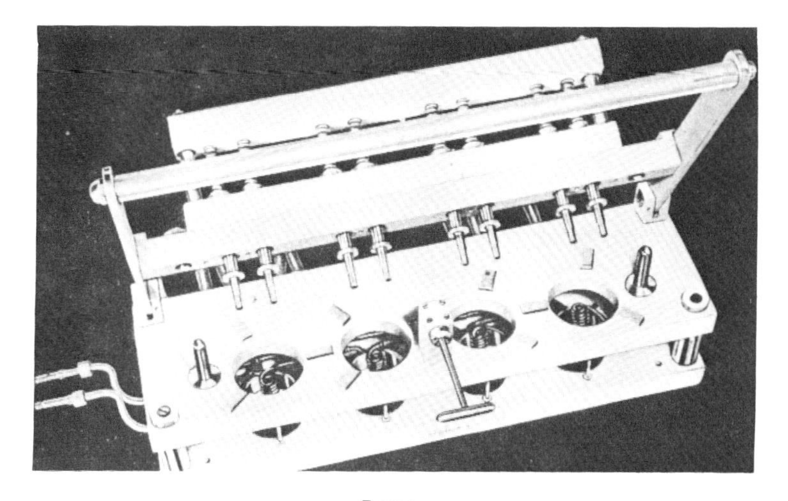

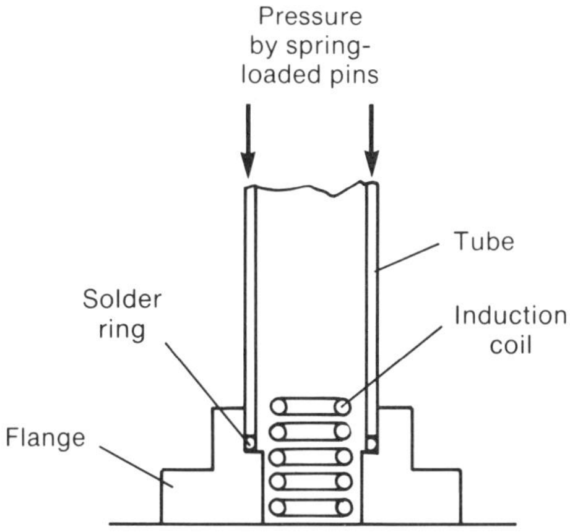

Fig. 10.17. Multiple-position fixture for soldering of four assemblies at once; pressure pads ensure proper component positioning during induction soldering; the sketch at bottom illustrates positions of induction coil and solder ring (from J. Libsch and P. Capolongo, *Lepel Review*, Vol 1, No. 5, p 1)

MATERIALS HANDLING IN OTHER INDUCTION HEATING PROCESSES

Two other induction heating processes in which material handling plays an important role are welding and melting. The fabrication of high-frequency induction welded tubing has been discussed in Chapter 6 on design guidelines. For this product, process design and material handling must be considered together. Induction welding can also be used for manufacture of structural shapes. Among the advantages of this process are the ability to form parts whose individual sections are relatively thin and the ability to make structural members whose components are of different metals. The former attribute is especially attractive in comparison with conventional hot rolling methods, in which rolling loads increase dramatically with increases in reduction and decreases in section thickness.

An arrangement for continuous welding of H sections is shown in Fig. 10.19; similar setups can be envisioned for other shapes such as T's, channels, and various asymmetrical shapes. As shown in the figure, slit strip is fed into the line from a series of uncoilers. Several series of rolls orient the strip. Before the strip enters the welding station, the edges of the web might be upset or another operation carried out on some or all of the individual strips. Some systems are also flexible enough to enable H sections with thick flanges to be fabricated. For such a product, the two coils supplying flange material are removed from the line and individual pieces of flange metal are fed automatically, one immediately after another, into the weld station by a magazine-type device. After passing through the welding area, the structural member is scarfed, cooled, cold straightened, and cut to length.

In conventional induction melting in air, the type and weight of charge materials and the furnace cycle time influence the size of magnets, cranes,

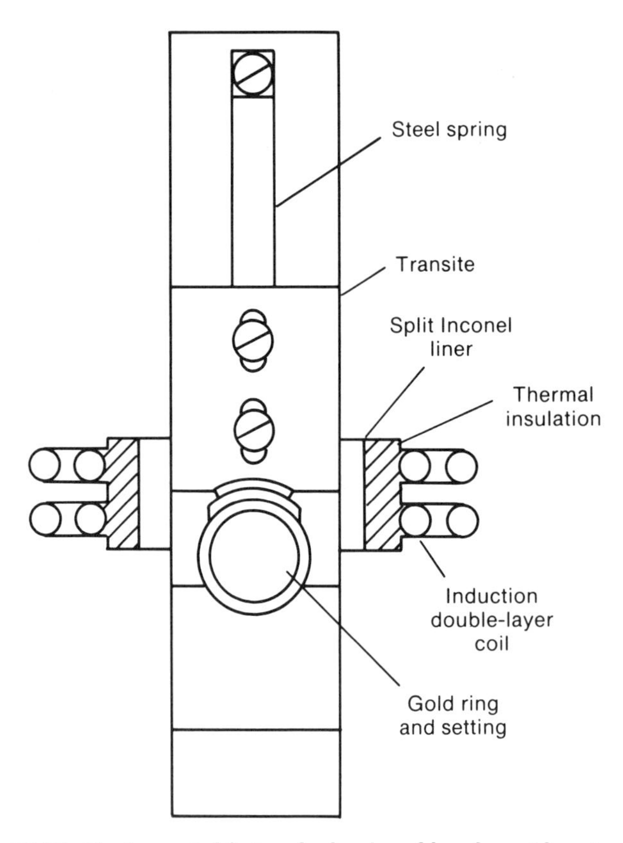

Fig. 10.18. Hand-operated fixture for brazing of jewelry settings to gold rings by induction methods (from J. Libsch and P. Capolongo, *Lepel Review*, Vol 1, No. 5, p 1)

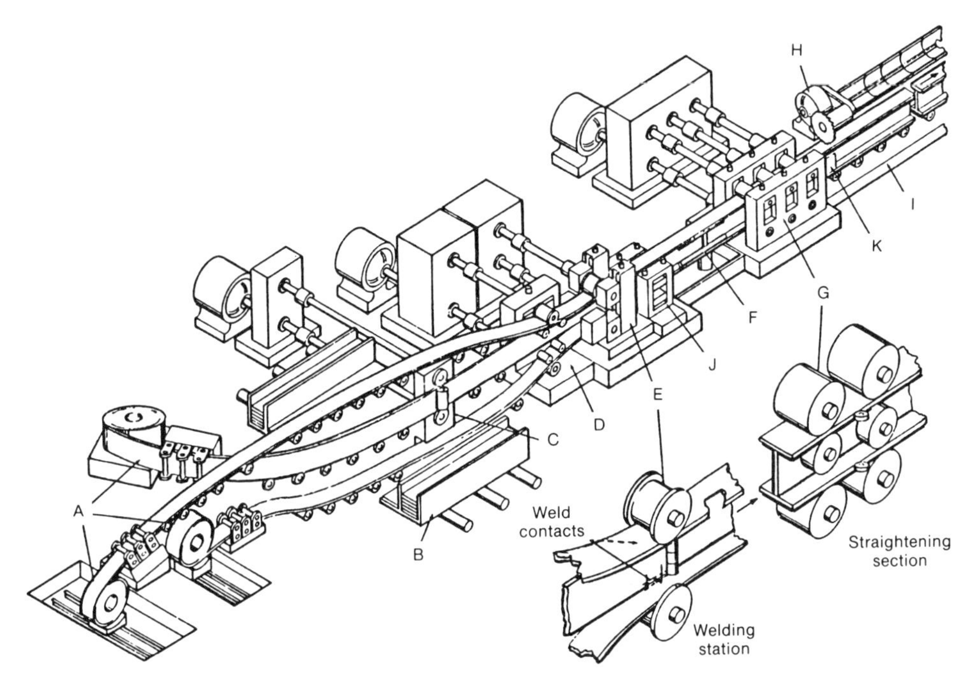

A–uncoilers and flatteners. B–cut flange feeder. C–web upsetter. D–flange prebender. E–welder, one each side. F–cooling zone. G–straighteners, longitudinal and flange. H–cutting saw. I–runout and takeaway. J–scarfing. K–flange joint finder.

Fig. 10.19. Schematic illustration of facility utilizing high-frequency welding for fabrication of structural members (source: Thermatool)

hoists, and charging buckets that are required for materials-handling purposes. A simple large charge may require very large and expensive handling equipment that might be in use only a few minutes each hour. It is therefore preferable to charge induction furnaces with two or more buckets which can be stored until the furnace is ready to accept them.

As discussed by Alexander,* there are several major problems associated with the charging of either coreless or channel furnaces. These problems comprise the following:

- The risk of refractory damage due to heavy metals hitting the furnace lining. Very heavy solid charge materials should be placed carefully into the furnace bath and usually require handling by either an open magnet or some form of sling or caliper device to ensure that they are lowered into the molten metal rather than dropped. The same is true if the furnace is empty. Heavy scrap should be lowered to the bottom carefully to prevent

*A. P. Alexander, "Charge Materials Handling," Modern Casting Technical Report No. 734, American Foundrymen's Society, Des Plaines, IL, February, 1973.

damage to the refractory lining. To prevent wash of the lining when charging a furnace with molten metal, a layer of dry, solid scrap should be placed in the bottom first.

- Metal splashing due to materials being dropped into a molten metal bath. Because of this hazard, very little charging is done by hand or by operators in cranes directly over an induction furnace. Use of a tilting bucket on the end of a mobile crane is the preferred method. Typically, this involves loading of scrap into a chute which leads down to an opening in the furnace. An alternative device is the drop-bottom bucket, consisting of a heavy shell with a lifting yoke at the top and petal-type or clamshell doors at the bottom. The doors are opened by the crane operator or by gravity when the bucket is placed on the furnace.
- Explosions due to water or oil entrapped in the charge material and carried below the surface. Preheating of the charge to a temperature between 260 and 315 °C (500 and 600 °F) eliminates this problem.

To avoid these and other problems, conveyors and special drying equipment for scrap are being used increasingly in induction furnace charging operations. Conveyors may be of the vibrating type, the flat-stroke type, or the belt type. Charge materials can be spread out uniformly and charged at a given, constant rate to prevent the furnace from being overfilled. Rotary barrel driers for preheating and drying of scrap prior to loading onto a conveyor are also increasing in popularity.

ROBOT DESIGN

With the increase in factory automation, the use of robots in induction heating is becoming more widespread. The main elements of a robot are generally a pick-and-place mechanism and its associated controls.* These controls can be a part of the robot package itself (programmable robot) or can be derived from software contained in a remote or integrated CAM (computer-aided manufacturing) system. The main advantages of programmable robots include easy setup and the ability to unload parts from a variety of devices such as conveyors, in-line bar feeders, etc.; such devices are used in machining, inspection, and various types of induction heating and heat treating cells. In contrast, parts must be palletized in CAM-controlled robots. This disadvantage is often outweighed by high production rates, easy start-up, and good reliability. The main features of robot design, irrespective of whether it is programmable or CAM controlled, are its drive mechanism, its arm construction and motions, and its precise method of control and programming. These features are discussed by Faber and Welch and by Bobart.†

---

*M. R. Faber and G. P. Welch, *Industrial Heating*, Vol 53, No. 7, July, 1986, p 24.

†G. F. Bobart, "Part Handling Systems and Process Control," Lesson 8, ASM Induction Heating Homestudy Course, ASM International, Metals Park, OH, 1986.

Robot Drive Mechanisms

Robot drives are generally of three types: pneumatic, hydraulic, and electric. The pneumatic type is best suited for fast cycle rates in the processing of light loads. In contrast, hydraulically driven robots are most often utilized for heavier loads (e.g., induction billet heating) which do not require rapid part motions; the use of such robots may involve special start-up provisions to bring the hydraulic fluids to the proper operating temperature. Electrically driven robots are being used increasingly because they can handle moderately heavy loads at high speeds with good part-positioning accuracy.

Robot Tooling Systems and Motions

All robots use one or more arms in order to handle parts within a prescribed area. Often, the system consists of one, two, or three arms; the selection is based on application requirements*:

- One-arm robots usually have no rotational capability and are used in low-production situations.
- Two-arm robots generally are used in medium-to-high-production applications, particularly in induction heat treatments involving integral or submerged quenching.
- Three-arm robots are used in high-production multistage heating/heat treating or in heat treating using multiple zones. A schematic illustration of a three-arm robot with 90° wrists and gripper rotation is shown in Fig. 10.20.

Often the end "effector" of a robot arm is designed with several grippers or hands in order to load and unload several parts at a time. Hands are usually custom designed for the application and can be cam, cylinder, or electrically actuated. The gripping action is typically achieved by stationary, moveable, or self-aligning jaws or fingers; by electromagnetic pickup plates; or by vacuum cups.

The application also determines the kind of motion that the robot arm (or wrist) must perform. These motions may rely on cylindrical, spherical, revolute (jointed), or rectangular (Cartesian) coordinate systems.

Control and Programming of Robots†

Control techniques for robots include nonservo-controlled point-to-point, servo-controlled point-to-point, and servo-controlled continuous-path systems. The first of these is the simplest kind of system, utilizing relay logic control and limit switches for part location.

The other two (servo-controlled) systems use feedback devices such as opti-

---

*M. R. Faber and G. P. Welch, *Industrial Heating*, Vol 53, No. 7, July, 1986, p 24.

†G. F. Bobart, "Part Handling Systems and Process Control," Lesson 8, ASM Induction Heating Homestudy Course, ASM International, Metals Park, OH, 1986.

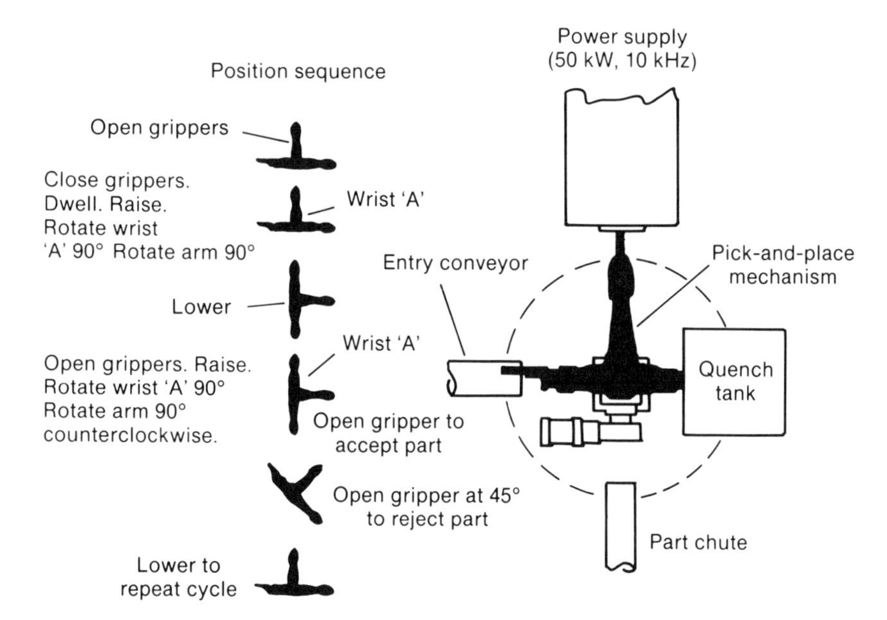

Fig. 10.20. Robot design including three arms, 90° wrists, and gripper rotation (from M. R. Farber and G. P. Welch, *Industrial Heating*, Vol 53, No. 7, July, 1986, p 24)

cal encoders, resolvers, tachometers, and potentiometers to maintain the velocity and position of each axis and thereby to provide a feedback control signal. In point-to-point servo systems, control is programmed by storing discrete points in the robot's memory, which are then used subsequently to carry out the desired sequence of motions. A hand-held pendant with separate buttons for each axis of motion is normally used to "teach" the program to the robot.

Continuous-path servo-controlled systems are similar to point-to-point systems except that more extensive data storage is involved, usually in a memory storage system such as a cassette or floppy disk. Another difference lies in the method by which the robot motion is taught. In continuous systems, the end of the arm is grasped and led through a pattern of motions while the control system continuously samples and records feedback data from the axis-position sensors. Most often, applications requiring high speeds and/or control of five or more axes of motion (e.g., induction heat treating and brazing) employ continuous-path servo-controlled systems.

Chapter 11

Special Applications of Induction Heating

In a large number of applications, induction heating is utilized to raise the temperature of a metal prior to forming or joining, or to change its metallurgical structure. In most cases, the induction process is chosen over other methods of heating because of special features such as:

- The ability to provide surface or localized heating
- The ease of control of the heating rate and heating cycle
- The ability to obtain very high processing temperatures
- The ability to heat the workpiece inside a nonconductive vessel which holds a protective atmosphere
- High energy efficiency and good process repeatability.

Recognizing that these are rather unique properties, induction heating should be viewed, therefore, as a specialized heating technique rather than as a *metal* heating process *per se*. In this chapter, some of the special applications of induction heating which make use of the above features are summarized. These include applications in the plastics, packaging, electronics, glass, and metal-finishing industries.

INDUCTION HEATING APPLICATIONS IN THE PLASTICS AND RUBBER INDUSTRIES

Applications of induction heating in the plastics industry include bonding or forming of plastics, coating of metals, and salvage operations. Some typical examples are discussed below.

Bonding and Forming of Plastics

Many plastics such as polyethylene cannot be bonded using conventional adhesives. However, they can be joined by being heated and flowed under

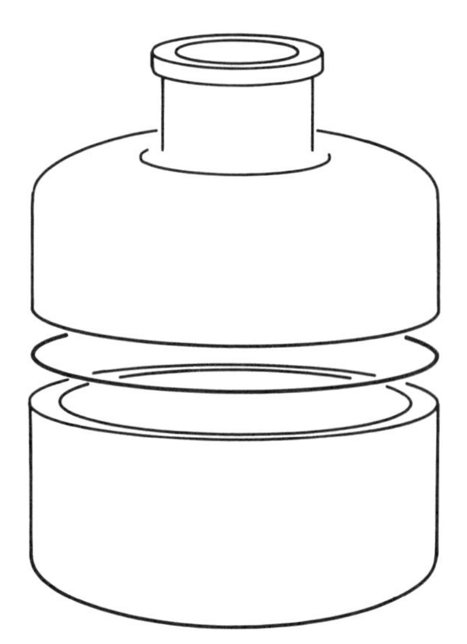

Fig. 11.1. Polyethylene assembly with a steel ring insert that is joined using induction heating (source: Lepel Corp.)

localized heating conditions. To accomplish this, a susceptor (Chapter 9), which becomes a part of the final assembly, is often used. For example, insertion of a wire ring in a special cavity of the assembly permits the use of induction for local heating of the area to be joined. Toilet-tank floats have been made in this manner (Fig. 11.1) by using a steel wire ring inserted where the two float halves join. Placing the wire ring in an induction field causes it to heat rapidly. This in turn melts the adjacent rims of the plastic hemispheres, causing them to form a plastic-to-plastic bond. In this technique, proper joint design ensures that the metal ring is fully encompassed by the plastic at the completion of the operation, precluding subsequent rusting of the wire. If it is desired to heat the plastic more slowly, a gap can be left between the ends of the metal ring.

In a similar approach, the application of localized heat to thermoformable plastics may be employed to create a mechanical bond. Typically, the tang of a metal kitchen utensil such as a knife or a fork (Fig. 11.2) is heated to the flow temperature of the plastic handle to which it is to be joined. The heated tang is then inserted in the predrilled or premolded handle, melting the adjacent material and forming a strong mechanical bond on cooling. If holes or ridges are provided on the tang, greater mechanical strength can be obtained through the ability of the plastic to flow into these areas. Twisted wire-handle brushes, similar to bottle brushes, are assembled in this manner as well. The plastic is deformed on contact with the heated wire, causing it to match the ridges in the twisted wire handle.

At other times, the metal susceptor does not remain a part of the final assembly but is used solely as a means of supplying heat to a plastic preform which is to be shaped. As an example, catheters for medical treatment are

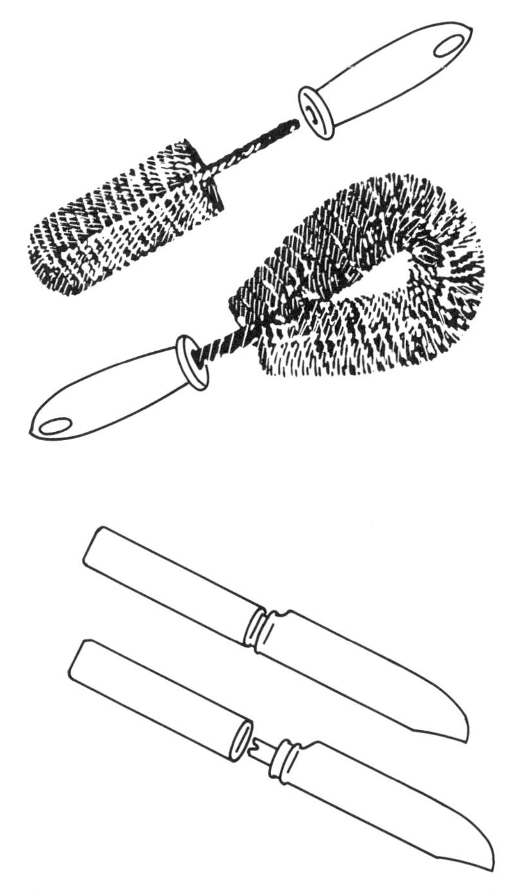

Fig. 11.2. Application of induction heating for attaching knife and brush handles (source: Lepel Corp.)

formed of inert plastic tubing and must have specific shapes at their ends so that they can be coupled to intravenous or similar types of apparatus. In these instances, a stainless steel forming die is fabricated with internal channels for water cooling (Fig. 11.3). In operation, the die is heated by an induction coil on an automatic forming machine. When the die is raised to the forming temperature of the plastic, the tubing is mechanically pressed against it. Thereby, the tubing assumes the shape of the die. The induction power is turned off and the die water cooling system is activated to cool the plastic and thus set it in its new shape. Because these dies are normally rather small, most machines of this type heat and form multiple parts simultaneously.

In some instances, it is not possible to insert a metal ring or preform into an assembly where the deformation is carried out. The technique described above for forming of catheters can then be utilized with a *movable* die. In this case, the die is heated by induction. When it is at proper temperature, it is then moved through the coil and pressed into the assembly to deform the plastic. Induction heating is preferred to resistance heating of the die because of

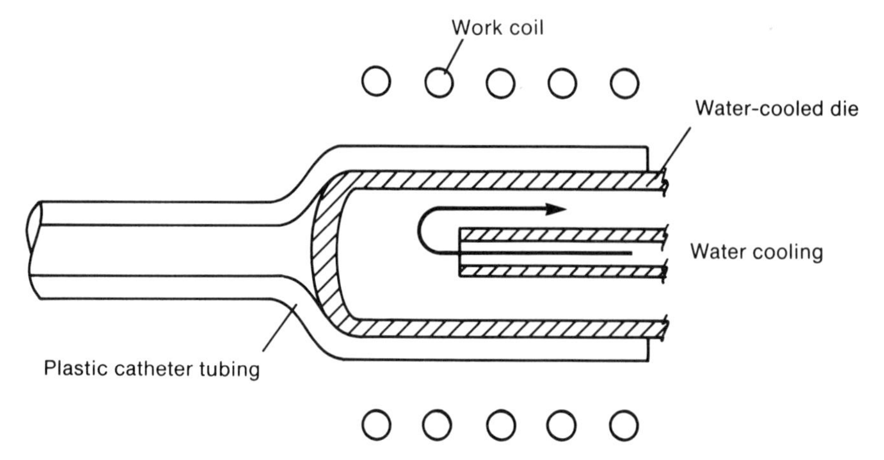

Fig. 11.3. Schematic illustration of induction heating method used to form catheter ends on tubing

its capability for rapid heating. Unlike resistance, it can provide the die with a high recovery heating rate for high-production applications. Typically, glass lenses are staked in a lens tube (Fig. 11.4) using a die that is slightly larger than the ID of the plastic tube.

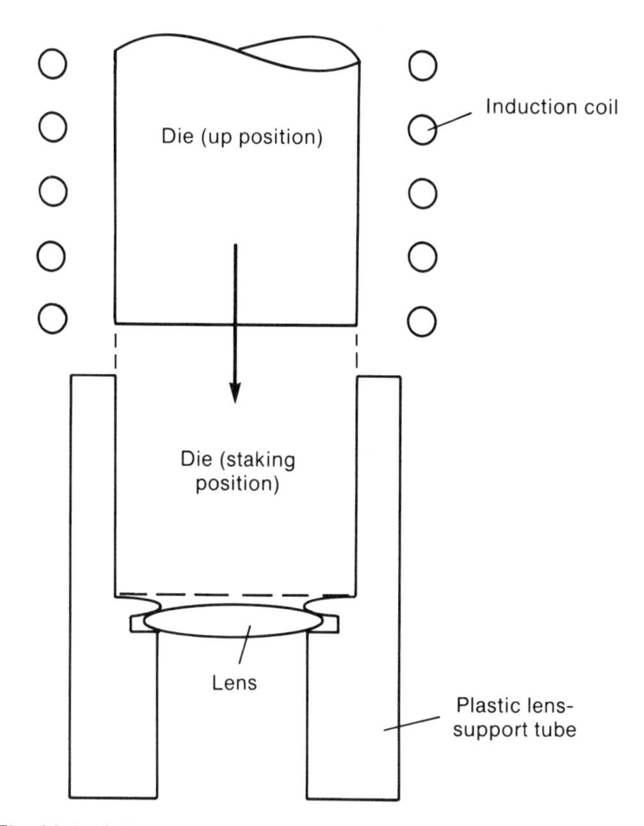

Fig. 11.4. Schematic illustration of induction technique used for staking of lenses in plastic tubes

Plastic Coatings

Application of plastic coatings (epoxy, vinyl, nylon, teflon, etc.) on metals results in parts with attractive combinations of strength and wear resistance, thermal and electrical insulation, corrosion resistance, and aesthetic qualities. Typically, the coating material is in powder form and is suspended in a fluidized bed by an air-flow system (Fig. 11.5). The material to be coated is raised to the temperature necessary for reflow of the coating material by induction methods, and, as it is passed through the bed, the powder melts on the surface of the part. The thickness of the coating is dependent on the metal temperature among other variables. When induction heating is used in such applications, only the surface of the metal need be brought to temperature, as compared with standard oven techniques, in which the entire mass is heated; thus, efficiency is greatly improved.

During coating of thin materials, temperature is lost in transporting the material from the oven to the bed containing the plastic. Because the induction coils being water cooled are at low temperature, they can therefore be placed directly in the bed. The material is then heated in the coating medium with no subsequent heat loss during transport, and coating can be closely controlled. Again, if localized coating is required, as on the handles of hand tools, scissors, etc., the heat and thus the coating can be restricted to a specific area.

Thermosetting materials can also be applied electrostatically to the base metal, as is done in the manufacture of coated anchor chains. Unlike the methods described above, the dry powder is given a negative charge as it exits

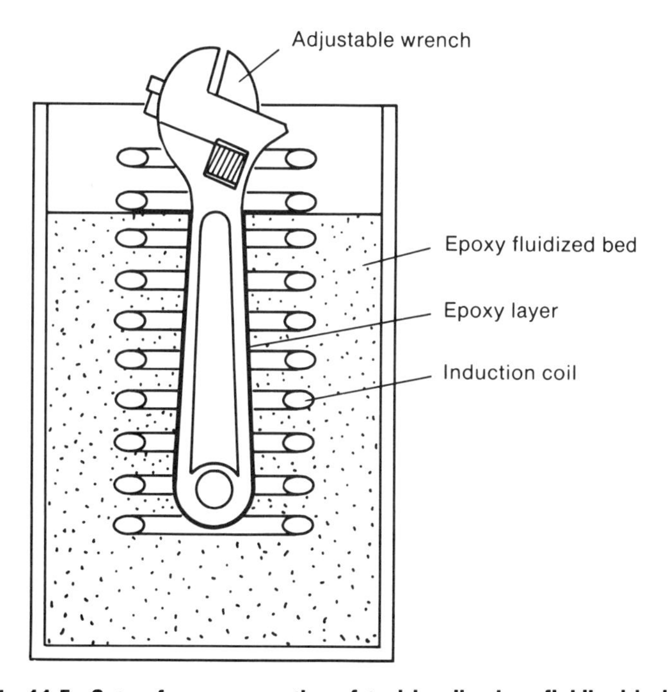

Fig. 11.5. Setup for epoxy coating of tool handles in a fluidized bed using induction heating (source: Lepel Corp.)

Fig. 11.6. Photograph of a processing line for induction curing of a polyimide film on wire (source: Lepel Corp.)

from a spray gun and is attracted to the positively charged metal being coated. The part is passed through an induction heating coil of proper frequency with a suitable power density, which results in heating of the powder and excellent flow characteristics.

In each instance, the thermal conductivity of the coating plays an important role in the process. Because heat is transferred to the powder by conduction from the metal surface, the extent of heat transfer from the substrate to the powder decreases as the powder thickness increases. Coating thickness is therefore limited by the properties of the plastic itself.

Similar techniques are employed in insulating electrical conductors using polyemide film or tape. The high-temperature resistance of the polyemide film, which is applied in the form of high-integrity tape heat sealed to the conductor, permits thinner insulation than was previously possible. In this case, because the material will not bond directly to copper, one or both sides of the tape are coated with a fluorocarbon resin to provide a heat-sealable surface. The over-all arrangement of equipment for continuous processing of insulated conductors is shown in Fig. 11.6 and 11.7. It consists of equipment for continuously transporting the conductor and tape; an arrangement for mechanically and chemically cleaning the conductor; dual-lead tape-wrapping machines; an RF induction generator with infrared temperature control; and

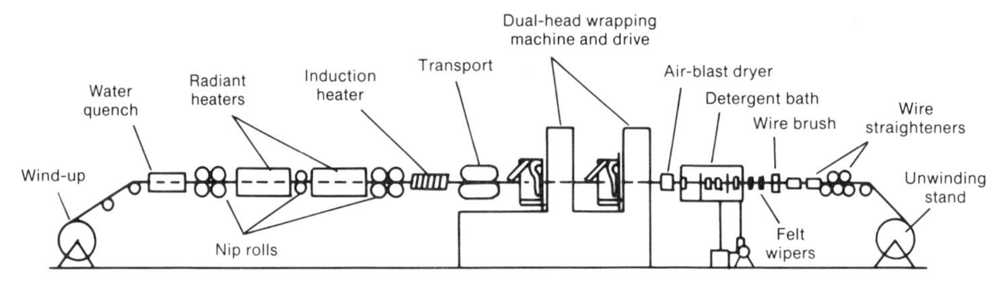

Fig. 11.7. Schematic diagram of the processing line for induction curing shown in Fig. 11.6 (source: Lepel Corp.)

radiant heaters for final curing, if desired. A 40-kW RF generator operating at 450 kHz heats the wire conductor to approximately 300 °C (570 °F) at a speed of 5 m/min (15 fpm). The infrared radiation pyrometer focused on the conductor at the exit end of the work coil continuously monitors the temperature of the wire. It can thus be used to control the power of the induction generator to maintain a specific processing temperature.

Salvage Operations

When coated materials have exhausted their useful life, the base-metal structure may still have considerable value and can be salvaged. Printing rolls, vibration eliminators, and doffer disks on cotton machinery are typical of assemblies that can be usefully and economically salvaged with induction heating. For example, Fig. 11.8 depicts a coil that is used to heat 152-cm- (60-in.-) diam printing rolls for this process.

In practice, the part is placed in a work coil (Fig. 11.9), and the heat produced solely on the surface of the metal breaks the bond at the interface

Fig. 11.8. Induction heating arrangement used to strip 152-cm (60-in.) printing rolls (source: American Induction Heating Corp.)

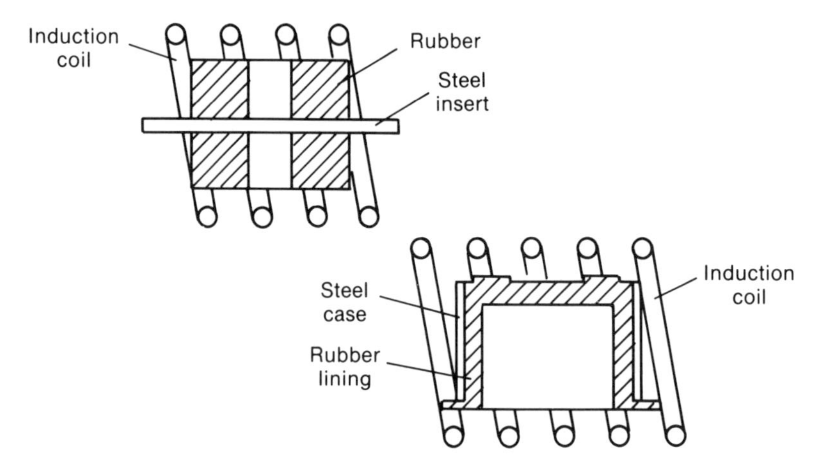

Fig. 11.9. Typical setup for stripping rubber by an induction heating method (source: Lepel Corp.)

between the materials. The parts can then be easily separated. Short processing times prevent appreciable heating and, when rubber is the second material, deterioration of the nonconductive coating is minimal, allowing both parts to be saved for reuse.

In the case of printing or drive rolls, the assembly can be heated on a progressive basis, and the coated cover can be split with a skiving knife following the coil. The covering material is thus easily removed, leaving an essentially clean surface that can be recoated. Rubber linings on brass, aluminum, and steel parts can all be reclaimed with this process.

BONDING APPLICATIONS OF INDUCTION HEATING

Several common applications of induction heating in the area of joining include adhesive bonding and setting of metallic-particle gaskets. Adhesive bonding relies on the localized heating of a metallic substrate to achieve melting and curing of an adhesive. In a typical example, shafts are bonded directly to squirrel-cage rotors in the manufacture of small motors (Fig. 11.10 and 11.11). An adhesive with a high-temperature, short-time curing cycle is coated on the shaft which is inserted in the fixture. The laminated steel rotors are then placed in position and the assemblies are lifted into a six-place single-turn coil. The rotors are heated, thus setting the adhesive and precisely locating the shaft with respect to the rotor. The mechanical position is now set; however, the full curing time of the adhesive is somewhat longer. Nevertheless, the large thermal mass of the rotor continues to cool slowly after this operation, thus maintaining sufficient heat to fully cure the adhesive as the parts progress to the next operation.

In the manufacture of copying machines, plastic components are adhesively bonded to aluminum rotors in a similar operation. Operated by a robot, the

Fig. 11.10. Photograph of multiplace induction coil used for adhesive bonding of rotors to shafts (source: Lindberg Cycle-Dyne, Inc.)

system uses a single RF generator operating two alternate remote tanks and bonds both sides of the assembly in approximately 12 s. A similar approach is utilized in the automotive industry where insulation panels are bonded to body parts. The induction coil is on flexible leads. The adhesive to be cured is preplaced on the stamped metal panels. The insulation is put in place, and

Fig. 11.11. Photograph of rotor-and-shaft assemblies being induction heated during an adhesive bonding operation (source: Lindberg Cycle-Dyne, Inc.)

the adhesive is activated and then cured by induction heating of the metal panel at the bond location. In addition, hot melt adhesives can be preplaced with this technique, allowing them to be located in an earlier operation and then activated when bonding is desired.

Joining of plastic assemblies using metallic-particle gaskets is similar to plastic bonding using metal rings, described earlier in this chapter. The major difference in the present case is that a gasket formed of the base plastic is used rather than a solid metal susceptor (Fig. 11.12 and 11.13). However, the gasket material does contain a dispersion of ferrous particles which serve essentially the same purpose. When placed in the induction field, the ferrous particles heat, in turn melting the plastic gasket material in order to complete the bond. The number, size, and dispersion of metallic particles in the gasket material are closely controlled. The operating frequency of the induction generator is generally 3 to 5 MHz because of the small size of the particles. Since the current through the coil at these high frequencies is not great, the

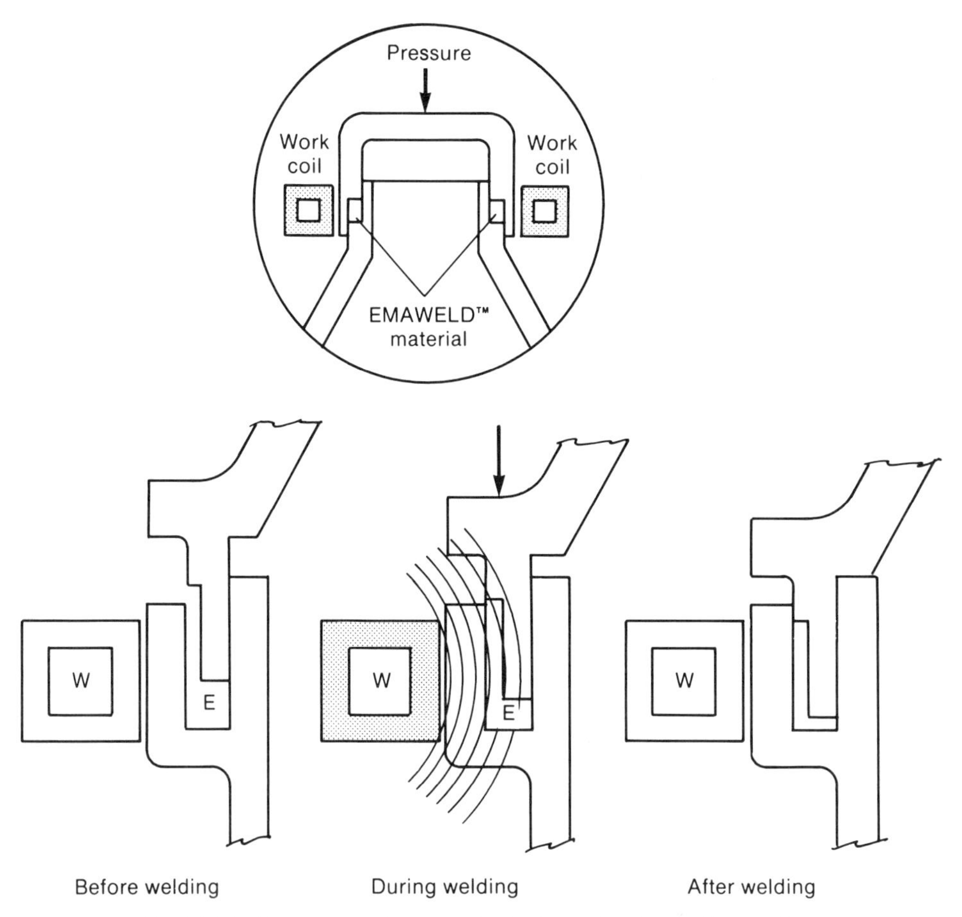

Fig. 11.12. Illustration of a typical plastic-to-plastic welding application using a metallic-particle gasket (source: Emabond, Inc.)

flux field is comparatively weak. The coil must therefore be designed to be as close to the joint as practical while maintaining minimum over-all coil inductance.

INDUCTION CAP SEALING AND PACKAGING

The food, drug, and chemical industries require easily assembled tamper-proof seals that assure users of their products that no contamination has taken place between the manufacturing site and the place of purchase. Induction cap sealing is a unique process in which an aluminum disk or seal is bonded to a product container. These seals can be used with most caps and corresponding containers to provide hermetic bonds suitable for use with foods, drugs, beverages, alkalis, most acids, oils, organic solvents, and corrosion inhibitors in liquid, powder, or pellet form. These seals meet the requirements of the Food and Drug Administration and can be used on containers made of glass, polyethylene, PVC, and most other thermoplastics, and are suitable for use with screw-type, snap-on, child-resistant, and other custom closures.

Like most conventional lining materials, such inner seals are supplied in coiled strips or special shapes, and are handled by conventional cap-lining equipment. Most have the same basic components: a pulpboard backing, a wax coating, aluminum foil, and a heat-sealable polyester film coating. This four-part disk is inserted in the product cap. The container is filled, and the cap is tightened. The capped container is then passed through a medium- or high-frequency induction field (Fig. 11.14). The field penetrates the cap and heats only the outer edge of the foil, melting the polyester coating beneath it. As the product leaves the induction field, the coating cools, bonding the

Fig. 11.13. Typical parts made with metallic-particle gaskets (source: Emabond, Inc.)

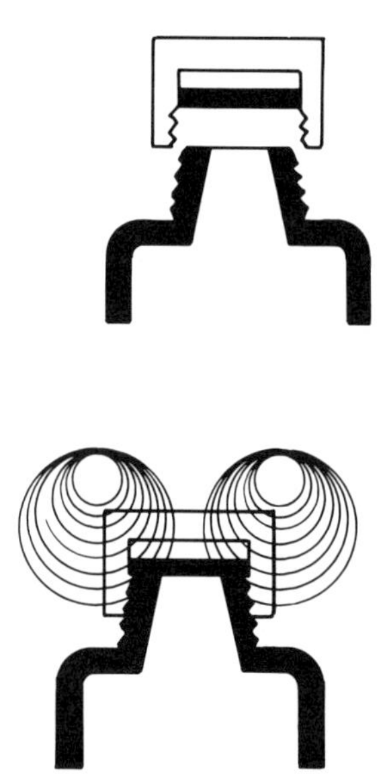

Fig. 11.14. Schematic illustration of a typical cap-sealing application in the packaging industry (source: Lepel Corp.)

foil to the lip of the container. When the cap is removed, the waxed pulpboard remains in the cap to permit resealing after the foil seal has been removed or punctured. The consumer removes the closure and then peels off the foil seal to gain access to the product.

Cap-sealing operations are usually continuous, utilizing a channel coil in a conveyorized processing system. However, low-production operations can be accomplished using single-turn coils at the ends of flexible leads. In some cases—particularly with large-diameter seals, for which the heat toward the middle of a channel coil might be a problem—the part can be indexed on a conveyor and the coil lowered into position for the bonding operation.

Tubes for toothpaste, hair conditioner, etc. are also formed from laminated materials on automated systems (Fig. 11.15). In these operations, however, the sheet material is wrapped on a mandrel with the end of the sheet overlapping the front. A narrow sealing band is required, which could be produced with a split inductor. However, a single-turn hairpin loop utilizing a powdered-ferrite core, as shown in Fig. 11.16, is also highly effective. In operation, the side of the coil is insulated and is mechanically pressed against the seal area. This places the two sides of the seal in close physical contact while minimizing the coupling distance from the coil. Sealing times on the order of 0.1 s are achieved with a 3-kW generator operating at 450 kHz.

INDUCTION HEATING APPLICATIONS IN THE ELECTRONICS INDUSTRY

The ease of control and the localized heating capabilities of induction heating are widely used in the electronics industry. The manufacture of semicon-

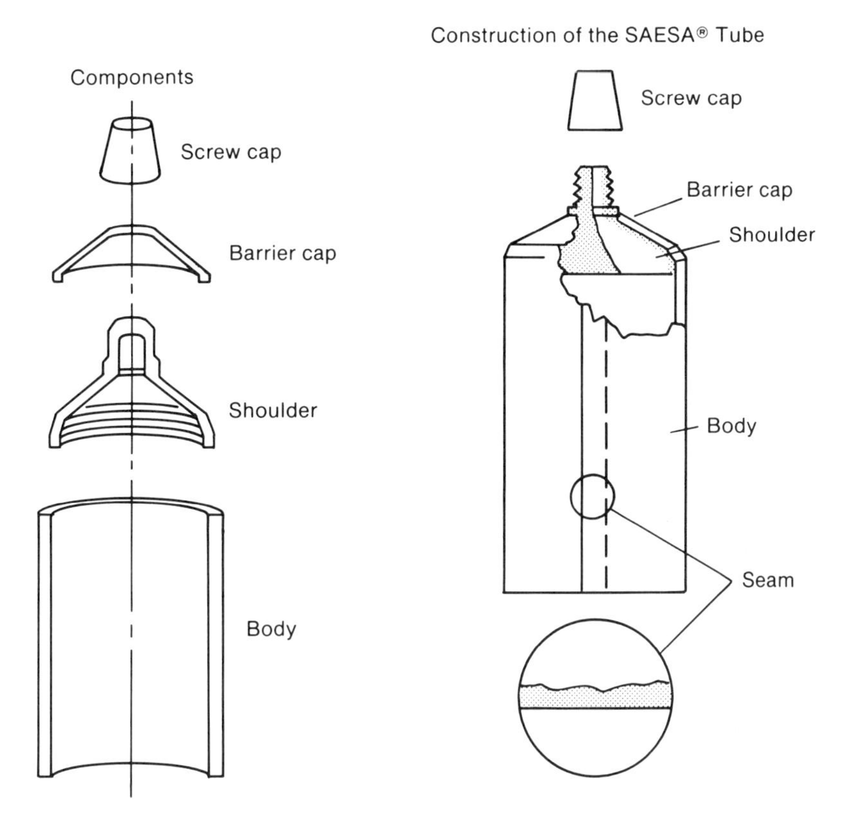

Fig. 11.15. Construction of toothpaste tubes that are assembled using high-frequency induction heating (source: SAESA/Goodway Tools Corp.)

ductor materials, silicon solar cells, and vacuum tubes are but a few of the products which make use of this heating technique. These and other applications are described next.

Zone Refining of Semiconductors

During manufacture of semiconductors, induction heating is used for refining, doping, and processing of crystalline materials. The high resistivities of semiconductor materials such as germanium, silicon, and gallium arsenide preclude direct induction heating, however. The use of susceptors, together with the ability to provide close temperature control by induction, provide specific benefits in the electronics industry.

In the processing of germanium and silicon, for example, the material must first be refined to remove impurities. The technique utilized to perform this function is called zone refining. In this method, the material is incrementally melted while being gradually moved through an induction coil, thus producing a narrow molten zone. The impurities tend to remain in the narrow molten zone, much as slag is formed in conventional melting. Thus they are driven to one end of the bar as it proceeds through the coil, and the molten zone stays within the confines of the coil. Successive passes through the induction coil further refine the product. In operation, a multiturn coil, with sections counterwound to provide repetitive narrow zones (Fig. 11.17a), produces the same effect as multipass heating but in a single pass. The degree of refining is related to the number of passes through the molten zone, within practical limits. Purities achieved have been reported to be within one part per billion of electrically active impurities.

In processing of germanium at 950 °C (1740 °F), the material is generally carried through the coil in a carbon susceptor or boat. During zone refining of silicon (or gallium arsenide), however, the carbon from a graphite susceptor or boat can contaminate the product. Hence, a process known as float-zone refining (Fig. 11.17b) is utilized. In this technique, a balance between the temperature of the molten zone in the rod and its surface tension must be maintained; thus, no boat is required to support the molten zone. In float-zone refining, the material can be passed through the coil, or the coil can be moved along the rod. The molten zone, at 1450 °C (2640 °F), in effect concentrates the impurities at the ends of the rod. Because of the extremely high resistivity of silicon, it cannot be heated with 450-kHz current until it has been raised in temperature. If a frequency in excess of 3 MHz is used, the material may be heated directly. However, preheating of the silicon by a resistance method can be used to permit power-supply operation at 450 kHz. Due to the fact that control of zone length is critical, and proper surface tension on the molten zone must be maintained, automatic closed-loop control of temperature and power is used for this process.

Growth of Semiconductor Single Crystals

After an electronic material has been refined, a single crystal of the product must often be grown. The material is placed in a graphite crucible (Fig. 11.17c), with a quartz liner in the case of silicon and gallium arsenide. Impurities are added to the crucible in controlled amounts to produce the

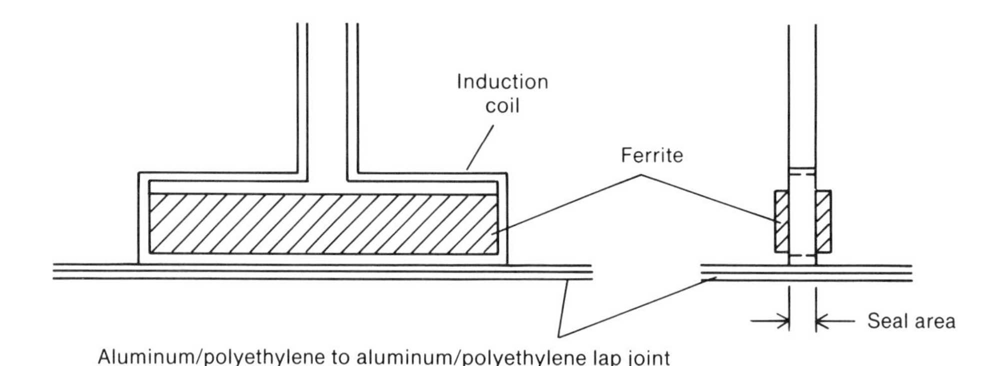

Fig. 11.16. Bonding of aluminum/polyethylene joints (such as for toothpaste tubes) with a high-frequency induction coil and ferrite flux modifier

desired electrical characteristics. The material is brought to its melting temperature, and a seed crystal of the proper orientation is inserted into the melt, rotated, and slowly withdrawn. The molten element solidifying on the end of the seed crystal continues the formation of the single crystal. Temperature, speed of seed withdrawal, and rotation of the crucible are all critical to the process and control both the diameter of the crystal and the dispersion of impurities. The process is temperature controlled on a programmed basis and is performed in either an atmosphere or a vacuum. An infrared temperature monitor or a sapphire rod thermocouple system is used to sense and control the temperature.

Germanium crystals can also be grown by the zone-leveling method, which employs essentially the same equipment as that used in the germanium zone-refining method, with a single, oriented, impure seed added to one end of the

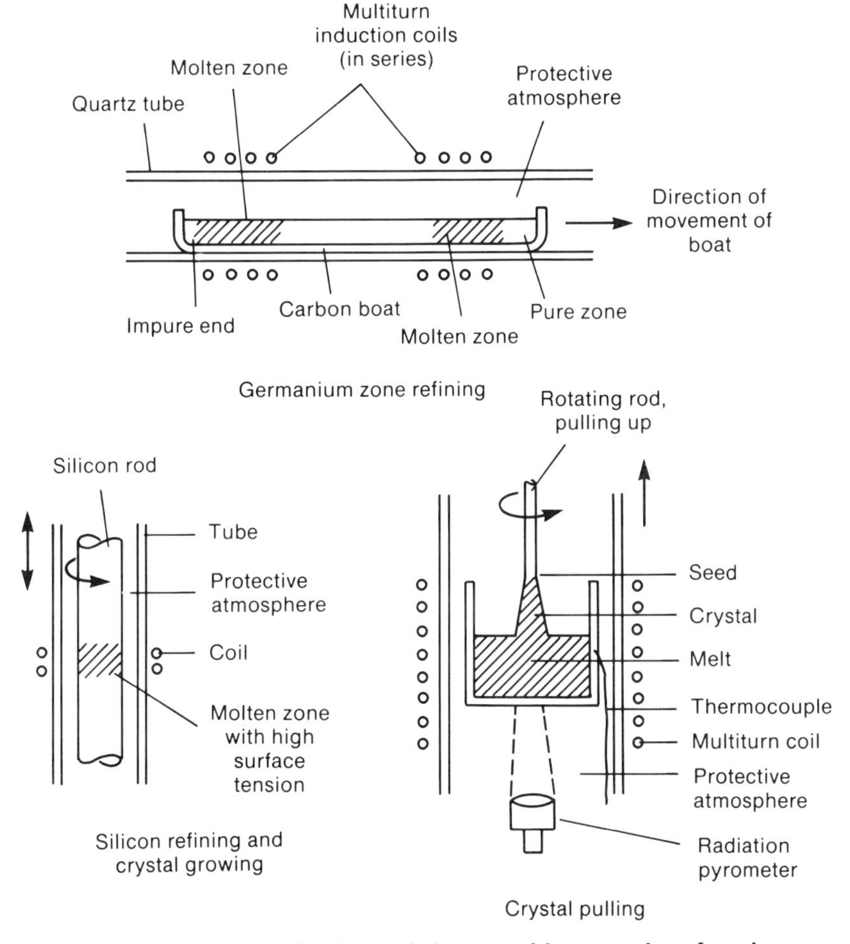

Fig. 11.17. Induction heating techniques used in processing of semiconductors (from J. Davies and P. Simpson, *Induction Heating Handbook*, McGraw-Hill, Ltd., London, 1979)

graphite boat. The rest of the boat contains purified germanium; as the boat is pulled through an induction coil, a single crystal of the desired purity is formed.

The float-zone refining technique for silicon, also called the Czochralski crystal growing method, is another common technique. Essentially it is a float-zone method. However, one end of the rod is a perfect crystal of the proper purity. Moving the rod through the coil, or the coil along the rod, in the proper direction and at the proper rate causes the material from the molten zone to solidify and assume the proper crystalline orientation.

Epitaxial Deposition

After a semiconductor crystal has been formed, it is subsequently cut into wafers, which are then lapped and polished. Using photographic techniques, multiple exposures of the desired circuitry are produced on the wafers. These circuits are then etched into the substrate surface. Frequently, an epitaxial layer of silicon is then deposited on the etched areas, or a reactive material is deposited by CVD (chemical vapor deposition) methods.

Epitaxial deposition involves heating of the wafer by placing it on a graphite susceptor within a quartz tube. The temperature is then raised to approximately 1350 °C (2460 °F), and silicon vapor is passed over the wafer. The silicon is deposited on the wafer in the etched areas and grows in a crystalline structure.

Originally, epitaxial deposition was performed in a vertical reactor (so called because of the direction of gas flow) using a disk susceptor with a pancake coil (Fig. 11.18a). Because this is essentially a batch process, it was desired to design a system by which larger numbers of wafers could be processed economically. The first step in this direction was the horizontal reactor (Fig. 11.18b). In this system, a graphite susceptor which typically measures approximately 10.2 by 76.2 by 1.3 cm (4 by 30 by 0.5 in.) is used. The major design considerations are as follows:

- The frequency (450 kHz) of the power supply is chosen on the basis of the thickness of the susceptor for most effective heating. To prevent field cancellation, the susceptor thickness must be at least three times the reference depth. It is desirable to keep this thickness to a minimum consistent with efficient heating in order to minimize the cooling time following processing. However, some thickness greater than the reference depth is required to minimize thermal gradients in the substrate (wafer).
- Power requirements are not determined by the mass of the susceptor as much as by radiation losses from its free surface. At 1350 °C (2460 °F), this radiation loss is 40 W/cm^2 (250 W/in.^{2}) for a total loss of approximately 69 kW for a susceptor of the above dimensions. An additional 10% allowance for loss to the hydrogen gas must also be made.

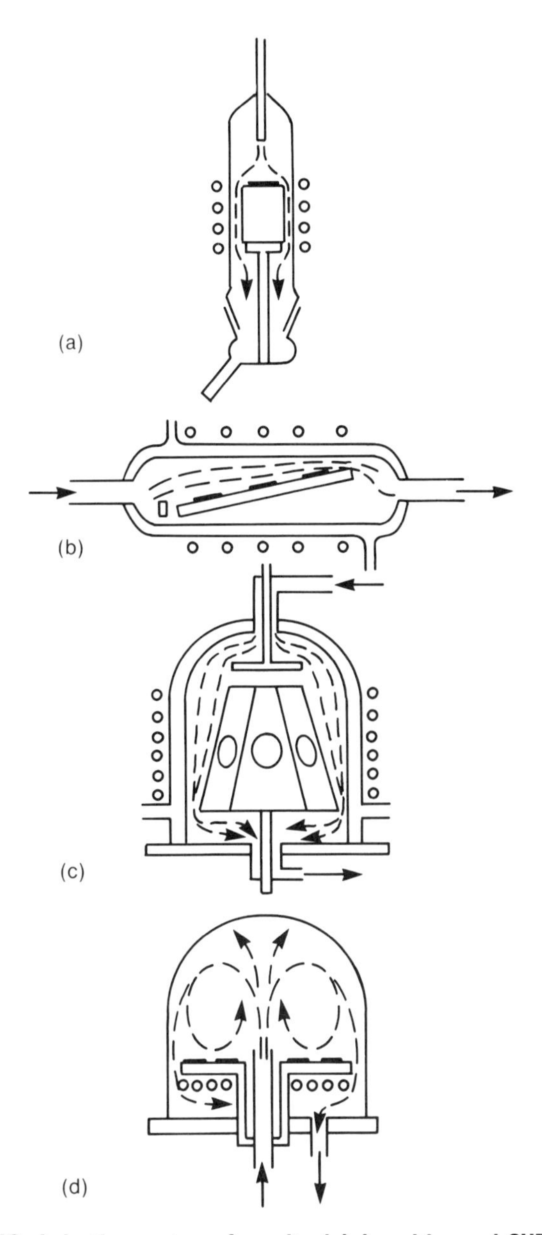

Fig. 11.18. Induction systems for epitaxial deposition and CVD coating of semiconductors (from S. Berkman, V. S. Ban, and N. Goldsmith, *Heteroepitaxial Semiconductors for Electronic Devices*, G. W. Cullen and C. C. Wang, eds., Springer-Verlag, New York, 1978, p 264)

A further increase in the number of wafers that can be processed at any one time has been achieved with the design of the barrel or drum reactor (Fig. 11.18c). Here, the wafers are mounted in pockets on the outer surface of a cylindrical susceptor. The coil surrounds the susceptor, which is rotated for uniformity. The susceptor is somewhat conical to equalize gas flow across the over-all surface. With this technique, it is possible to treat the susceptor wall (for induction purposes) as a portion of a tube; thus, a frequency of only 10 kHz satisfactorily heats a susceptor of this size without the problem of field cancellation.

Because more power is drawn from an induction generator while bringing the susceptor to temperature, a large portion of the heating capacity is left unused. One means of more fully utilizing a motor-generator system is by employing multiple heating stations which can utilize the excess power. However, this is not possible with most solid-state power supplies, which can operate only one heating station at a time.

Production of Silicon Solar Cells

Conventional methods of growing silicon crystals for solar cells are typically batch processes. However, the normal crystal-growing technique, involving subsequent wafer slicing and polishing, is too costly for practical applications. The development of a continuous method of producing inexpensive crystals has been sought as a means of lowering the cost of this material for solar cells. Continuous ribbon growth using induction as a power source is thus used as a way of greatly reducing the cost of such substrate materials (Fig. 11.19). In this method, material is melted in a graphite crucible using induction as a basic heat source. Molten silicon is drawn through a graphite die by capillary action. Solidifying silicon is pulled upward from the die by a belt. A similar process eliminates the die by growing a single, continuous dendritic crystal that is shaped like a dog bone. This crystal is thicker at the sides than in the middle, which constitutes a mechanically stronger shape.

Hermetic Sealing and Salvage of Electronic Components

Hermetically sealed enclosures are often necessary for protection of electronic components. Hermetic sealing generally calls for a continuous solder seal at the periphery of a metal package. However, it is necessary that the contents be protected from excessive heat while the cover is being soldered to the assembly. Conventional soldering techniques, which use conduction to seal the part, are slow, allowing considerable time for heat transfer. Accordingly, by the time the seal is made, the package contents generally attain a temperature approaching that of the solder and may therefore be damaged.

Induction provides rapid, localized heating that can be introduced at all areas of the joint simultaneously, thus greatly reducing heat transfer to the contents of the package. Figure 11.20 shows a typical assembly of a case and a header that is soldered quickly in a single operation. The coil is designed to place the greatest amount of heat into the larger mass of the header. A case vent is provided to allow the escape of heated air as it expands in the case.

If this were not provided, the heated air would vent through the molten solder, preventing an optimal peripheral seal. At the completion of the soldering operation, the vent may be used to introduce a protective atmosphere before it is sealed. The soldering of the periphery of a semiconductor flat pack can also be done using induction heating. Because the area to be heated is greatly restricted, a metallic susceptor is heated by the induction coil. The susceptor is then pressed against the flat pack to effect a seal. The die (susceptor) is designed to contact the assembly only at the periphery of the cover.

Sometimes it is necessary to open a hermetically sealed assembly for repair or removal of the contents. The metal case may be easily destroyed if a torch or iron is used, because only one seal area can be heated at a time. By use of an induction coil, however, the assembly can be heated throughout the full joint area with the same advantages (rapid, even heating) realized in the initial sealing operation. Under these conditions, the assembly can be safely opened, and the case can generally be salvaged for reuse.

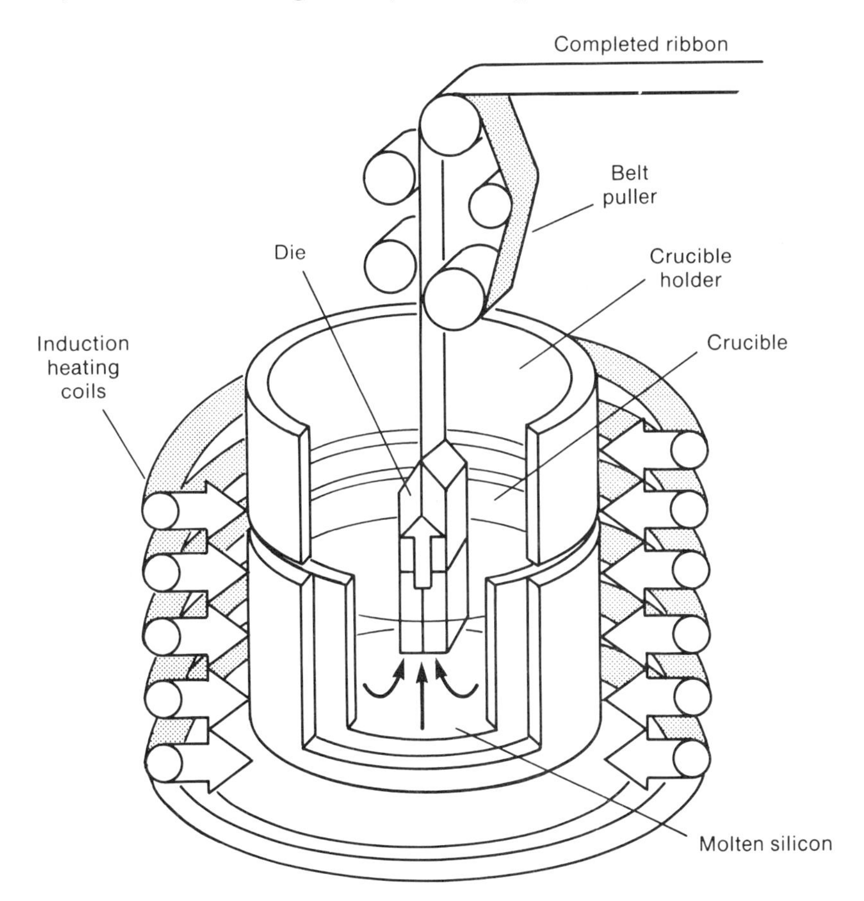

Fig. 11.19. Illustration of the use of induction heating in growth of continuous silicon ribbon (source: Mobil Tyco Corp.)

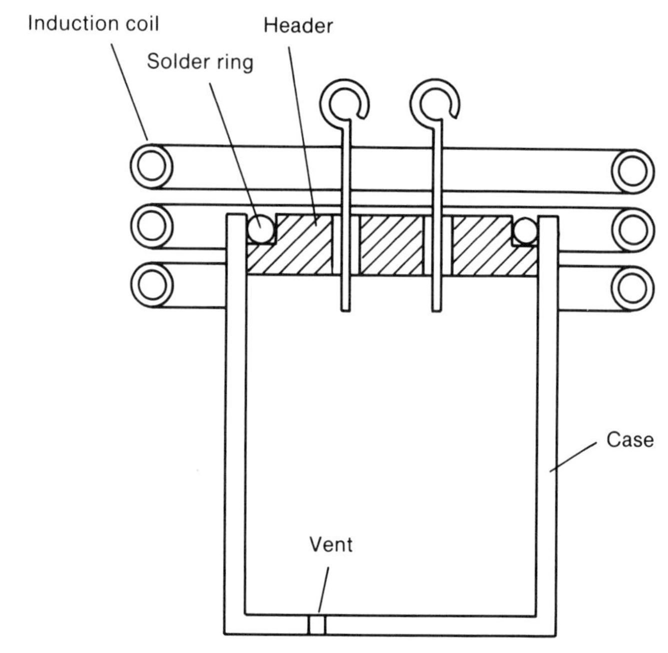

Fig. 11.20. Hermetic solder sealing of electronic case assemblies (source: Lepel Corp.)

Manufacture of Vacuum Tubes

Induction heating is also widely used in the manufacture of TV and cathode-ray tubes. It is utilized in outgassing and band sealing (a shrink-fitting process).

The metallic parts of the tubes are fabricated in an air environment, and thus oxygen becomes entrapped in the metal. The heat produced by the filaments in such tubes during operation would release this oxygen, thereby oxidizing the metal and reducing service life. Outgassing during production is thus required. Typically, during outgassing, each tube rests on a rolling cart which proceeds through a long oven. The cart contains its own vacuum pumping system which exhausts the air in the tube as it proceeds through the oven. The cart also carries a work coil which surrounds the electron-gun structure in the tube. This coil is connected by flexible leads and brushes to a length of buswork which in turn is connected to an induction heating power supply. The metallic parts contained in the tube are induction heated to 850 °C (1560 °F) as it travels through the oven, and the entrapped oxygen is driven off. When the tube exits the oven, it is "tipped off" or sealed to maintain the vacuum.

After the tube has been sealed, a slight amount of oxygen still remains in the tube envelope. A small amount of material which rapidly oxidizes at high temperature, usually barium, is incorporated as part of the assembly in each tube. This is called a "getter." In a subsequent operation, the getter is heated

Fig. 11.21. Photograph of a getter flashing system used in production of television tubes (source: Lindberg Cycle-Dyne, Inc.)

by coupling it to a high-frequency induction coil (Fig. 11.21). When the getter is flashed (i.e., brought to its flashpoint temperature by the induced field of the coil), it oxidizes at a rapid rate, "getting" the remaining oxygen in the tube.

In some types of cathode-ray tubes, the face of the tube, for reasons of simplicity in manufacture, is separate from the neck and "funnel." It is necessary to join these components by shrink fitting a steel band around the juncture of the two parts. The band is heated with an induction coil, creating sufficient expansion to enable it to be slipped over the assembly. The resultant compressive forces on the glass, as the band cools to its original dimensions, provide an airtight, mechanically strong seal.

INDUCTION HEATING APPLICATIONS IN THE GLASS INDUSTRY

Applications of induction heating in the glass industry involve joining of glass to metal, repair of glass tanks, lens blocking, and manufacture of fiber optics.

Glass-to-Metal Sealing

Glass-to-metal seals are often required in the fabrication of components such as connectors. Problems associated with dissimilar rates of thermal expansion and contraction, leading to cracking, have been solved through the use of an intermediate material called Kovar®, which can be fused to the glass and soldered to the base-metal assembly. Induction is used to provide even heating around the joint between the two materials so that the expansion and contraction are uniform. A lead-tin alloy is generally used as a low-temperature solder, and a liquid noncorrosive flux or reducing atmosphere removes the surface oxides. In some instances, where assemblies require slow cooling to prevent cracking due to residual stresses, the parts to be assembled are mounted on a graphite susceptor whose thermal mass retards the heat dissipation on cooling.

Glass Melting

In manufacture of glass for production of optical lenses and special glass components, the molten glass must be kept free of contaminants. For this reason, it is generally melted in a platinum crucible. It is also important that the glass be kept free of the products of combustion. Hence, the platinum crucible is induction heated. Because platinum has a low resistivity, low frequencies in the range of 1 to 3 kHz can be used with very thin platinum crucibles, reducing costs considerably.

In some melting operations, induction is also utilized to control the flow of glass from the "spigot" of a melting tank. There is a relatively narrow temperature range over which the glass begins to flow freely. By utilizing a platinum or molybdenum "spigot" within the coil of a low-power induction heater, the spigot temperature can be regulated to start or stop glass flow, much as a stopcock controls the flow from a pipe.

Lens Blocking

When optical lens blanks have been received from a casting system, they must be ground to specific contours depending on their ultimate application. This operation is performed by mounting the blank(s) on a cast iron or steel block which is used as a fixed base in grinding or polishing equipment. This block is coated with pitch. The pitch is then heated to a soft condition, utilizing an induction coil to heat the metal base. The blanks are locked in place as the base cools and the pitch solidifies. This heating operation generally takes a matter of minutes as compared with the long oven soaking times previously required. Depending on the mass of the base, power supplies for this application range from small RF generators to multiple blocking stations operated from a central motor-generator or solid-state supply.

Fiber-Optic Manufacture

Production of fiber-optic waveguides for telecommunications has become a major technology. It is estimated that most of the world's communications will be carried on optical fibers by the year 2000.

In its simplest form, an optical waveguide is a glass tube that has been coated on its inner or outer surface with an oxide layer that prevents light from passing through its walls. This tube is then drawn into a fiber as small as 125 μm (5 mils) in diameter. In one process, the glass preform, approximately 25 to 50 mm (1 to 2 in.) in diameter, is filled with a gaseous form of germanium or silicon. An induction power supply operating at 2 to 5 MHz is connected to a work coil that is passed at a controlled rate over the preformed tube. A gas plasma is created which oxidizes approximately 85% of the germanium or silicon, evenly distributing the coating over the inner surface of the preform and producing a light pipe or light conductor. The treated tube is then placed in a vertical chamber known as a draw tower (Fig. 11.22), where it is maintained in either an inert atmosphere or a vacuum. Here, the glass is heated by radiation from an induction heated susceptor to a temperature at which it becomes electrically conductive (815 °C, or 1500 °F). The susceptor is removed, and an RF generator operating at 2 to 5 MHz then couples its energy directly into the now-conductive glass, raising it to the required final temperature which, depending on the material, can be as high as 2500 °C (4530 °F). The tubing is drawn from the narrow molten zone at a controlled rate to the final diameter to produce the finished optical waveguide.

INDUCTION HEATING APPLICATIONS IN STEEL FINISHING

Steel sheet products are often coated or painted for aesthetic reasons or to prevent corrosion. Induction heating is widely used in such applications to reflow (or smooth) an electrolytically deposited coating or to cure paint.

Fig. 11.22. Draw tower used in manufacture of optical fibers (source: Astro Industries, Inc.)

Tin Reflow

In the manufacture of tinplate, tin is electrolytically deposited on steel sheet. It is desirable to "flow-brighten" the tin coating by heating it to a temperature above its melting point (230 °C, or 450 °F). This causes the tin surface to melt and flow into a uniform, tightly adhering layer. Pinhole voids in the plating are also reduced by this means.

The tin-reflow operation is generally performed with wide coils wound around the strip as shown in Fig. 11.23. Due to the fact that the strip is usually of thin-gage material (0.18 to 0.36 mm, or 0.007 to 0.014 in.), RF heating is generally used. However, for thicker materials, 10-kHz heating has been utilized. Strip in widths up to 1.5 m (60 in.) has been processed by this technique. Strip speeds can reach 610 m/min (2000 fpm).

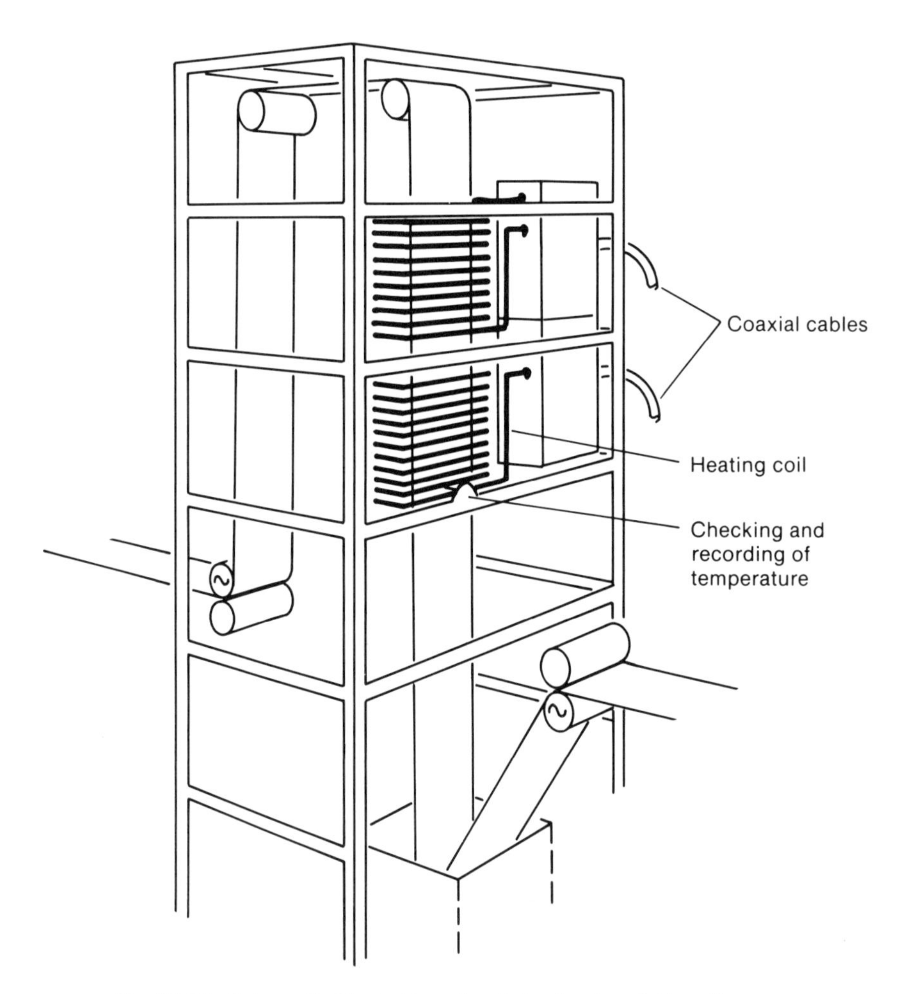

Fig. 11.23. Schematic illustration of an induction-based, tin-reflow system (from J. Davies and P. Simpson, *Induction Heating Handbook*, McGraw-Hill, Ltd., London, 1979)

Paint Curing

In recent years, paints that cure almost instantly when heated to an elevated curing temperature have come into wide use. Coatings of this type are especially suitable for strip-type products which require the material to be rolled into coils immediately after painting. Typical of these products are steel tape measures (Fig. 11.24). Steel strapping and banding of the types used on packing crates and in related packaging applications are also now largely protected by this type of coating.

The use of induction heating produces heat at the interface between the metal and the coating, as compared with the surface curing performed in ovens. In a conventional oven, the outer surface of the coating is heated and cured first, trapping the volatile compounds between the surface and the base

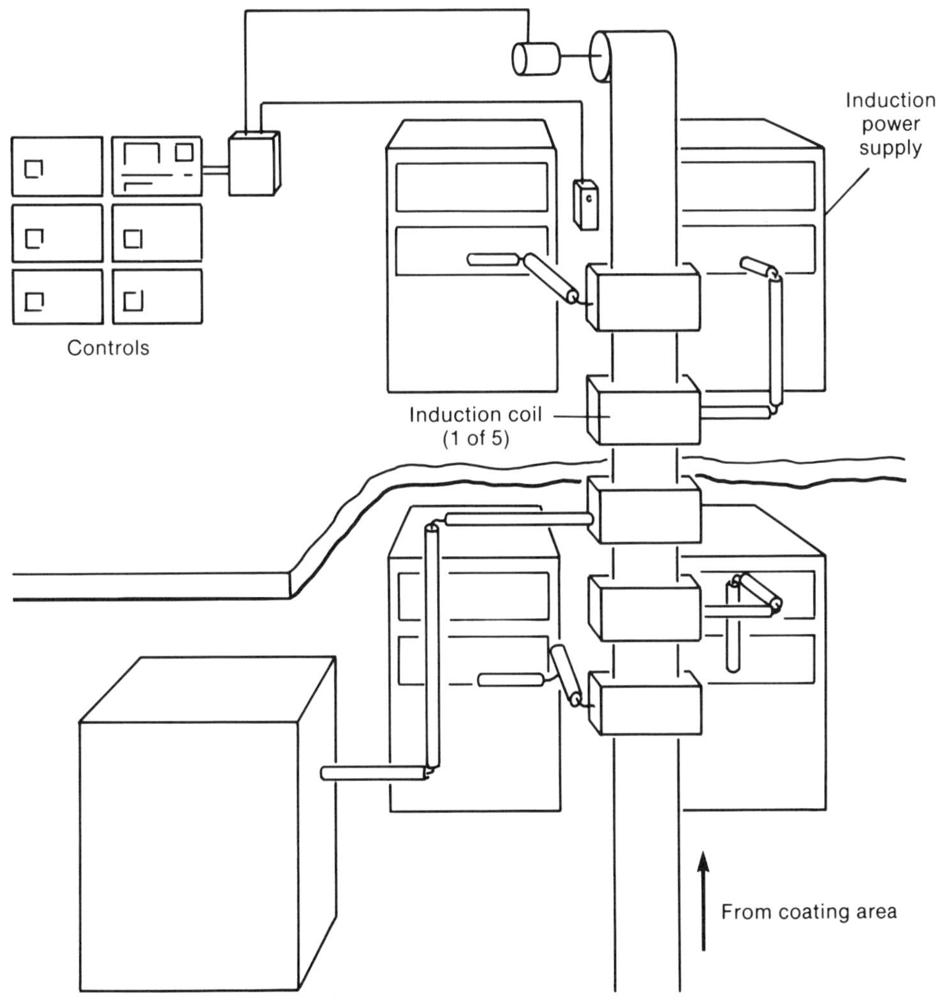

Fig. 11.24. Schematic illustration of an induction-based paint-curing line (source: H. Hanser)

metal. In extreme cases, the coating spalls, or porosity, which weakens the coating, is developed. By contrast, the induction heating technique cures from the inside out, eliminating volatile elements through the uncured coating and achieving a much stronger bond between coating and base material.

In a typical high-speed induction curing line, the heating portion of the system occupies approximately 3 m (10 ft) of line length. Material moves at a maximum speed of approximately 180 m/min (600 fpm). Line speeds are limited only by the mechanical capabilities of the facilities. Because the only heat produced in the process is that which occurs in the metal, working conditions are cool and quiet, and maintenance costs are negligible. Automatic line-temperature feedback control adjusts the heating-equipment power level to maintain a constant curing temperature regardless of line speed. Scrap due to improper curing is virtually eliminated. By-products of the induction curing process are limited to vapors and solvents which are evaporated from the paint during the process.

In a typical system for curing of paint on wide (1.5-m, or 60-in.) steel strip, the material moves vertically through the coater and the induction heating section. The latter consists of two induction coils, each controlled independently. This gives the line the needed flexibility to cope with the variety of coatings the plant handles. Typically, most of the solvents are evaporated when the strip passes through the first coil; the second coil heats the strip to curing temperature. The solvents are collected by a blower system and ducted to a unit on the plant roof. This unit mixes large amounts of ambient air with the solvent-saturated mixture in the required ratio, which can be adjusted. Some of the solvents and particulate matter condense and collect at the bottom of an inner shell from which they are removed. Upon exit from the induction line, the strip is air cooled and then passed over water-cooled rolls prior to coiling at a temperature close to ambient.

VESSEL HEATING

In recent years, line-frequency induction heating of vessels has come into use in the chemical, paint, plastics, and related industries. The greatest use of such heating methods is probably in the manufacture of synthetic resins. Typical holding capacity and power requirements for this application are given in Table 11.1.

In a typical setup, a coil wound around the outside of the metal vessel (Fig. 11.25) induces low-voltage, high-current flow within the vessel wall when energized at line frequency. The wall acts as a single, short-circuited secondary turn, causing the vessel to heat, which in turn rapidly transfers heat to the vessel contents. Any tank can be successfully heated by induction provided it is constructed from a metallic or electrically conductive material. In line-frequency heating, the tank is generally constructed from 6-mm (0.25-in.) or thicker mild steel plate which may be internally clad with stainless steel or

Table 11.1. Typical sizes and ratings of induction units used for vessel heating (source: Cheltenham Induction Heating, Ltd.)

Vessel working capacity		Typical vessel diameter		Nominal temperature		Typical heater rating,
L	gal(a)	mm	in.	°C	°F	kW
5	1	200	8	300	570	2
45	10	380	15	300	570	10
115	25	530	21	300	570	20
225	50	610	24	300	570	30
450	100	760	30	300	570	50
1,150	250	1000	40	300	570	100
2,300	500	1375	54	300	570	150
3,500	750	1525	60	300	570	200
5,000	1100	1830	72	300	570	300
7,000	1500	1980	78	300	570	375
8,500	1800	2150	84	300	570	450
10,000	2200	2300	90	300	570	550
12,000	2600	2440	96	300	570	600
14,000	3000	2590	102	300	570	675

(a) Imperial gallons.

glass as process requirements dictate. Solid stainless steel vessels or vessels with external jackets can also be efficiently heated.

With induction heating of vessels, it is possible to achieve input power densities much higher than those of other heating systems. Inputs on the order of 60 kW/m^2 (20,000 Btu/ft$^2 \cdot$h) are easily obtained. The only limitation is the absorption capacity of the product or process. By comparison, the maximum operational power density of electrical-resistance heaters is usually limited to approximately one-tenth that of induction systems to reduce element burnout. Another advantage of induction heating is the fact that heat is generated uniformly within the vessel wall. There are no hot spots which can cause local overheating of the contents or deterioration of the tank. In addition, there are no entry and exit temperature gradients as are normally encountered in vessels heated by noninduction methods. Induction heating also results in minimal thermal inertia because only the vessel wall and contents are at elevated temperatures. This in turn gives rapid control response with minimum temperature overshoot. Because the vessel wall is the heat source in the induction heated system, the over-all temperature gradient is significantly lower than with other methods. For example, the typical maximum temperature differential between the batch and the vessel wall is 20 °C (35 °F) while heating to temperature and 5 °C (10 °F) during holding at the set point. Coupled with the low thermal inertia, this allows extremely close control of process temperature.

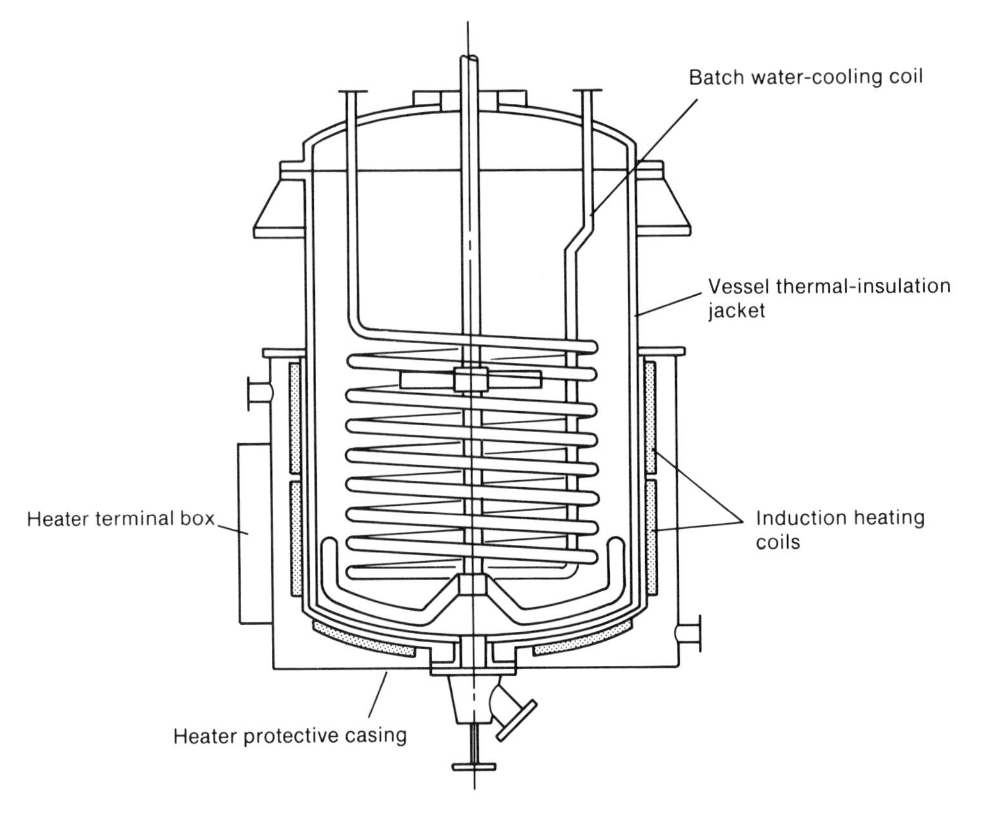

Fig. 11.25. Schematic illustration of vessel heating using low-frequency induction (source: Cheltenham Induction Heating, Ltd.)

APPLICATION OF INDUCTION HEATING FOR VACUUM PROCESSES

The processing of many special metals and semiconductors often requires an atmosphere free of oxygen or other gases which might act as contaminants. Accordingly, induction heating in vacuum has progressed from a laboratory curiosity to a production process.

The area of greatest concern in vacuum processing generally is related to the creation of an ionized gas or plasma due to the voltages encountered with induction. Therefore, it is desirable that the applied voltage at the incoming terminals of the vacuum feedthrough be limited to a maximum of 200 V. In addition, because the voltage generated in the workpiece is directly related to the frequency, there is a predisposition to create a plasma or arcing between turns at higher power-supply frequencies. Thus, good practice dictates use of the minimum frequency consistent with the application. Further, when backfill gases are used, selection should be based on the maximum ionization voltage as well as the appropriateness of the atmosphere for the process.

Fig. 11.26. Two views of a small vacuum induction melting furnace (source: Vacuum Industries, Inc.)

Vacuum Melting

Probably the most basic vacuum induction process involves metal melting. Figure 11.26 illustrates a small laboratory setup utilizing a 20-kW, 10-kHz power supply capable of melting 2.3 kg (5 lb) of iron or steel. For melting of nonferrous materials, graphite crucibles are common. The crucible is held by means of insulating material in the work coil which is connected to the power supply by a rotary vacuum feedthrough assembly called a coax. This enables the crucible to be tilted manually when the melt has been made, thus pouring the metal into a mold which is also in the chamber. Provision is also made for the addition of alloying materials to the crucible during the melting process. A remotely removable cover on the crucible conserves heat during the melting operation. Some larger systems feature air locks so that the crucible may be charged with additional material for melting without breaking the vacuum seal. In addition, some systems contain additional air locks so that preheated molds can be introduced into the main chamber, filled with molten material, and then removed, also without breaking the vacuum seal.

Directional Solidification

Directional solidification is another popular vacuum melting process. It is frequently used in the manufacture of superalloy blades for jet engines (Fig. 11.27). In this technique, the mold is supported on a pedestal whose movement is closely controlled. The mold rests vertically in the pouring position within an induction heating coil whose power supply is separate from that of the induction melter. When the mold is filled from the crucible, the pedestal slowly lowers the unit out of the mold heating coil. Then, similar to the process of crystal growth for semiconductors, the slow controlled cooling creates a single crystal or columnar grain structure in the shape of the mold—i.e., a turbine engine blade. The creep strength of this structure at the high service temperatures which it undergoes is superior to that of the normally cast or forged product.

Levitation Melting

Minimal contamination of melted metal is obtained by a melting technique known as levitation melting, which is frequently carried out in vacuum. Gen-

Fig. 11.27. Photograph of a vacuum induction furnace used for controlled solidification of metals (source: Vacuum Industries, Inc.)

erally it is used for producing small quantities of electrically conductive materials while suspending them electromagnetically. It offers a rapid means of preparing melts of reactive or high-purity metals without the danger of crucible contamination in either a vacuum or an inert atmosphere. Levitation and melting are accomplished in a specially shaped induction coil; casting is performed subsequently by turning off the power and allowing the melt to fall into a mold.

When a conductor is placed in an induction field, the induced current heats the conductor. It also creates an opposing magnetic flux that tends to push the object into a region of lower field strength—i.e., out of the field (or coil). This force may be computed using the Lorentz equation. If the magnetic field is uniform, there is no net force on an object that is placed in it. A field gradient is needed to provide a lifting force. To create a strong magnetic field and a strong field gradient, it is customary to use a conical induction coil such as that shown in Fig. 11.28. Note the bucking plate located above the induction coil. Induced current created in the bucking plate causes an opposing magnetic field, distorting the magnetic flux and creating a stable pocket of balanced magnetic forces for the levitated sample.

When given enough current, most coil configurations capable of providing a field gradient will levitate a sample. A conical coil with five to seven turns of small-diameter tubing provides a stable support for molten charges. An extra turn of tubing outside the lower turn of the levitation coil increases the strength and gradient of the magnetic field and thus the lifting force exerted on the specimen. One or two reversed turns or a bucking plate are used over the coil for horizontal stability. The coil shown in Fig. 11.29 pro-

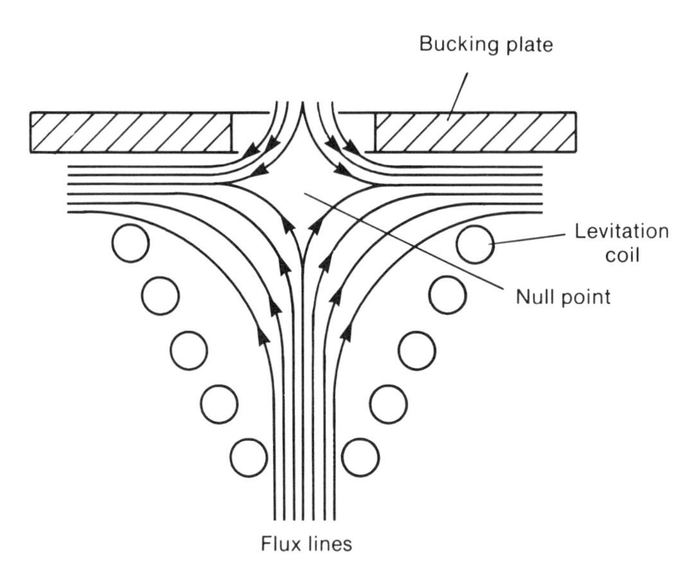

Fig. 11.28. Induction coil used for levitation melting of metals (source: Lepel Corp.)

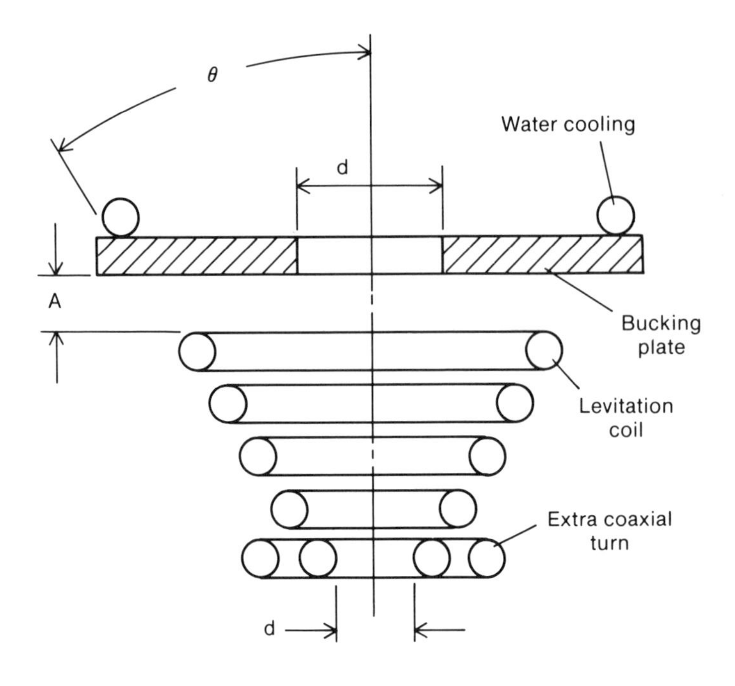

Typical parameters:
θ — 30 to 45°
d — 9.5 to 19 mm (3/8 to 3/4 in.)

A — 1.6 to 9.5 mm (1/16 to 3/8 in.)
Coil frequency, ~250 kHz
Coil current, to 650 A

Fig. 11.29. Induction-coil design parameters used in levitation melting of metals (source: Lepel Corp.)

vides a good electrical match and good levitation in a 10-kW system. Copper tubing 3.2 mm (0.13 in.) in diameter provides the best compromise between the high turn density needed for maximum magnetic field strength and field gradient and the current-carrying capacity which is limited by cooling-water flow rate.

Levitation melting systems have been used successfully for melting aluminum, copper, iron, chromium, and their alloys, primarily to produce high-purity materials free of crucible contamination. Typical melts of 5 to 15 g (0.18 to 0.53 oz) have been melted and cast in less than 1 min. Melting in vacuum or in a protective atmosphere is completed as desired.

In addition to melting of reactive materials, uses of levitation melting include:

- Rapid alloy preparation
- Gas-metal reaction kinetic studies

- Solidification studies
- Liquid slag and refractory reaction studies
- Melt purification and homogenization.

Levitation melting has also been performed aboard the space shuttle, providing an opportunity to work in an almost perfect vacuum and thus produce ultrapure materials for testing. In addition, working in a weightless condition permits the creation of special alloys that cannot be produced under the influence of the earth's magnetic and gravitational fields because of the high densities of their components. The properties of these alloys are presently being examined.

Chapter 12

Economics

Induction heating affords a rapid, efficient technique for producing localized or through heating in a wide range of industries. The economics as well as the technical feasibility of induction heating should be important considerations prior to investing in such a system. A number of different cost elements enter into the analysis. These include equipment and energy costs, production lot size and ease of automation, material savings, labor, and maintenance. Each of these factors is discussed below. There are also several secondary benefits such as those related to improved working conditions and special process capabilities which impact productivity but which are not readily quantifiable except on a case-by-case basis.

In the discussion that follows, the cost elements of induction heating are compared with those of its main rival in large process heating applications—namely, the gas-fired furnace.

COST ELEMENTS OF INDUCTION HEATING

Equipment Costs

Induction power supplies, which constitute the major cost items in induction systems, typically cost $2\frac{1}{2}$ to 3 times as much as gas-fired furnaces of equal capacity. Gas furnaces with recuperators (used to preheat combustion air) have higher efficiencies than conventional furnaces and cost about one-half as much as comparable induction equipment. Both induction power supplies and gas furnaces have very long service lives. Because of this, amortization of both types of equipment is readily spread over 5 to 10 years or even longer. Thus, induction heating equipment will cost more, but the cost can usually be offset by increases in productivity and by energy and materials savings.

The absolute cost of induction generators varies with frequency and power rating. For a given power rating it increases with frequency, and for a given frequency, it scales linearly with power. In terms of 1983 dollars, typical costs range from $100 to $200 per kilowatt for 60-to-1000 Hz generators, from $200 to $300 per kilowatt for 3-to-10 kHz generators, and from $300 to $400 per kilowatt for radio-frequency (RF) generators. A typical gas-fired furnace costs $6000 to $7000 per million Btu/h capacity. This corresponds to approximately $25 per kilowatt. Frequently, furnaces are only about one-fourth as efficient as induction heating units; hence a furnace comparable to an induction system, in terms of useful energy supplied to the part, would cost about $100 per kilowatt.

Energy Costs

Energy costs are determined by the base fuel cost and the over-all efficiency of the heating system. On the average, electricity costs about 3.5 times as much as natural gas in recent years (Fig. 12.1). However, heating-system efficiencies can mitigate this difference in fuel cost. Efficiency is the ratio of energy used for actual heating to the energy supplied to the system. For induc-

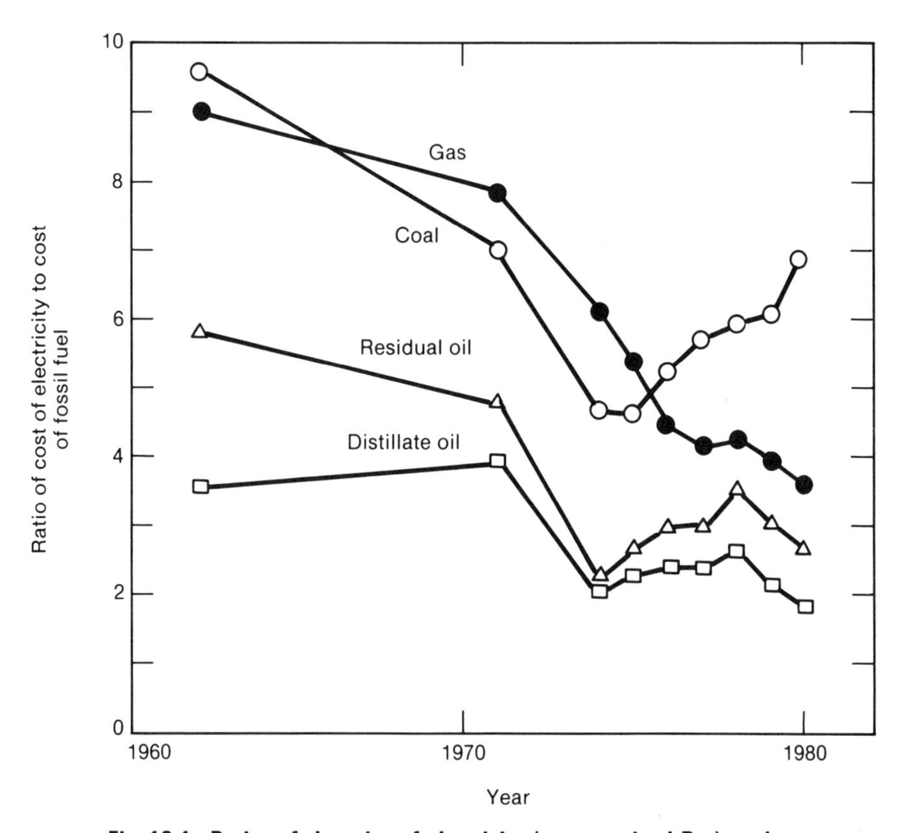

Fig. 12.1. Ratios of the price of electricity (per contained Btu) to the prices of the indicated fossil fuels (per contained Btu) paid by manufacturers (from C. C. Burwell, "Industrial Electrification: Current Trends," Research Memorandum ORAU/IEA-83-4 (M), Institute for Energy Analysis, Oak Ridge Associated Universities, Oak Ridge, TN, February, 1983)

tion heating, system efficiencies of 55 to 85% are typical. For gas furnaces, in which heat losses through doors, walls, etc. are common, typical efficiencies are 15 to 25%; with special insulation methods, which, of course, increase the furnace cost, it is reported that these efficiencies can be increased to perhaps 40 to 50%. In any case, it is obvious that the higher efficiencies of induction units can often offset the high cost of electricity.

Besides the energy used during the actual processing of material, the energy required during idle periods should also be taken into account. With induction systems, start-up is instantaneous. In addition, standby power (needed during short delays in production) is rather small, amounting to much less than 1 kVA for line-frequency and solid-state equipment. For RF tube-type power supplies, filament power is generally shut off when delays exceed 30 min. In comparison with induction units, furnaces must be idled at fairly high temperatures even when not in use during off shifts. Thus, considerable energy is still used during these periods. The effects of utilization level on energy costs for several means of heating prior to forging are illustrated in Fig. 12.2. In this plot, data are given for a typical motor-generator induction system as well as for a direct-resistance heating method. Not shown is the trend for solid-state induction power supplies, which lie between "whole-piece" induction heaters and direct-resistance units.

Production Lot Size and Ease of Automation

Because a given induction coil can often be used only for a given part and heating application, induction heating is almost always selected for large production runs of a given product for which coil changes are not necessary. Such processes are best carried out on an automated basis, which can lead to substantial economic benefits.

As shown in previous chapters, the major factors inherent in the induction process (i.e., time, temperature, power, and coil/part placement) all combine to provide a repeatable, uniform process. Control of these factors enables the system to repeat the particular operations within specified process parameters on a consistent basis. Product uniformity and quality therefore are independent of changes in operator capability. It is apparent that time, temperature, and output power can be monitored and controlled within close limits for any operation. Therefore, the equipment can be simply integrated into an automated handling system once the part/coil location has been precisely determined. Process time can then usually be adjusted by means of power control or coil design to permit operations to mesh within allowable constraints. In addition, the lack of undesirable heating of the surroundings simplifies integration directly into a production line without the need to employ off-line heat processing. This in turn greatly reduces handling costs. Moreover, quality-control and inspection techniques may be incorporated directly into the induction process to monitor the results and maintain process requirements.

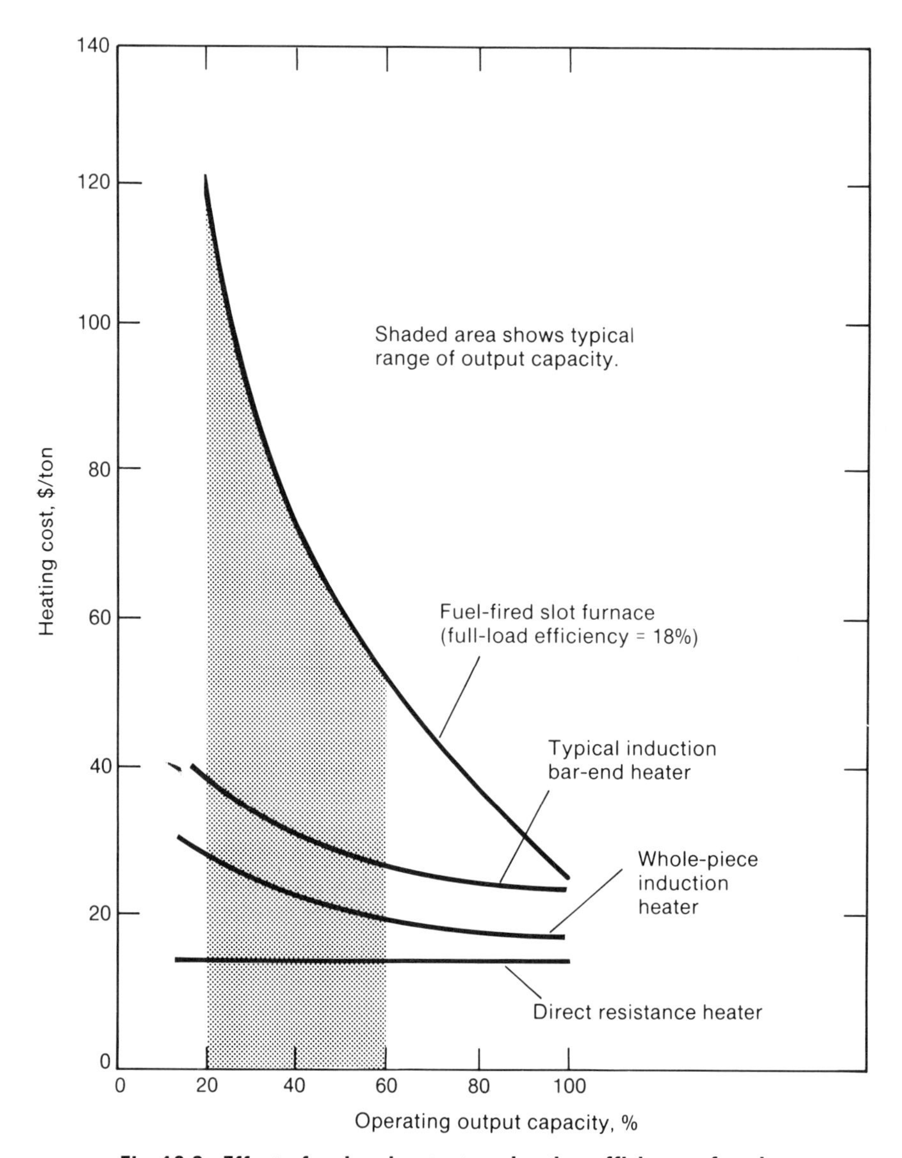

Fig. 12.2. Effect of reduced output on heating efficiency of various methods as reflected in cost per ton of product (from A. Perkins, *Metallurgia*, Vol 48, December, 1981, p 568)

Scale and Scrap Losses

Scale and scrap losses are of particular importance in an economic analysis because a large part of total production cost is tied up in the workpiece material itself. Often, the material cost constitutes as much as one-half or more of the net cost of the finished product. Because of this, material loss, as scale or scrap, is a prime factor in any economic analysis.

Scaling occurs during heating of steels prior to forming or during heat treating processes. In heating of steel billets to hot working temperatures in air using gas-fired furnaces, scaling often amounts to 2 to 4% of total part weight. At hardening temperatures, thc amount of scalc formation is somewhat less, around 1 to 2% for furnaces. In contrast, the shorter heating times characteristic of induction heating lead to considerably less scale, usually no more than one-fourth of the amount formed during furnace heating processes. In forging installations, scale formation is also a serious consideration with regard to die wear. It has been shown that major increases in die life and thus reduced die costs result from the reduction in scale generated during induction, as opposed to furnace, heating.

Scrap losses due to improper heating are usually higher in gas-fired furnace operations. In furnace-based processes, thermal cracking and distortion are the usual causes of part rejection. Such defects cause scrap losses of 1 to 2% for steels. Scrap losses associated with induction heating are typically around 0.25 to 0.5%. Thus, considering both scale and scrap losses, induction heating could result in materials savings of as much as 3 to 4%.

Labor Costs

Induction systems require less labor for both operation and maintenance than do comparable furnace-based heat treating systems. This is because the widespread use of reliable solid-state power generators, and the fast response and controllability of induction heaters, make these systems very amenable to automation. On the other hand, furnace heat treating processes normally require a considerable amount of labor for loading, unloading, and handling of the steel to be processed. In addition to requiring fewer operators, automation also allows the use of less-skilled operators with a minimal amount of training. The use of automatic as opposed to manual operation in a single-shot induction application has reduced labor requirements as much as 50 to 80% in some instances. Taking all these factors into account, a general rule of thumb for cost-analysis purposes is that induction heat treating usually requires one-half the labor needed for furnace processes with an equivalent production rate.

Maintenance Costs

Induction-based heating systems normally require considerably less maintenance than gas-fired furnaces. Total system rebuilding is not a typical requirement for induction equipment as it can be for gas furnaces. An annual allowance of approximately 2 to 3% of the installation cost should be budgeted for repair of coils and other maintenance. When mechanical handling or poor maintenance procedures decrease coil life, this figure can approach 4%. By contrast, the refractory linings of gas furnaces are subject to severe service conditions and may deteriorate rapidly, particularly when furnace temperatures are cycled. The cost of relining is not inconsiderable, and lost production time during the repair operation can become an expensive factor.

Other Cost Elements

Other cost factors that may enter into a cost analysis concern the quality of the final product. For example, much less decarburization of steel surfaces occurs during induction hardening than during longer furnace-hardening cycles. Reduced decarburization translates into lower costs for finish machining (to remove the decarburized layer) and lower material losses during such final machining. In addition, because distortion is generally lower in induction-based processes, straightening and final machining of induction heated parts are usually less costly than for identical furnace-heated or furnace heat treated parts.

TYPICAL COST COMPARISONS

The above (and other pertinent, application-dependent) factors can be employed to estimate the economic feasibility of candidate induction heating processes. Several typical examples are given below to illustrate the methodology.

Heating of Steel Billets Prior to Forging

Cost analyses for the induction heating of steel billets prior to forging have been quoted many times in the literature. Invariably, it is concluded that induction heating for steel forging is very attractive. Examples of such analyses are given in Tables 12.1 and 12.2.

The first example (Table 12.1) is for an induction system operating at 52% efficiency and a gas-fired furnace at 17.5% efficiency. Neglecting the capital investment, it can be seen that the only item for which induction poses a greater expense is energy. With the more rapid increase in gas prices since 1980 (the most recent year for which costs are estimated in the table), the differential has decreased somewhat. It can also be noted that substantial savings in labor and in scrap and scale losses are realized through the use of induction, leading to lower over-all operating costs. Such lower costs can aid in the amortization of the induction equipment in an equal or shorter period of time in a high-production shop.

The second example cost analysis (Table 12.2) results in similar conclusions. In this example, the net energy cost includes the effects of the base fuel price, the thermal efficiency of the equipment, and the expense of idling the equipment when not in use. The bottom-line operating-cost difference of $300,000 per year suggests a reasonable payback period for the induction system, whose estimated initial investment cost was $600,000.

Heating of Nonferrous Billets Prior to Forging/Extrusion

Cost data for forging/extrusion of nonferrous metals are not as abundant as those available for steel. In one analysis,* the costs of preheating alumi-

*J. Stott, *Metallurgia and Metal Forming*, Vol 44, No. 5, May, 1977, p 212.

Table 12.1. Hot forging operating costs that differ between induction and natural gas (from N. W. Lord, R. P. Ouellette, and P. N. Cheremisinoff, *Advances in Electric Heat Treatment of Metals*, Ann Arbor Science Publishers, Ann Arbor, MI, 1981)

Item	1977 cost(a)	1980 cost(b)
Induction		
Energy	$8.34	$12.70
Production labor	0.93	1.25
Maintenance labor	0.33	0.44
Scale loss	1.75	2.33
Scrap loss	1.75	2.33
TOTAL	$13.10	$19.05
Natural Gas		
Energy	$5.93	$9.20
Production labor	1.87	2.49
Maintenance labor	0.67	0.89
Scale loss	7.00	9.34
Scrap loss	7.00	9.34
TOTAL	$22.47	$31.26

(a) Cost per ton of product in 1977 dollars. (b) Cost per ton of product in 1980 dollars.

num billets prior to extrusion using induction and gas furnaces were estimated. The analysis showed the cost of the former method to be about twice that for the latter when the cost elements of fuel, thermal efficiency, and capital equipment were included. Because aluminum does not scale, this cost driver was not included for aluminum extrusion. Despite the apparent cost penalty for induction, however, it was estimated that the excess cost of induction heating could be recovered if a production time savings of only 5 min per shift could be realized. In fact, a time savings of 30 min per shift was realized by using induction. In addition, the analysis demonstrated that the difference in equipment cost between the induction unit and the gas furnaces could be offset by the substantially smaller floor space required by the induction equipment.

Heat Treating of Steel

Analyses similar to those for preheating prior to forging may be performed in order to compare the costs of induction and furnace heat treatment of steel. For through heat treatment, the analyses* demonstrate that induction-based processes are quite competitive.

---

*S. L. Semiatin and D. E. Stutz, *Induction Heat Treatment of Steel*, American Society for Metals, Metals Park, OH, 1986.

Table 12.2. Economic comparison between induction and gas-fired furnace heating of forging billets(a) (from J. Skelton, private communication, Battelle Columbus Division, Columbus, OH, September, 1984)

Item	Induction installation	Gas-fired furnace
Installation cost	$600,000	$200,000
Heating efficiency	60%	15%
Annual energy cost	$720,000	$540,000
Scale loss	1/2%	2%
Scrap loss	1/4%	1%
Annual scrap and scale loss cost	$150,000	$600,000
Labor requirement	1 operator; 1/4 maintenance	2 operators; 1/2 maintenance
Annual labor cost	$60,000	$120,000
Total annual operating cost	$930,000	$1,230,000

(a) Hot forging of steel. Annual throughput, 30,000 tons. Annual hours of operation, 4000. Value of raw material, $500/ton. Value of product scrap, $1000/ton. Labor cost, $12/h. Energy cost: electricity, $0.06/kW·h; gas, $4/MBtu.

For surface-hardening applications, induction appears to hold a considerable cost advantage over other established methods. These other methods include salt-bath nitriding, gas nitriding, and carburizing. Using pricing data from commercial heat treaters, it has been estimated that the over-all costs of the various processes are in the ratio of 0.11 to 1.75 to 8 to 2.5 for induction, salt-bath nitriding, gas nitriding, and carburizing methods of case hardening.* Thus, in applications in which geometry and production volume allow the use of induction, it is the preferred surface-hardening method from a cost standpoint.

Several recent technologies have emerged that could compete with induction surface hardening on a cost basis. These are laser and electron-beam surface hardening. Generally these processes are employed to obtain rather shallow case-hardened depths (0.05 cm or 0.02 in., or less). In these instances, and for complex geometries for which induction coils are difficult to design, these emerging technologies may be less expensive.† Nevertheless, in situations in which case depths of 0.05 to 0.10 cm (0.02 to 0.04 in.) are required, induction systems which produce very high power inputs per unit of surface area (i.e., "high-intensity" induction setups) can compete effectively because of lower capital costs, higher productivity, less maintenance, and lower floor-space requirements.

---

*F. H. Reinke, "The Different Applications of Induction Heat Treatment with Particular Respect to Energy Demand," *Proc. World Electrotechnical Conference*, Moscow, June 21–25, 1977.

†R. Creal, *Heat Treating*, Vol 14, No. 9, September, 1982, p 21.

Table 12.3. Comparison of energy requirements for high-frequency welding and arc welding of carbon steel(a) (source: Thermatool Corp.)

Process(b)	Thickness, cm (in.)	Arc volts	Arc amps	Power, kW	Speed, m/h (ft/h)	Energy, kW·h per 30 m or 100 ft
FCAW.....	0.32 (0.125)	28	325	9.1	60 (200)	4.6
HFRW	0.32 (0.125)	...	...	275(c)	4,025 (13,200)	2.1
FCAW.....	0.64 (0.25)	28	475	13.3	40 (125)	10.6
SAW	0.64 (0.25)	33	900	29.7	40 (130)	22.8
HFRW	0.64 (0.25)	...	...	555(d)	3,600 (11,820)	4.7
SAW	1.27 (0.50)	34	1100	37.4	25 (90)	41.6
HFRW	1.27 (0.50)	...	...	1100(e)	2,560 (8,400)	13.2

(a) Data for arc welding are from *Welding Handbook*, 7th Edition, Vol 2, W. H. Kearns, ed., American Welding Society, Miami, 1978, pp 177, 179, and 210. Data for high-frequency welding are from Thermatool Corp. (b) FCAW = flux-cored arc welding. HFRW = high-frequency resistance welding (similar to high-frequency induction welding). SAW = submerged-arc welding. (c) Input-power requirement for 160-kW welder output capacity. (d) Input-power requirement for 300-kW welder output capacity. (e) Input-power requirement for 600-kW welder output capacity.

Tube Welding

The use of high-frequency induction and resistance heating for tube welding also appears very attractive from a cost viewpoint. Not only is the cost per unit of product considerably less with these techniques than with submerged-arc welding (SAW) and fluxed-cored arc welding (FCAW), but production rates are more than an order of magnitude greater (Table 12.3), thus giving rise to substantially lower labor costs.*

*H. N. Udall, "Energy Saving Through High Frequency Electric Resistance Welding," unpublished paper, Thermatool Corp., Stamford, CT, 1984.

Index